Triangular Arrays with Applications

Triangular Arrays with Applications

Thomas Koshy

OXFORD
UNIVERSITY PRESS
2011

OXFORD
UNIVERSITY PRESS

Oxford University Press, Inc., publishes works that further
Oxford University's objective of excellence
in research, scholarship, and education.

Oxford New York
Auckland Cape Town Dar es Salaam Hong Kong Karachi
Kuala Lumpur Madrid Melbourne Mexico City Nairobi
New Delhi Shanghai Taipei Toronto

With offices in
Argentina Austria Brazil Chile Czech Republic France Greece
Guatemala Hungary Italy Japan Poland Portugal Singapore
South Korea Switzerland Thailand Turkey Ukraine Vietnam

Published by Oxford University Press, Inc.
198 Madison Avenue, New York, NY 10016

www.oup.com

Library of Congress Cataloging-in-Publication Data
Koshy, Thomas.
Triangular arrays with applications / Thomas Koshy.
p. cm.
Includes bibliographical references and index.
ISBN 978-0-19-974294-3
1. Pascal's triangle. 2. Number theory. 3. Combinatorial analysis.
4. Binomial coefficients.
I. Title.
QA241.K687 2011
511′.1—dc22 2010009157

Dedicated to
My granddaughter Asha Grace Thomas

Contents

List of Symbols

Symbol	Meaning
$\{x, y, z\}$	set consisting of the elements x, y, z
$x \in A$	x belongs to set A
$\emptyset$	empty set
$\|A\|$	number of elements in set A
$A \cup B$	union of the sets A and B
$A \cap B$	intersection of the sets A and B
$\{x, y^2, z^3\}$	multiset $\{x, y, y, z, z, z\}$
$\{s_n\}_{n=1}^{\infty} = \{s_n\}$	sequence with general term s_n
$\sum_{i=k}^{i=n} a_i = \sum_{i=k}^{n} a_i$	$a_k + a_{k+1} + \cdots + a_n$
$\sum_i \sum_j a_{ij}$	$\sum_i \left(\sum_j a_{ij} \right)$
$\prod_{i=k}^{i=n} a_i = \prod_{i=k}^{n} a_i$	$a_k a_{k+1} \cdots a_n$
$n!$	$n \cdot (n-1) \cdots 3 \cdot 2 \cdot 1$, where $0! = 1$
$(2n-1)!!$	$1 \cdot 3 \cdots (2n-1)$
$\|x\|$	absolute value of x
$\lfloor x \rfloor$	greatest integer $\leq x$
$\lceil x \rceil$	least integer $\geq x$
$\lg x$	$\log_2 x$
$\delta_{n,r}$	Kronecker delta

LHS	left-hand side
RHS	right-hand side
PMI	principle of mathematical induction
$a\|b$	a is a factor of b
$a \nmid b$	a is not a factor of b
$a^n \|\| b$	$a^n\|b$, but $a^{n+1} \nmid b$
(a, b)	greatest common factor of a and b
$[a, b]$	least common multiple of a and b
$\leftarrow$	assignment operator
α	golden ratio $(1+\sqrt{5})/2$
$A = (a_{ij})_{m \times n}$	$m \times n$ matrix A whose ijth element is a_{ij}
$a \approx b$	a is approximately equal to b
$a \equiv b \pmod{m}$	a is congruent to b modulo m
$a \bmod b$	remainder when a is divided by b
$\overline{AB}$	line segment AB
$\widehat{AB}$	arc with endpoints A and B
$v - w$	edge between vertices v and w
K_n	complete graph with n vertices
$P(n, r)$	number of permutations of n things taken r at a time
$\binom{n}{r}$	$\frac{n!}{r!(n-r)!}$
$\binom{-n}{r}$	$(-1)^r\binom{n+r-1}{r}$
$\binom{p/q}{r}$	$\frac{p(p-q)(p-2q)\cdots[p-(r-1)q]}{q^r\, r!}$
$\binom{-p/q}{r}$	$(-1)^r\frac{p(p+q)(p+2q)\cdots[p+(r-1)q]}{q^r\, r!}$
$(a_1, a_2, \cdots, a_n)$	n-tuple
$f = \begin{pmatrix} a_1 & a_2 & \cdots & a_n \\ b_1 & b_2 & \cdots & b_n \end{pmatrix}$	permutation f such that $f(a_i) = b_i$
$\mathbf{Z}_n$	additive group of integers modulo n
F_n	nth Fibonacci number
L_n	nth Lucas number
t_n	$n(n+1)/2$
T_n	tetrahedral number $n(n+1)(n+2)/6$
$[x]_n$	descending factorial $x(x-1)\cdots(x-n+1)$
$< x >_n$	ascending factorial $x(x+1)\cdots(x+n-1)$
$s(n, r)$	Stirling numbers of the first kind
$S(n, r)$	Stirling numbers of the second kind
t_n^*	$t_n t_{n-1} \cdots t_2 t_1$

$\left\langle {n \atop r} \right\rangle$	Eulerian coefficient
$\left[{n \atop r} \right]$	$\dfrac{t_n^*}{t_r^* t_{n-r}^*}$
$\binom{n}{i_1, i_2, \cdots, i_k}$	multinomial coefficient
■	end of an example, proof, theorem, or corollary

Preface

Mathematics is the most beautiful and most powerful creation of the human spirit.
Stefan Banach (1892–1945)

A Byproduct of My Fascination

This book is the result of my fascination for triangular arrays of positive integers. My interest in them got deeper and wider after writing two earlier books, *Fibonacci and Lucas Numbers with Applications* (Wiley, 2001) and *Catalan Numbers with Applications* (Oxford, 2009). Both contain numerous such mathematical objects. Over the years, my interest has taken me on a long and curious mathematical journey, searching for triangular arrays from different sources. This book, a compendium of numerous such arrays, is the product of such a rewarding journey.

This unique work brings together a great deal of material that is found in journal articles, but not in existing textbooks. The examples and applications come from a variety of areas: combinatorics, geometry, number theory, calculus, and graph theory.

Triangular arrays pop up quite often in discrete mathematics, especially in combinatorics, probability, and number theory. For example, the first such array most of us encounter early in our study of mathematics is undoubtedly the well-known Pascal's triangle. Although it was named after the French mathematician Blaise Pascal (1623–1662), it was known in China and India at least two centuries earlier. Pascal's triangle appears in numerous situations and has hundreds of intriguing properties, as seen in Chapters 6–8. Although known for several centuries, it has kept its beauty and charm. It continues to surprise the mathematical community with its ubiquity and fascinating properties.

Triangular arrays are certainly a great source of fun and excitement for both curious amateurs and seasoned professionals. They often contain many hidden treasures, providing numerous rewarding opportunities for us to explore, experiment, conjecture, and prove or disprove conjectures.

Intended Audience

The book is written for a broad audience, ranging from mathematically curious amateurs to mathematically sophisticated professionals. It includes high school students and their teachers, as well as undergraduate and graduate students in both mathematics and computer science. The book provides ample opportunities for the readers to experiment, explore, and create group discussions, projects, and senior capstone experiences. These opportunities could even result in new examples, as well as new applications and properties, which would extend the mathematical horizon. This book is an excellent source of topics for student research.

Organization and Coverage

This book is divided into 29 small chapters. Like *Catalan Numbers with Applications*, it is a unique product: it is the only book to date that deals with a wealth of arrays, found in numerous journals which are not within reach of many. This book brings that work together in an accessible format.

Pascal's triangle has far more properties than any other triangular array; consequently, two chapters have been devoted to it. One of the truly intriguing jewels of Pascal's triangle is its relationship with Mersenne numbers, Tower of Brahma, and Hanoi graphs.

The chapter on miscellaneous arrays includes a number of unrelated arrays, each interesting in its own right. Some of them are contributions of Angelo S. DiDomenico of Milford, Massachusetts, who has a passion for creating such arrays and extracting interesting properties from them.

The book ends with a chapter on multinomial arrays, based on the multinomial theorem, which is an extension of the binomial theorem. As a byproduct of this endeavor, we get an opportunity to see Catalan numbers again, as well as the related Motzkin numbers.

Prerequisites

I have tried to ensure that most of the presentation is within reach of most amateurs. However, a sound knowledge of Precalculus mathematics is essential. So is a reasonable amount of mathematical sophistication. For example, a reader would need a good working background in recursion, binomial coefficients, binomial theorem, congruences, polygonal and pyramidal numbers, generating functions, induction, basic calculus, introductory graph theory, fundamental counting principles, proof techniques, as well as a lot of patience, for exploring and enjoying the beauty of these mathematical stars.

Historical Background

A historical perspective has been incorporated into this material, to put a human face on every development, wherever possible. I have made every effort to cite the name of each contributor as well as relevant dates. I apologize in advance for any omission, either partial or full; it is wholly unintentional.

Mathematical Symbols and Notations

Unlike spoken languages, mathematics is a precise, concise, and elegant language. It has its own symbolism and vocabulary. In the interest of clarity, preciseness, and unambiguousness, a glossary of symbols used in the book is given on the inside front cover, for quick and easy reference.

References

I have had the privilege of referring to an extensive literature, a vast array of over 300 articles, problems, and solutions. Needless to say, they have played a central role in the development of the book.

Acknowledgments

Finally, I remain indebted to a number of people who have played a significant role in fine-tuning the project. Without their support, cooperation, and encouragement, this arduous task would not have come to fruition.

To begin with, I am grateful to the following reviewers for their enthusiasm, boundless support, and constructive suggestions, and for tolerating my indiosyncrasies: Ralph P. Grimaldi at Rose-Hulman Institute of Technology and Fredric T. Howard at Wake Forest University. My sincere thanks to Angelo S. DiDomenico for his passion for triangular arrays, and for sharing some and their properties with me over the years.

My gratitude also goes to my colleague, Zhenguang Gao, for creating the spectacular computer-generated designs in Figures 8.3, 18.2, and 27.2; to Margarite Landry for her extraordinary editorial skills; and to Phyllis Cohen, Editor of Mathematics and Science at Oxford University Press for her support, enthusiasm, promptness, and attention to details.

Thomas Koshy
Framingham, Massachusetts
August 21, 2010

If I have been able to see farther, it was only
because I stood on the shoulders of giants.
Sir Isaac Newton (1643–1727)

Triangular Arrays with Applications

1

Fundamentals

Triangular arrays of integers are a great source of fun and excitement, for both seasoned mathematicians and enthusiastic amateurs. A cornerstone of number theory and hence of discrete mathematics, they provide fascinating opportunities for experimenting, exploring, conjecturing, and establishing or disproving conjectures.

We begin our study of triangular arrays with a brief introduction to some fundamental concepts and notations, which will facilitate later discussions. We invite you to refer to the author's *Elementary Number Theory with Applications* for a detailed study of the various topics and the various proof techniques.

Although we are interested only in integers, there are many instances when we will need to deal with rational and real numbers. Floor and ceiling functions are two useful number-theoretic functions that enable us to transform such numbers into integers. Needless to say, they have many applications to discrete mathematics and computer science.†

Floor and Ceiling Functions

The *floor* of a real number x, denoted by $\lfloor x \rfloor$, is the greatest integer $\leq x$. The *ceiling* of x, denoted by $\lceil x \rceil$, is the least integer $\geq x$. These two notations and the names *floor* and *ceiling* were introduced by Kenneth Eugene Iverson in the early 1960s. Both notations are slight variations of the original greatest integer notation [x].

† See, for example, the author's *Discrete Mathematics with Applications*.

Thus $\lfloor x \rfloor$ is the nearest integer to the left of x on the real line, whereas $\lceil x \rceil$ is the nearest integer to the right of x, as Figure 1.1 shows. The floor of x rounds down x, whereas the ceiling of x rounds it up.

For example, $\lfloor 3.456 \rfloor = 3$, $\lfloor -3.456 \rfloor = -4$, $\lceil 3.456 \rceil = 4$, and $\lceil -3.456 \rceil = -3$.

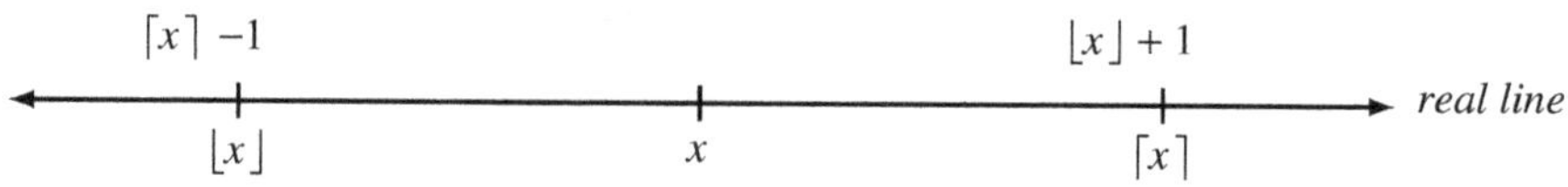

Figure 1.1 The Floor and Ceiling Functions

The floor function $f(x) = \lfloor x \rfloor$ and the ceiling function $g(x) = \lceil x \rceil$ are also known as the *greatest integer function* and the *least integer function*, respectively.

The Internal Revenue Service lets taxpayers have the option of entering actual amounts on their tax returns or using the *whole-dollar method*; but they cannot mix the two methods. In the whole-dollar method, the amount entered for x dollars is given by $\lfloor x + 0.5 \rfloor$.

The following example presents a simple everyday application of the ceiling function.

Example 1.1 (*The post-office function*) In 2010, the US postage rate for a first-class letter of not more than one ounce was 44 cents. The rate for each additional ounce or fraction thereof, up to 11 ounces, was an additional 17 cents. Then the postage $p(x)$, in dollars, of a first-class letter of weight x ounces is given by $p(x) = 0.44 + 0.17\lceil x - 1 \rceil$, where $0 < x \le 11$.

For instance, the postage of a first-class letter weighing 5.7 ounces is $p(5.7) = 0.44 + 0.17\lceil 5.7 - 1 \rceil = \$1.29$. ■

A few properties of the floor and ceiling functions are listed in the following theorem.

Theorem 1.1 Let x be any real number and n any integer. Then:

1) $\lfloor n \rfloor = n = \lceil n \rceil$
2) $\lceil x \rceil = \lfloor x \rfloor + 1$, where x is not an integer.
3) $\lfloor x + n \rfloor = \lfloor x \rfloor + n$
4) $\lceil x + n \rceil = \lceil x \rceil + n$
5) $\left\lfloor \frac{n}{2} \right\rfloor = \frac{n-1}{2}$, where n is odd
6) $\left\lceil \frac{n}{2} \right\rceil = \frac{n+1}{2}$, where n is odd
7) $\left\lfloor \frac{n^2}{4} \right\rfloor = \frac{n^2-1}{4}$, where n is odd
8) $\left\lceil \frac{n^2}{4} \right\rceil = \frac{n^2+3}{4}$, where n is odd
9) $\left\lfloor \frac{n}{2} \right\rfloor + \left\lceil \frac{n}{2} \right\rceil = n$, where n is odd
10) $\lceil x \rceil = -\lfloor -x \rfloor$ ■

Next, we turn to the summation and product notations used throughout this book.

The Summation Notation

Sums, such as $a_k + a_{k+1} + \cdots + a_m$, can be rewritten in a concise way using the *summation symbol* Σ (the upper case Greek letter *sigma*): $\sum_{i=k}^{i=m} a_i = a_k + a_{k+1} + \cdots + a_m$. The summation notation was introduced in 1772 by the French mathematician Joseph Louis Lagrange (1736–1813), for practical purposes.

The variable i is the *summation index*, and the values k and m are the *lower* and *upper limits*, respectively, of the index i. The "$i =$" above the $\sum$ is omitted, when there is no confusion. Thus

$$\sum_{i=k}^{i=m} a_i = \sum_{i=k}^{m} a_i = \sum_{k}^{m} a_i$$

The index i is just a *dummy variable*; its choice has *no* bearing on the sum. For example,

$$\sum_{i=k}^{m} a_i = \sum_{j=k}^{m} a_j = \sum_{t=k}^{m} a_t$$

The following results are very useful in evaluating finite sums. These results can be established easily, using the principle of mathematical induction (PMI).

Theorem 1.2 Let n be any positive integer, c any real number, and $a_1, a_2, \cdots, a_n$ and $b_1, b_2, \cdots, b_n$ any two number sequences. Then:

1) $\sum_{i=1}^{n} c = nc$

2) $\sum_{i=1}^{n} (ca_i) = c\left(\sum_{i=1}^{n} a_i\right)$

3) $\sum_{i=1}^{n}(a_i + b_i) = \sum_{i=1}^{n} a_i + \sum_{i=1}^{n} b_i$

4) $\sum_{i=1}^{n} a_i = \sum_{i=1}^{k} a_i + \sum_{i=k+1}^{n} a_i$

5) $\sum_{i=1}^{n} a_i = \sum_{j=k}^{n+k-1} a_{j-k+1}$

6) $\sum_{i=1}^{n}\left(\sum_{j=1}^{i} a_{ij}\right) = \sum_{j=1}^{n}\left(\sum_{i=j}^{n} a_{ij}\right)$

These results can be extended to any integral lower limit. ■

For example,

$$\sum_{i=-1}^{3}(3i^2 + 4) = 3\left(\sum_{i=-1}^{3} i^2\right) + 4\left(\sum_{i=-1}^{3} 1\right) = 65$$

The following summation formulas will come in handy in our later discussions. They also can be established using PMI.

1) $\sum_{i=1}^{n} i = \frac{n(n+1)}{2}$

2) $\sum_{i=1}^{n} i^2 = \frac{n(n+1)(2n+1)}{6}$

3) $\sum_{i=1}^{n} i^3 = \left[\frac{n(n+1)}{2}\right]^2$

4) $\sum_{i=1}^{n} ar^{i-1} = \frac{a(r^n-1)}{r-1}, \quad r \neq 1$

We now turn to the product notation.

The Product Notation

Using the *product symbol* $\prod$ (the uppercase Greek letter *pi*), the product $a_k a_{k+1} \cdots a_m$ is written as $\prod_{i=k}^{i=m} a_i$. Thus $\prod_{i=k}^{i=m} a_i = a_k a_{k+1} \cdots a_m$. As before, the "$i =$" above the $\prod$ is omitted, when its absence causes no confusion. Thus $\prod_{i=k}^{i=m} a_i = \prod_{i=k}^{m} a_i = \prod_{k}^{m} a_i$. Once again, i is a dummy variable.

For example,

$$\prod_{i=2}^{5}(2i^3 - 1) = (2 \cdot 2^3 - 1)(2 \cdot 3^3 - 1)(2 \cdot 4^3 - 1)(2 \cdot 5^3 - 1) = 25{,}140{,}285$$

We now turn to the division algorithm, which is often used to establish the correctness of a division problem.

The Division Algorithm

Suppose we divide an arbitrary integer a (*dividend*) by a positive integer b (*divisor*). We then get a unique *quotient* q and a unique *remainder* r such that $a = bq + r$, where $0 \leq r < b$. This is formally stated as follows.

Theorem 1.3 (*The Division "Algorithm"*) Given any integer a and a positive integer b, there exist unique integers q and r such that $a = b \cdot q + r$, where $0 \leq r < b$. ■

For example, when $a = 23$ and $b = 5$, $q = 4$ and $r = 3$: $23 = 5 \cdot 4 + 3$. But when $a = -23$, $q = -5$ and $r = 2$: $-23 = 5 \cdot (-5) + 2$. (Remember that the remainder is *never* negative.)

Although this theorem has been historically called an algorithm, it is certainly *not* an algorithm; it does not tell us how to compute the quotient and the remainder when a is divided by b. They are found by the usual long division method.

If $r = 0$, then b is a *factor* of a. We then write $b|a$; otherwise, we write $b \nmid a$. For example, $3|6$, but $3 \nmid 7$.

The Euclidean Algorithm

The Euclidean algorithm is named after the Greek mathematician Euclid (*c.* 330–275 B.C.), who included it in Book VII of his extraordinary work, *The Elements*. The Euclidean algorithm enables us to compute the greatest common divisor (gcd) of any two positive integers a and b; it is denoted by (a, b). Let $a \geq b$. If $a = b$, then $(a, b) = a$; so we assume that $a > b$. The algorithm hinges on the fact $(a, b) = (b, r)$, where r is the remainder when a is divided by b.

Let $r_0 = b$. Then by the successive application of the division algorithm, we have:

$$\begin{aligned} a &= q_0 r_0 + r_1, & 0 &\leq r_1 < r_0 \\ r_0 &= q_1 r_1 + r_2, & 0 &\leq r_2 < r_1 \\ r_1 &= q_2 r_2 + r_3, & 0 &\leq r_3 < r_2 \\ &\vdots \\ r_{n-2} &= q_{n-1} r_{n-1} + \boxed{r_n}, & 0 &\leq r_n < r_{n-1} \\ r_{n-1} &= q_n r_n + 0, \end{aligned}$$

where $r_0 > r_1 > r_2 > \cdots > r_n \geq 0$. Since the remainders are nonnegative, and are getting smaller and smaller, this sequence must terminate with remainder $r_{n+1} = 0$. Then $(a, b) = r_n =$ last non-zero remainder.

For example, let $a = 5076$ and $b = 1976$. Then:

$$\begin{aligned} 5076 &= 2 \cdot 1976 + 1124 \\ 1976 &= 1 \cdot 1124 + 852 \\ 1124 &= 1 \cdot 852 + 272 \\ 852 &= 3 \cdot 272 + 36 \\ 272 &= 7 \cdot 36 + 20 \\ 36 &= 1 \cdot 20 + 16 \\ 20 &= 1 \cdot 16 + \boxed{4} \leftarrow \text{gcd} \end{aligned}$$

So $(5076, 1976) = 4$.

By reversing the steps in the algorithm, it follows that the gcd (a, b) can be expressed as a *linear combination* of a and b. For example, $4 = 109 \cdot 5076 - 280 \cdot 1976$. In general, $4 = 5076(109 - 1976t) + 1976(-280 + 5076t)$, where t is an arbitrary integer.

More generally, the gcd of the positive integers $a_1, a_2, \cdots, a_n$ is the least positive integer that is a linear combination of $a_1, a_2, \cdots, a_n$.

Base-b Expansions

Using the division algorithm, we can successfully transform a decimal integer to any other base $b \geq 2$. This is a consequence of the following result.

Theorem 1.4 Let b be any integer ≥ 2. Then every positive integer N can be expressed uniquely in the form $N = n_k b^k + n_{k-1} b^{k-1} + \cdots + n_1 b + n_0$, where $n_0, n_1, \cdots, n_k$ are nonnegative integers less than b, $n_k \neq 0$, and $k \geq 0$. ■

The expression $N = n_k b^k + n_{k-1} b^{k-1} + \cdots + n_1 b + n_0$ is the *base-b expansion* (or *representation*) of the integer N. We often rewrite this as $N = (n_k n_{k-1} \cdots n_1 n_0)_b$. The powers b^k, b^{k-1}, $\cdots$, b, and 1 are the *place values* of the base-b digits $n_k, n_{k-1}, \cdots, n_1$, and n_0, respectively. Thus:

$$\begin{aligned} N &= (n_k n_{k-1} \cdots n_1 n_0)_b \\ &= n_k b^k + n_{k-1} b^{k-1} + \cdots + n_1 b + n_0 \end{aligned}$$

If the base is not explicitly given, it is assumed to be ten. For example, $234 = 234_{\text{ten}}$.

When $b = 2$, the expansion is called the *binary expansion*. Then each coefficient a_i is either 0 or 1; they are called *b*inary dig*its* (or *bits*). For example, $10011_{\text{two}} = 1(2^4) + 0(2^3) + 0(2^2) + 1(2) + 1(2^0) = 19$.

Although converting a non-decimal integer to base ten can be accomplished fairly easily and quickly, conversion in the reverse direction is a bit involved. By Theorem 1.4, all we need to do is express the decimal integer as a sum of multiples of powers of the base b, beginning with the largest power of b, then collect the various coefficients in the correct order; always remember to account for all missing powers of b.

For example, $49 = 1(2^5) + 1(2^4) + 0(2^3) + 0(2^2) + 0(2^1) + 1(2^0) = 110001_{\text{two}}$.

Fortunately, there is an elegant algorithm for such a conversion, which is also based on the division algorithm: divide the decimal integer and its successive quotients by b and then pick the remainders in the reverse order.

The following example illustrates this algorithm.

Example 1.2 Express 49 in base two.

Solution Using the division algorithm, we have:

$$\begin{aligned} 49 &= 24 \cdot 2 + \boxed{1} \\ 24 &= 12 \cdot 2 + \boxed{0} \\ 12 &= 6 \cdot 2 + \boxed{0} \\ 6 &= 3 \cdot 2 + \boxed{0} \\ 3 &= 1 \cdot 2 + \boxed{1} \\ 1 &= 0 \cdot 2 + \boxed{1} \end{aligned} \quad \uparrow \text{ read up}$$

Reading up the remainders in the reverse order, we get $49 = 110001_{\text{two}}$. ■

As this example demonstrates, the algorithm terminates when we arrive at a zero quotient.

Next, we pursue two important and closely linked classes of positive integers: Mersenne primes and even perfect numbers.

Mersenne Primes

Numbers of the form $2^m - 1$ were investigated extensively by the French mathematician and Franciscan monk Marin Mersenne (1588–1648). Such numbers are called *Mersenne numbers*. Because they play a crucial role in cryptology, a significant amount of energy and time have been spent finding larger and larger Mersenne primes. Today, the largest known (Mersenne) prime is $M_p = M_{43112609} = 2^{43112609} - 1$, discovered in 2008 by a team of mathematicians at the University of California, Los Angeles; its decimal value is about 13 million digits long.

The infinitude of Mersenne primes has not yet been proved. Establishing it continues to be one of the many outstanding problems in number theory.

Interestingly, Mersenne primes are closely related to the class of *perfect numbers*. The term *perfect number* was coined by the *Pythagoreans*, the followers of the Greek mathematician and philosopher Pythagoras (*c*. 572–500 B.C.). They observed that 6 equals the sum of its proper factors: $6 = 1 + 2 + 3$; 6 is a perfect number. The next two perfect numbers are 28 and 496. The discovery of perfect numbers is sometimes credited to the Greek mathematician Nichomachus (*c*. 100 A.D.).

We now formally define a perfect number.

Perfect Numbers

A positive integer n is a *perfect number* if the sum of its proper factors equals n; that is, if the sum of its positive factors equals $2n$.

The first six perfect numbers are 6, 28, 496, 8128, 33550336, and 8589869056. The first four of these were known to the ancient Greeks. A quick study reveals two interesting properties: they are all even, and are of the form $2^{p-1}(2^p - 1)$, where both p and $M_p = 2^p - 1$ are prime. Euclid established that every such number is indeed a perfect number, as the next theorem confirms.

Theorem 1.5 (Euclid) If $2^p - 1$ is a prime, then $N = 2^{p-1}(2^p - 1)$ is a perfect number. ■

In 1757, about two millennia after Euclid made this discovery, the greatest Swiss mathematician, Leonhard Euler (1707–1783), proved that the converse of Euclid's theorem is also true, as the following theorem shows.

Theorem 1.6 (Euler) Every even perfect number is of the form $2^{p-1}(2^p - 1)$, where p and $M_p = 2^p - 1$ are prime. ■

Theorems 1.5 and 1.6 together characterize even perfect numbers. Since every even perfect number is of the form $2^{p-1}M_p$, it follows that the number of Mersenne primes equals that of even perfect numbers.

We do *not* yet know if there are odd perfect numbers. It has been shown, however, that an odd perfect number, if it exists, must be very large. In fact, there is serious doubt in the mathematical community that an odd perfect number even exists. So, stay tuned.

Next, we present the congruence relation.

Congruences

One of the most astounding and powerful relations in mathematics, especially in number theory, is the congruence relation. It was developed by the greatest German mathematician, Karl Friedrich Gauss (1777–1855), popularly known as the "prince of mathematics." Gauss presented the theory of congruences in his outstanding work, *Disquistiones Arithmeticae*, published in 1801, when he was only 24 years old.

We begin with a definition of the congruence relation.

Congruence Modulo *m*

Let m be an integer ≥ 2. Then an integer a is *congruent to* an integer b *modulo m* if $m \mid (a-b)$. Symbolically, we then write $a \equiv b \pmod{m}$; m is the *modulus* of the congruence relation. If a is not congruent to b modulo m, then we write $a \not\equiv b \pmod{m}$.

For example, since $5 \mid (43-23)$, $43 \equiv 23 \pmod{5}$; likewise, $33 \equiv 5 \pmod{7}$. But $6 \nmid (37-4)$; so $37 \not\equiv 4 \pmod{6}$.

The next theorem summarizes some fundamental properties of the congruence relation. We omit their proofs in the interest of brevity.

Theorem 1.7 Let a and b be arbitrary integers and m any integer ≥ 2. Then:

- $a \equiv b \pmod{m}$ if and only if $a = b + km$ for some integer k.
- $a \equiv a \pmod{m}$ (*Reflexive property*).
- If $a \equiv b \pmod{m}$, then $b \equiv a \pmod{m}$ (*Symmetric property*).
- If $a \equiv b \pmod{m}$ and $b \equiv c \pmod{m}$, then $a \equiv c \pmod{m}$ (*Transitive property*).
- $a \equiv b \pmod{m}$ if and only if a and b leave the same remainder when divided by m.
- r is the remainder when a is divided by m if and only if $a \equiv r \pmod{m}$, where $0 \leq r < m$.
- If $a \equiv b \pmod{m}$ and $c \equiv d \pmod{m}$, then $a + c \equiv b + d \pmod{m}$ and $ac \equiv bd \pmod{m}$. ■

Next, we present two celebrated results from the theory of congruences: Wilson's theorem and Fermat's little theorem.

Theorem 1.8 (*Wilson's theorem*) If p is a prime, then $(p-1)! \equiv -1 \pmod{p}$. ■

For example, let $p = 11$. Then $10! = 3628800 \equiv -1 \pmod{11}$.

Although John Wilson (1741–1793) had conjectured this result, he never provided a proof. The first proof was furnished by Lagrange. In fact, the German mathematician Gottfried Leibniz (1646–1716) knew the result as early as 1682, although he did not publish it.

Next, we present Fermat's little theorem.

Fermat's Little Theorem

In 1640, the French mathematician Pierre de Fermat (1601–1665) conjectured the following result: *If p is a prime and $p \nmid a$, then $p \mid \left(a^{p-1} - 1\right)$*. Although Fermat never provided a proof of this result, it has come to be known as *Fermat's little theorem*.

Theorem 1.9 (*Fermat's Little Theorem*) Let p be a prime and a any integer such that $p \nmid a$. Then $a^{p-1} \equiv 1 \pmod{p}$. ■

For example, $13^4 \equiv 3^4 = 81 \equiv 1 \pmod 5$, as expected.

Fermat's little theorem can in fact be extended to all positive integers a, as the next theorem shows. Because its proof is short, we will include it.

Theorem 1.10 Let p be a prime and a any positive integer. Then $a^p \equiv a \pmod{p}$.

Proof: If $p \nmid a$, then, by Fermat's little theorem, $a^{p-1} \equiv 1 \pmod{p}$. So $a^p \equiv a \pmod{p}$.

If $p \mid a$, then $a \equiv 0 \pmod{p}$. So $a^p \equiv 0 \pmod{p}$ and $0 \equiv a \pmod{p}$. Then $a^p \equiv a \pmod{p}$, by transitivity. ■

This theorem has an interesting byproduct, as the next corollary† shows.

Corollary 1.1 Let p be a prime, and a and b be arbitrary integers. Then $(a+b)^p \equiv a^p + b^p \pmod{p}$.

Proof: By Theorem 1.10, $(a+b)^p \equiv a+b \equiv a^p + b^p \pmod{p}$. ■

†A *corollary* is a result that follows from a theorem as a byproduct.

More generally, $(a+b)^{p^n} \equiv a^{p^n} + b^{p^n} \pmod{p}$ for every positive integer n. This can be confirmed using PMI.

Next, we turn to one of the most elegant and useful problem-solving techniques: recursion. It is so powerful that it is used very often in discrete mathematics and computer science.

Recursion

Suppose we would like to solve a problem and it can be defined in terms of a simpler version of itself. So the given problem can be solved if the simpler version can be solved. Such a definition is an *inductive definition*.

An inductive definition of a function f consists of three steps:

1. The *basis clause* defines a few initial values $f(a), f(a+1), f(a+2), \cdots,$ $f(a+k-1)$. Equations which define such initial values are *initial conditions*.
2. The *recursive clause* provides a formula to compute $f(n)$ from preceding functional values. Such a formula is a *recurrence relation* (or a *recursive formula*).
3. The *terminal clause* guarantees that only the values obtained by steps 1 and 2 are valid values of the function. (For convenience, the terminal clause is often omitted.)

In general, in a *recursive definition* of f, $f(n)$ may be defined using the values $f(k)$, where $k \neq n$, so *not* all recursive definitions are inductive.

Thus, the recursive definition of f consists of a finite number of initial conditions and a recurrence relation.

The next four examples illustrate the recursive definition.

Example 1.3 Consider the *factorial function* $f(n) = n!$, where $f(0) = 1$ and $n \geq 0$. When $n \geq 1$, $n! = n(n-1)!$; so $f(n) = n \cdot f(n-1)$. Thus the factorial function can be defined recursively as follows:

$f(0) = 1$ $\leftarrow$ initial condition

$f(n) = n \cdot f(n-1),\ n \geq 1$ $\leftarrow$ recurrence relation ■

Next, we present the well-known handshake problem.

Example 1.4 (*The handshake problem*) Suppose there are n guests at a party, where $n \geq 1$. Each person shakes hands with everybody else exactly once. Let $h(n)$ denote the number of handshakes made by the guests. Define $h(n)$ recursively.

Solution Clearly, $h(1) = 0$. So let $n \geq 2$. Let X be one of the guests. Then there are $n - 1$ guests remaining; by definition, they make $h(n - 1)$ handshakes. Now X can shake hands with each of the $n - 1$ guests; they yield $n - 1$ new handshakes. So the total number of handshakes made equals $h(n - 1) + (n - 1)$.

Thus $h(n)$ can be defined recursively as follows:

$$h(1) = 0 \qquad \leftarrow \text{initial condition}$$
$$h(n) = h(n - 1) + (n - 1), n \geq 2 \qquad \leftarrow \text{recurrence relation}$$ ■

Next, we turn to the pizza problem. It is an interesting close relative of the handshake problem. We will revisit it in Chapter 6.

Example 1.5 (*The pizza problem*) Find the maximum number of pieces f_n a pizza can be cut into with n cuts, where $n \geq 0$.

Solution Clearly, $f_0 = 1$. With one cut c_1, we get two pieces; so $f_1 = 2$ (see Figure 1.2). A second cut c_2 crosses the previous cut c_1 at exactly one point; so it cuts each piece into two, adding two more pieces (see Figure 1.3). Thus $f_2 = f_1 + 2 = 4$. Now we make a third cut c_3. It cuts the previous cuts c_1 and c_2 in two points; in other words, c_3 divides c_1 and c_2 into three parts. Each region divides an existing piece into two, yielding three new pieces (see Figure 1.4); so $f_3 = f_2 + 3 = 7$.

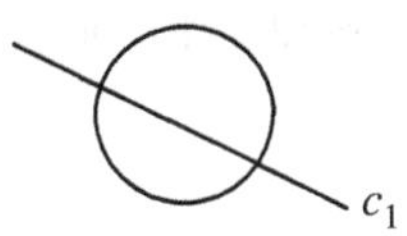

Figure 1.2 One Cut

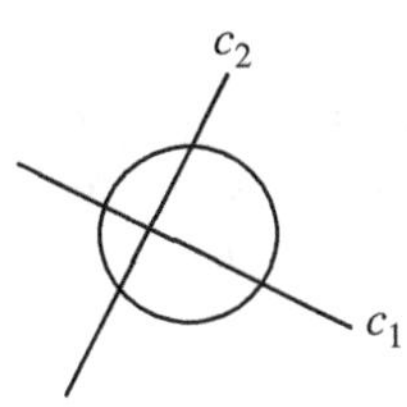

Figure 1.3 Two Cuts

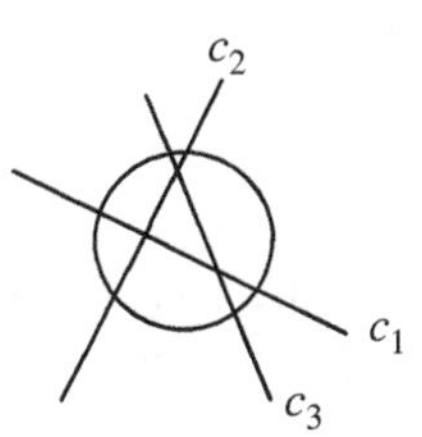

Figure 1.4 Three Cuts

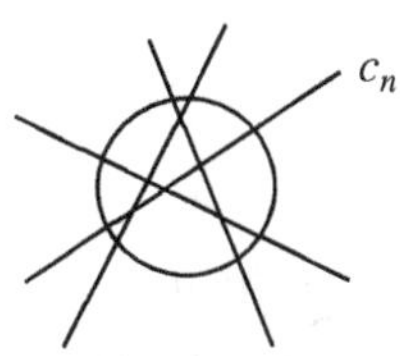

Figure 1.5 n Cuts

More generally, suppose cuts $c_1, c_2, \cdots, c_{n-1}$ divide the pizza into a maximum of f_{n-1} pieces. Now make one more cut c_n (see Figure 1.5). Since we would like to maximize the number of pieces, c_n must cross the cuts $c_1, c_2, \cdots, c_{n-1}$ at new points and must be divided by them into n segments. Each segment divides an existing piece into two smaller pieces, contributing n more pieces, so $f_n = f_{n-1} + n$. Thus f_n can be defined recursively as follows:

$$f_0 = 1$$
$$f_n = f_{n-1} + n, \quad n \geq 1$$

Using iteration, an explicit formula for f_n can be conjectured:

$$\begin{aligned} f_n &= f_{n-1} + n \\ &= f_{n-2} + (n-1) + n \\ &\vdots \\ &= f_0 + 1 + 2 + \cdots + n \\ &= 1 + 1 + 2 + \cdots + n \\ &= 1 + \frac{n(n+1)}{2} \end{aligned}$$

This formula can be confirmed using induction. ■

The following example, also pursued in Chapter 6, derives a formula for the maximum number of non-overlapping regions formed inside a circle by joining n distinct points on it. It was proposed around 1950 by L. Moser of the University of Alberta, Canada. The solution is based on the one given in 1951 by A. V. Boyd and M. J. Glencross of the University of the Witwatersrand, Johannesburg, South Africa.

Example 1.6 (*The circle problem*) Let g_n denote the maximum number of non-overlapping regions formed in the interior of a circle by the chords joining n distinct points on it. Define g_n recursively.

Solution It follows from Figure 1.6 that $g_1 = 1, g_2 = 2, g_3 = 4, g_4 = 8$, and $g_5 = 16$. So we are tempted to conjecture that $g_n = 2^n$ for every $n \geq 1$. Unfortunately, the pattern takes a detour when $n = 6$; it can be shown that $g_6 = 31$, *not* 32.

More generally, let $P_1, P_2, \cdots, P_n$ be n points on a circle such that no three chords joining them meet inside the circle; see Figure 1.7. Now choose a new point X on arc $\overset{\frown}{P_{n-1}P_n}$, and draw every chord $\overline{XP_r}$, where $1 \leq r \leq n$; $\overline{XP_r}$ should not pass through the point of intersection of any two chords. The chord $\overline{XP_r}$ introduces a new region inside the circle.

Now consider the chord $\overline{XP_r}$, where $1 \leq r \leq n-1$. It intersects each of the chords joining one of the r points $P_n, P_1, \cdots, P_{r-1}$ to one of the $n-1-r$ points $P_{r+1}, P_{r+2}, \cdots, P_{n-1}$. There are $r(n-1-r)$ such intersections. Since each intersection corresponds to a new region, they yield $r(n-1-r)$ new regions.

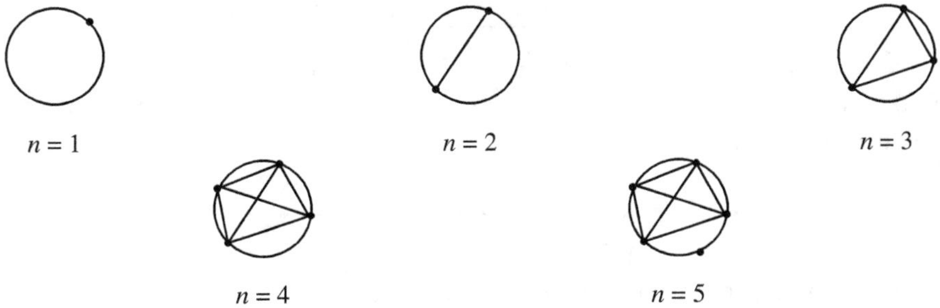

Figure 1.6

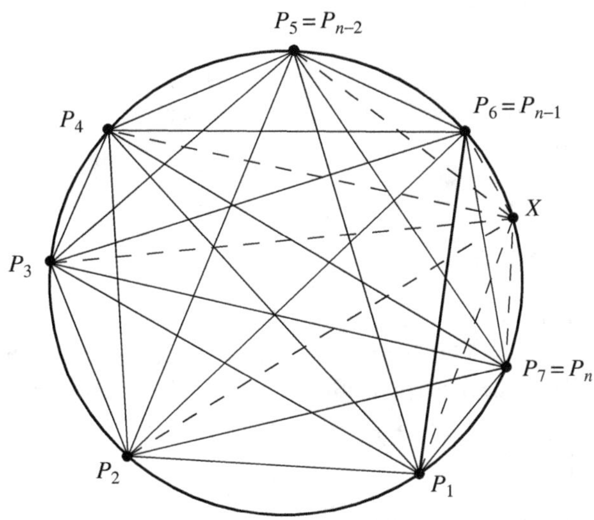

Figure 1.7

So $\overline{XP_r}$ produces a total of $1 + r(n - 1 - r)$ new regions in the interior, where $1 \leq r \leq n - 1$.

So the number of new regions formed by the selection of the $(n + 1)$st point X equals:

$$\begin{aligned}1 + \sum_{r=1}^{n-1} [1 + r(n - 1 - r)] &= \sum_{r=1}^{n} 1 + (n - 1)\sum_{r=1}^{n-1} r - \sum_{r=1}^{n-1} r^2 \\ &= n + (n - 1) \cdot \frac{(n - 1)n}{2} - \frac{(n - 1)n(2n - 1)}{6} \\ &= n + \frac{n(n - 1)}{2}\left[(n - 1) - \frac{(2n - 1)}{3}\right]\end{aligned}$$

$$= n + \frac{n(n-1)(n-2)}{6}$$

But, by definition, the number of new regions equals $g_{n+1} - g_n$. Thus, we have the recurrence relation

$$g_{n+1} - g_n = n + \frac{n(n-1)(n-2)}{6}$$

Consequently, g_n can be defined recursively as follows:

$$g_1 = 1$$

$$g_n = g_{n-1} + n - 1 + \frac{(n-1)(n-2)(n-3)}{6}, \quad n \geq 2 \qquad \blacksquare$$

For example, $g_6 = g_5 + 5 + \frac{5 \cdot 4 \cdot 3}{6} = 16 + 5 + 10 = 31$.
Next, we pursue two classes of figurate numbers, triangular numbers and tetrahedral numbers.

Figurate Numbers

Figurate numbers are positive integers that can be represented by geometric patterns. They are thought to have been discovered by the Pythagoreans. In any case, in 1665, the French mathematician and philosopher Blaise Pascal (1623–1662) wrote a book on them, *Treatise on Figurate Numbers*.

Triangular Numbers

If you have been to a bowling alley, then you know that there are 10 pins in bowling and initially they are placed in a triangular shape. Likewise, the 15 balls in the game of pool are also initially arranged in a triangular array. Both numbers, 10 and 15, are triangular numbers.

More generally, a *triangular number* is a positive integer that can be represented by a triangular shape. The nth triangular number is denoted by t_n, where $n \geq 1$. The first four triangular numbers are $t_1 = 1, t_2 = 3, t_3 = 6$, and $t_4 = 10$. Figure 1.8 represents them pictorially.

Figure 1.8 The First Four Triangular Numbers

Since the ith row in such a triangular array contains i dots, it follows that

$$t_n = \sum_{i=1}^{n} i = \frac{n(n+1)}{2}$$

For example, $t_{36} = \frac{36 \cdot 37}{2} = 666$, and $t_{49} = \frac{49 \cdot 50}{2} = 1{,}225$.

Since row n in the triangular array, except the first, contains n dots more than the previous row, it follows that t_n can be defined recursively:

$$t_1 = 1$$

$$t_n = t_{n-1} + n, \quad n \geq 2$$

For example, $t_{10} = t_9 + 10 = 45 + 10 = 55$.

Next, we take a look at the well-known carol *The Twelve Days of Christmas*, and see how it is related to triangular numbers.

Twelve Days of Christmas

On the first day of Christmas, my true love gave me a partridge in a pear tree. On the second day of Christmas, my true love gave me two turtle doves and a partridge in a pear tree. On the third day of Christmas, my true love sent me three French hens, two turtle doves, and a partridge in a pear tree. This pattern continues until the twelfth day.

This carol raises two interesting questions:

- Suppose the given pattern continues for n days. Find the number gifts g_n given on the nth day.
- Find the total number of gifts s_n given in n days.

Clearly, the number of gifts g_n given on the nth day is n more than g_{n-1}; so $g_n = g_{n-1} + n$, where $g_1 = 1$. Consequently,

$$g_n = \sum_{i=1}^{n} i = t_n = \frac{n(n+1)}{2}$$

Thus:

$$\begin{aligned} s_n &= \sum_{i=1}^{n} t_i = \sum_{i=1}^{n} \frac{i(i+1)}{2} \\ &= \frac{1}{2}\left(\sum_{i=1}^{n} i^2 + \sum_{i=1}^{n} i\right) = \frac{1}{2}\left[\frac{n(n+1)(2n+1)}{6} + \frac{n(n+1)}{2}\right] \\ &= \frac{n(n+1)}{12}[(2n+1)+3] = \frac{n(n+1)(n+2)}{6} \end{aligned}$$

Tetrahedral Numbers

Next, we pursue *tetrahedral numbers* T_n; they are positive integers that can be represented by tetrahedrons; they are obtained by taking successive sums of triangular numbers, as Figure 1.9 shows.

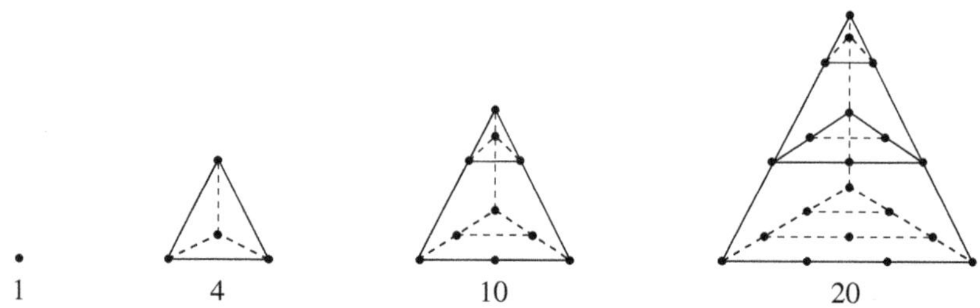

Figure 1.9 The First Four Tetrahedral Numbers

The first four tetrahedral numbers are:

$$T_1 = t_1 = 1 \qquad T_3 = t_1 + t_2 + t_3 = 1 + 3 + 6 = 10$$
$$T_2 = t_1 + t_2 = 1 + 3 = 4 \qquad T_4 = t_1 + t_2 + t_3 + t_4 = 1 + 3 + 6 + 10 = 20$$

It now follows that T_n can be defined recursively:

$$T_1 = 1$$
$$T_n = T_{n-1} + t_n, \quad n \geq 2$$

Since $T_n = T_{n-1} + t_n$, each tetrahedral number T_n can be obtained by adding the corresponding triangular number t_n to the previous tetrahedral number T_{n-1}. For instance, $T_5 = T_4 + t_5 = 20 + 15 = 35$ (see Table 1.1).

Table 1.1 Tetrahedral Numbers

n	1	2	3	4	5	6	...	n
t_n	1	3	6	10	15	21	...	$n(n+1)/2$
T_n	1	4	10	20	35	?	...	?

Since $T_n = \sum_{i=1}^{n} t_i$, it follows from the previous discussion that $T_n = \frac{n(n+1)(n+2)}{6}$. We will revisit both triangular and tetrahedral numbers in Chapter 6.

Next, we present a brief introduction to generating functions, which we will use in later chapters.

Generating Functions[†]

Generating functions are an important tool for solving recurrence relations and combinatorial problems. They were invented in 1718 by the French mathematician Abraham De Moivre (1667–1754).

Let a_i be any real numbers, where $i \geq 0$. Then the function

$$g(x) = a_0 + a_1x + a_2x^2 + \cdots + a_nx^n + \cdots \tag{1.1}$$

is the *generating function* for the sequence $\{a_n\}$; $g(x) = a_0 + a_1x + a_2x^2 + \cdots + a_nx^n$ is the generating function for the finite sequence $a_0, a_1, a_2, \cdots, a_n$.

For example,

$$f(x) = 1 + 3x + 6x^2 + \cdots + \frac{n(n+1)}{2}x^2 + \cdots$$

is the generating function for triangular numbers.

We emphasize that the RHS[*] of equation (1.1) is a *formal power series* in x. The quantity x is a dummy variable and the various powers x^n of x are just place-holders. Furthermore, we are *not* interested in the convergence of the series.

Next, we define the equality of generating functions in the same way you would expect.

Equality of Generating Functions

Consider the generating functions $f(x) = \sum_{n=0}^{\infty} a_nx^n$ and $g(x) = \sum_{n=0}^{\infty} b_nx^n$. Then $f(x) = g(x)$ if and only if $a_n = b_n$ for every $n \geq 0$.

A Frequently Used Generating Function

A frequently used generating function is

$$\frac{1}{1 - ax} = 1 + ax + a^2x^2 + \cdots + a^nx^n + \cdots$$

It can be verified by the long division method or cross-multiplication.

Generating functions (of the same variable x) are added and multiplied in exactly the same way that two polynomials (in x) are combined.

[†]For a detailed discussion, see I. Niven. Formal Power Series, *American Mathematical Monthly*, 76 (1969), 871–889.

[*]RHS and LHS are abbreviations of *right-hand side* and *left-hand side*, respectively.

Addition and Multiplication of Generating Functions

Let $f(x) = \sum_{n=0}^{\infty} a_n x^n$ and $g(x) = \sum_{n=0}^{\infty} b_n x^n$ be any two generating functions. Then:

$$f(x) + g(x) = \sum_{n=0}^{\infty} (a_n + b_n) x^n$$

$$f(x)g(x) = \sum_{n=0}^{\infty} \left(\sum_{r=0}^{n} a_r b_{n-r} \right) x^n$$

For example:

$$\frac{1}{(1-ax)^2} = \frac{1}{1-ax} \cdot \frac{1}{1-ax} = \left(\sum_{n=0}^{\infty} (ax)^n \right) \left(\sum_{n=0}^{\infty} (ax)^n \right)$$

$$= \sum_{n=0}^{\infty} \left(\sum_{r=0}^{n} 1 \cdot 1 \right) (ax)^n = \sum_{n=0}^{\infty} (n+1)(ax)^n$$

$$= 1 + 2ax + 3a^2x^2 + \cdots + (n+1)a^n x^n + \cdots$$

Similarly,

$$\frac{1}{(1-ax)^3} = \frac{1}{1-ax} \cdot \frac{1}{(1-ax)^2}$$

$$= 1 + 3ax + 6a^2x^2 + 10a^3x^3 + \cdots$$

In particular, we have:

$$\frac{1}{(1-x)^2} = 1 + 2x + 3x^2 + 4x^3 + \cdots$$

$$\frac{1}{(1-x)^3} = 1 + 3x + 6x^2 + 10x^3 + \cdots$$

Two Fundamental Counting Principles

We will often use the familiar addition and multiplication principles from combinatorics, where $|X|$ denotes the number of ways task X can be done.

Addition principle: Let A and B be any two mutually exclusive tasks. Then task A or B can take place in $|A| + |B|$ ways.

Multiplication principle: Let A and B be any two tasks. Then these two tasks can be done in $|A| \cdot |B|$ ways in that order.

Obviously, these two fundamental counting principles can be extended to any finite number of tasks.

Next, we present a brief introduction to permutations and combinations in combinatorics. The roots of combinatorics can be found in the ancient Indian writer Pingala's work on Vedic meters, *Chandas Shastra* (200 B.C.?).

Permutations and Combinations

In combinatorics we often deal with arrangements of objects. In fact, there are two kinds of an arrangement. An order where order matters is a *permutation*; otherwise, it is a *combination*. For example, every English word is a permutation (an ordered set), whereas every ordinary set (or committee) is a combination.

Suppose we would like to find the number $P(n, r)$ of ways n distinct objects can be distributed into r slots, where $n \geq r$. There are n choices for the first slot, $n-1$ choices for the second slot, and so on; there are $n-r+1$ choices for the rth slot:

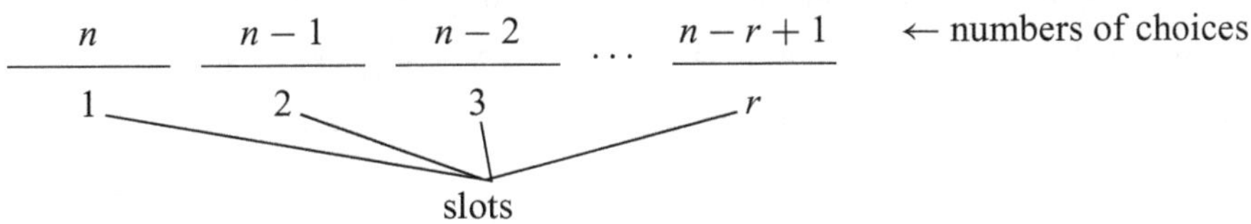

So, by the multiplication principle, $P(n, r) = n(n-1)\cdots(n-r+1) = \frac{n!}{(n-r)!}$. In particular, $P(n, n) = n!$; that is, there are $n!$ permutations of n distinct items. This was known in ancient India.

Using the multiplication principle, it now follows that there are $\frac{P(n,r)}{r!} = \frac{n!}{r!(n-r)!}$ combinations of n distinct items taken r at a time. For example, there are $\frac{5!}{3!(5-3)!} =$ 10 three-member subcommittees of a five-member committee.

The fact that there are $\frac{n!}{r!(n-r)!}$ combinations of n distinct things taken r at a time appears in Mahavira's *Ganita Sara Samgraha* and Bhaskara's *Leelavati*.

The next example is an interesting application of the *telescoping sum* $\sum_{i=1}^{n}(a_i - a_{i-1}) = a_n - a_0$.

Example 1.7 Show that $\sum_{k=1}^{n-1} k \cdot k! = n! - 1$.

Proof: Since $k \cdot k! = [(k+1) - 1]k! = (k+1)! - k!$, it follows that

$$\sum_{k=1}^{n-1} k \cdot k! = \sum_{k=1}^{n-1}[(k+1)! - k!] = [(n-1)+1]! - 1! = n! - 1 \qquad \blacksquare$$

Next, we present a delightful combinatorial interpretation of this summation formula.

A Combinatorial Interpretation

There are $n!$ permutations of the integers 1 through n. One of them is the natural permutation $123\cdots n$, where the integers occur in ascending order. Excluding this natural order, there are $n! - 1$ permutations left.

We will now count such permutations in a different way. Let $1 \leq k \leq n-1$. Consider the permutations in which the first $n-k-1$ integers occur in the natural order, but $n-k$ does *not* occur in position $n-k$; so the integers 1 through $n-k-1$ are followed by one of the integers $n-k+1, n-k+2, \cdots$, or n:

$$\underbrace{123\cdots(n-k-1)}_{(n-k-1)\text{ positions}} \quad \underbrace{}_{\substack{\uparrow \\ k \text{ choices}}} \quad \underbrace{}_{k \text{ positions}}$$

Since the $(n-k)$th position has k choices and the remaining k positions can be filled in $k!$ ways, there are $k \cdot k!$ such permutations. So, by the addition principle, the total number of such permutations also equals $\sum_{k=1}^{n-1} k \cdot k!$.

Thus

$$\sum_{k=1}^{n-1} k \cdot k! = n! - 1. \qquad \blacksquare$$

For example, let $n = 4$. Table 1.2 lists all $4! - 1 = 23$ permutations, where $4 - k$ is the first out-of-order integer and $1 \leq k \leq 3$.

Table 1.2

k	$4-k$	Desired permutations
1	3	1243
2	2	1324, 1342, 1423, 1432
3	1	2134, 2143, 2314, 2341, 2413, 2431
		3124, 3142, 3214, 3241, 3412, 3421
		4123, 4132, 4213, 4231, 4312, 4321

Finally, we present an intriguing occurrence of the combinatorial sum $(n+1)! - 1$. It appeared as a Leningrad Mathematical Olympiad problem in 1990.[‡]

[‡]D. Fomin and A. Kirichenko, *Leningrad Mathematical Olympiads 1987–1991*, MathPro Press, Mansfield, Ohio, 1994; R. Honsberger, *Mathematical Delights*, Math. Assoc. of America, Washington, D.C., 2004.

Example 1.8 Let F_n be the family of all nonempty subsets of the set $S_n = \{1, 2, 3, \cdots, n\}$, which do not contain consecutive integers. Let P_n be the set of products of elements in each subset. (Multiplication takes place only if a subset contains two or more elements.) Prove that the sum T_n of the squares of the elements in P_n equals $(n+1)! - 1$.

For example, $F_4 = \{\{1\}, \{2\}, \{3\}, \{4\}, \{1, 3\}, \{1, 4\}, \{2, 4\}\}$; $P_4 = \{1, 2, 3, 4, 3, 4, 8\}$; and $T_4 = 1 + 4 + 9 + 16 + 9 + 16 + 64 = 119 = 5! - 1$.

Proof (*by strong induction*): When $n = 1$, $S_1 = \{1\}$, $F_1 = \{\{1\}\}$, $P_1 = \{1\}$; so $T_1 = 1 = 2! - 1$. Thus, the formula works when $n = 1$. Similarly, it works when $n = 2$ also.

Now, assume it is true for all positive integers $i \le k - 1$, where $k \ge 3$: $T_i = (i+1)! - 1$. Let A be an acceptable subset of S_k.

Case 1 Suppose $k \notin A$. Then A is an acceptable subset of S_{k-1}. By our inductive hypothesis, it contributes a total of $k! - 1$ to the sum T_k.

Case 2 Suppose $k \in A$. Since A cannot contain consecutive integers, $k - 1 \notin A$. So A can be $\{k\}$ or an acceptable subset of S_{k-2} with k added to it.

Subcase 1 Suppose $A = \{k\}$. It contributes k^2 to the sum T_k.

Subcase 2 Suppose A is an acceptable subset of S_{k-2} with k added to it. By the inductive hypothesis, S_{k-2} contributes a total of $(k-1)! - 1$ to the sum of the squares; but, inserting k in each such acceptable subset, multiplying the elements, and squaring and then adding them increase this total by a factor of k^2; so the total now is $k^2[(k-1)! - 1] = k \cdot k! - k^2$.

Thus, by the addition principle, if $k \in A$, the total contribution to T_k equals $k^2 + (k \cdot k! - k^2) = k \cdot k!$.

Consequently, by cases 1 and 2, $T_k = (k! - 1) + k \cdot k! = (k+1)! - 1$. So the formula works when $n = k$ also.

Thus, by induction, the formula works for every positive integer. ■

2

Binomial Coefficients

Binomials are sums of two terms, and they occur often in mathematics. The binomial theorem provides a systematic way of expanding positive integral powers of binomials and is studied in Chapter 3. So we begin with a brief discussion of *binomial coefficients*, which are coefficients that occur in binomial expansions; they play a pivotal role in all our discussions to follow. The term *binomial coefficient* was introduced by the greatest German algebraist of the sixteenth century, Michel Stifel (1486–1567).

Binomial Coefficients

Let n and r be nonnegative integers. The *binomial coefficient* $\binom{n}{r}$ is defined as

$$\binom{n}{r} = \frac{n!}{r!(n-r)!} = \frac{n(n-1)\cdots(n-r+1)}{r!}$$

where $0 \le r \le n$. If $r > n$, then $\binom{n}{r}$ is defined as 0. Read $\binom{n}{r}$ as "n choose r" to be consistent with the *Latex* language. The bi-level parenthesized notation for binomial coefficient was introduced by the German mathematician and physicist Andreas von Ettinghausen (1796–1878) in his book *Die Combinatorische Analysis* (1826).

For example,

$$\binom{10}{3} = \frac{10!}{3!(10-3)!} = \frac{10!}{3!7!} = 120 \text{ and } \binom{3}{5} = 0$$

The earliest known occurrence of binomial coefficients can be found in a tenth-century commentary by the Indian mathematician Halayudha on Pingala's

Chandas Shastra; it also contains a discussion of Pascal's triangle (called *Meru Prastara*), studied in Chapter 6. Bhaskara (1114–1185?) gave a clear discourse on binomial coefficients in his classic work *Leelavati*, written in 1150.

It now follows from Chapter 1 that $\binom{n}{r}$ denotes the number of r-member subcommittees of an n-member committee. For example, we can draw exactly $\binom{10}{2} = 45$ distinct lines through 10 points in a plane, and $\binom{5}{3} = 10$ distinct triangles using five points in a plane, where no three points are collinear in both cases.

It now follows that an n-member committee has a total of $\sum_{r=0}^{n} \binom{n}{r}$ subcommittees; exactly, $\sum_{r=k}^{n} \binom{n}{r}$ of them have at least k members.

Next, we present an interesting geometric interpretation of the binomial coefficient $\binom{n}{r}$.

A Geometric Interpretation of $\binom{n}{r}$

Consider a city divided into blocks by avenues running east–west and streets running north–south. Each intersection (x, y) is a *lattice point*, meaning both x and y are integers. Suppose we would like to find the number of distinct paths from the origin O to the lattice point $A\,(n-r, r)$, where $n \geq r$. (Such a path is called a *lattice path*.) Figure 2.1 shows a lattice path from O to the point (5, 3); it can be represented by the permutation (or *word*) EENEENNE consisting of 5 Es and 3 Ns.

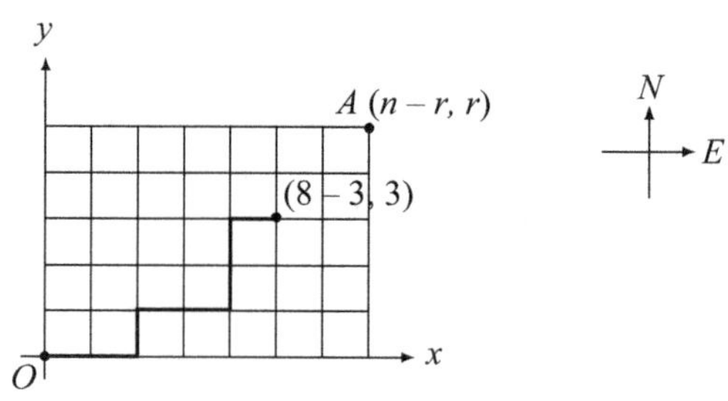

Figure 2.1 Lattice Path EENEENNE

More generally, every lattice path from O to A is a permutation of the $(n-r)+r = n$ letters of which $n-r$ letters are of one kind (E) and r letters of a second kind (N), and vice versa. The number of such permutations equals $\frac{n!}{(n-r)!r!} = \binom{n}{r}$ (see Theorem 5.3). Thus, there are exactly $\binom{n}{r}$ lattice paths from O to A.

Next we present an occurrence of the binomial coefficient in the theory of elementary symmetric functions (ESFs).

Elementary Symmetric Functions

Notice that:

$$(1 + ax)(1 + bx) = 1 + (a + b)x + abx^2$$

$$(1 + ax)(1 + bx)(1 + cx) = 1 + (a + b + c)x + (ab + ac + bc)x^2 + abcx^3$$

The coefficients $a + b + c$, $ab + ac + bc$, and abc, for example, manifest the interesting property that swapping any two letters does not impact the expression. For example, switching a and c in $ab + ac + bc$ yields the same expression. They are called the *elementary symmetric functions* on the letters a, b, and c. The *order* of such a function is the common degree of its terms. For example, the order of $ab + ac + bc$ is two.

Let $f(n, r)$ denote the ESF on n letters of order r. Clearly, $r \leq n$. Then:

$f(1,1)=a$	$f(2,1)=a+b$	$f(2,2)=ab$
$f(3,1)=a+b+c$	$f(3,2)=ab+ac+bc$	$f(3,3)=abc$
$f(4,1)=a+b+c+d$	$f(4,2)=ab+ac+ad+bc+bd+cd$	$f(4,3)=abc+abd+acd+bcd$

[You can confirm the functions $f(4, 1)$ through $f(4, 3)$ by considering the product $(1 + ax)(1 + bx)(1 + cx)(1 + dx)$.]

Notice that there are $\binom{n}{r}$ terms in $f(n, r)$, where $1 \leq r \leq n$ and $n = 4$. For example, $f(4, 2)$ contains $\binom{4}{2}$ terms.

More generally, there are exactly $\binom{n}{r}$ terms in $f(n, r)$; this is so, since an n-element set has $\binom{n}{r}$ r-element subsets.

It follows from the definition that $\binom{n}{0} = 1 = \binom{n}{n}$. You may confirm both.

Triangular and Tetrahedral Numbers Revisited

Interestingly, triangular numbers t_n and tetrahedral numbers T_n can be expressed as binomial coefficients:

$$t_n = \frac{n(n+1)}{2} = \binom{n+1}{2} \qquad T_n = \frac{n(n+1)(n+2)}{6} = \binom{n+2}{3}$$

The combinatorial expressions for both t_n and T_n will come in handy when we pursue the properties of Pascal's triangle in Chapter 6.

The Twelve Days of Christmas Carol Revisited

It also follows from our discussion of *The Twelve Days of Christmas carol* that the number of gifts sent on the nth day equals $\binom{n+1}{2}$, and the total number of gifts sent in the first n days equals $\binom{n+2}{3}$, where $n \geq 1$.

Often we will need to compute the binomial coefficients $\binom{n}{r}$ and $\binom{n}{n-r}$ in the same problem. It turns that $\binom{n}{r} = \binom{n}{n-r}$; this follows from the definition.

Combinatorial Identities

The following combinatorial identities also can be verified algebraically, where $0 \le r \le k \le n$:

$$\binom{n}{r} = \frac{n}{r}\binom{n-1}{r-1} \tag{2.1}$$

$$\binom{n}{r} = \frac{n}{n-r}\binom{n-1}{r} \tag{2.2}$$

$$\binom{n}{r} = \frac{n-r+1}{r}\binom{n}{r-1} \qquad \text{(Cardano's rule of succession)} \tag{2.3}$$

$$\binom{n}{k}\binom{k}{r} = \binom{n}{r}\binom{n-r}{k-r} \qquad \text{(Newton's identity)} \tag{2.4}$$

Such identities can often be interpreted combinatorially.* For example, consider identity (2.1). Suppose we would like to select an r-member subcommittee of an n-member committee. It can be selected in $\binom{n}{r}$ ways. The chair of such a subcommittee can be selected in r ways, so such a subcommittee can be selected in $r\binom{n}{r}$ ways.

On the other hand, suppose we first select the chair; this can be done in n ways. The remaining $r-1$ members of the subcommittee can be chosen from the $n-1$ people in $\binom{n-1}{r-1}$ ways. So the number of such subcommittees also equals $n\binom{n-1}{r-1}$ ways.

Thus $r\binom{n}{r} = n\binom{n-1}{r-1}$; this yields the desired identity.

Interestingly, identity (2.4) has an interesting byproduct:† If $0 < r \le k < n$, then $\binom{n}{r}$ and $\binom{n}{k}$ have non-trivial common factors. To confirm this, assume that they are relatively prime. Then it follows from the identity that $\binom{n}{r} | \binom{k}{r}$. But this is impossible, since $k < n$. Therefore, $\binom{n}{r}$ and $\binom{n}{k}$ cannot be relatively prime; so they have non-trivial common factors.

For example, clearly $\binom{12}{7} = 792 = 2^3 \cdot 3^2 \cdot 11$ and $\binom{12}{4} = 495 = 3^2 \cdot 5 \cdot 11$ have non-trivial common factors.

Identity (2.1) plays an important role in determining the parity of the binomial coefficient $\binom{n}{r}$‡ and in finding the number of odd numbers in row n of Pascal's triangle (see Theorem 4.8). Likewise, Cardano's rule (2.3) plays a pivotal role

*See Benjamin and Quinn's *Proofs That Really Count* for combinatorial interpretations of a number of combinatorial identities.

†This was discovered by P. Erdös and G. Szekeres in 1978.

‡See Benjamin and Quinn.

in identifying the rows of Pascal's triangle that consist of odd numbers only (see Theorem 4.7).

Binomial coefficients satisfy an interesting and useful recurrence relation, as the next theorem shows. It is called *Pascal's identity*, in honor of Pascal, although it was known to Pingala. Although its proof follows easily using basic algebra, we will give a combinatorial argument.

Theorem 2.1 (*Pascal's identity*) Let n and r be any positive integers, where $r \leq n$. Then

$$\binom{n}{r} = \binom{n-1}{r-1} + \binom{n-1}{r}$$

Proof: Suppose we would like to form an r-member subcommittee of an n-member committee. There are $\binom{n}{r}$ such subcommittees.

Assume that Alice belongs to the committee.

Case 1 Suppose Alice belongs to a subcommittee. Then there are $n - 1$ members left in the committee and $r - 1$ members in the subcommittee. There are $\binom{n-1}{r-1}$ such subcommittees.

Case 2 Suppose Alice does not belong to a subcommittee. An r-member subcommittee can be selected from the remaining $n - 1$ members in $\binom{n-1}{r}$ ways.

So, by the addition principle, the total number of r-member subcommittees equals $\binom{n-1}{r-1} + \binom{n-1}{r}$. Thus $\binom{n-1}{r-1} + \binom{n-1}{r} = \binom{n}{r}$. ■

The definition of the binomial coefficient $\binom{n}{r}$ appears to show a fractional appearance, so we might be tempted to assume that its value can be a fraction. Fortunately, this is not true, as the following theorem confirms. Its proof follows from Pascal's identity by the strong version of PMI.

Theorem 2.2 The binomial coefficient $\binom{n}{r}$ is an integer for every integer $n \geq 0$. ■

Theorem 2.2 has an interesting byproduct, as the following corollary reveals.

Corollary 2.1 The product of r consecutive integers is divisible by $r!$.

Proof: Clearly, it suffices to consider the product of r consecutive positive integers. Let n be the smallest of them. Since

$$\frac{n(n+1)\cdots(n+r-1)}{r!} = \frac{(n+r-1)!}{r!(n-1)!} = \binom{n+r-1}{r}$$

by Theorem 2.2, $\frac{n(n+1)\cdots(n+r-1)}{r!}$ is an integer; so $n(n+1)\cdots(n+r-1)$ is divisible by $r!$. ■

For example, $8! \mid 5\cdot6\cdot7\cdot8\cdot9\cdot10\cdot11\cdot12$ and $6! \mid (-6)(-7)(-8)(-9)(-10)(-11)$.

Consecutive Binomial Coefficients in the Ratio $1 : 2 : 3 \cdots k$

The following example, proposed in 1955 by C. W. Trigg of Los Angeles City College, identifies the consecutive binomial coefficients that are in the ratio $1 : 2 : 3 \cdots k$. The solution given is due to L. A. Ringenberg of then Eastern Illinois State College.

Example 2.1 Find the largest value of k such that the consecutive binomial coefficients $\binom{n}{r}$ are in the ratio $1 : 2 : 3 \cdots k$, and identify the corresponding binomial coefficients.

Solution Suppose that

$$\binom{n}{r} : \binom{n}{r+1} \cdots \binom{n}{r+k-1} = 1 : 2 : 3 \cdots k$$

Then

$$\binom{n}{r+s} : \binom{n}{r+s+1} = s+1 : s+2$$

where $0 \le s \le k-2$. So:

$$\frac{r+s+1}{n-r-s} = \frac{s+1}{s+2}$$

$$(n-r-s)(s+1) = (r+s+1)(s+2)$$

$$n(s+1) = r(2s+3) + 2(s+1)^2 \tag{2.5}$$

Setting $s = 0$ and $s = 1$, this yields the 2×2 linear system

$$n = 3r + 2$$

$$2n = 5r + 8$$

Solving, we get $r = 4$ and $n = 14$. Then

$$\binom{14}{4+k-2} : \binom{14}{4+k-1} = k-1 : k$$

This yields

$$\frac{4+k-1}{12-k} = \frac{k-1}{k}$$

Then $k^2 - 5k + 6 = 0$; so $k = 2, 3$. Thus, the largest value of k is 3, and the corresponding binomial coefficients are $\binom{14}{4} = 1001$, $\binom{14}{5} = 2002$, and $\binom{14}{6} = 3003$. Notice that $1001 : 2002 : 3003 = 1 : 2 : 3$.

Returning to equation (2.5), we see that *no* other pairs of values of s yield a larger number of consecutive binomial coefficients with the desired property.

For example, $s = 0$ and $s = 2$ yield the linear system

$$n = 3r + 2$$

$$3n = 7r + 18$$

Solving, we get $r = 6$ and $n = 220$. Then the condition

$$\binom{20}{6+k-2} : \binom{20}{6+k-1} = k - 1 : k$$

yields $k^2 - 6k + 8 = 0$; so $k = 2,\ 4$. The binomial coefficients corresponding to $k = 4$, are $\binom{20}{6}$, $\binom{20}{7}$, $\binom{20}{8}$, and $\binom{20}{9}$; they are *not* in the ratio 1:2:3:4. When $k = 2$, $\binom{20}{6} : \binom{20}{7} = 38760 : 77520 = 1 : 2$.

Thus, the largest value of k that works is $k = 3$; this occurs when $n = 14$ and $r = 4$.

Next, we pursue an interesting property of four consecutive binomial coefficients.

Four Consecutive Binomial Coefficients

Let w, x, y, and z be four consecutive binomial coefficients. Then z can be expressed in terms of the other three, as A. Shaumyan, a student at Southern Connecticut State University, showed in 1995.

To see this, let $w = \binom{n}{r-1}$, $x = \binom{n}{r}$, $y = \binom{n}{r+1}$, and $z = \binom{n}{r+2}$. Then

$$\frac{x}{w} = \frac{n+1}{r} - 1,\ \frac{y}{x} = \frac{n+1}{r+1} - 1,\ \text{and}\ \frac{z}{y} = \frac{n+1}{r+2} - 1$$

From the first two equations, we have:

$$\frac{1+\frac{x}{w}}{1+\frac{y}{x}} = 1 + \frac{1}{r}$$

$$r = \frac{wx + wy}{x^2 - wy}$$

$$n + 1 = \left(1 + \frac{x}{w}\right) r$$

$$= \frac{xy + x^2}{x^2 - wy}$$

Substituting for $n + 1$ and $r + 2$ in the third equation yields the desired result:

$$z = \frac{2wy^2 + xy^2 - x^2y}{2x^2 + wx - wy}$$

Hermite's Divisibility Properties

We now pursue two divisibility properties of binomial coefficients, discovered by the prominent French algebraist Charles Hermite (1822–1901).

Theorem 2.3 (Hermite) Let $m, n \geq 1$. Then

$$\frac{m}{(m, n)} \mid \binom{m}{n} \tag{2.6}$$

$$\frac{m-n+1}{(m+1, n)} \mid \binom{m}{n} \tag{2.7}$$ ■

Both proofs employ the Euclidean algorithm; we omit them in the interest of brevity.[†]

Hermite's divisibility properties yield three interesting byproducts, as the next corollary shows.

Corollary 2.2

1) The binomial coefficient $\binom{2n}{n}$ has even parity, where $n \geq 1$.
2) Let p be a prime. Then $\binom{p}{r} \equiv 0 \pmod{p}$, where $1 \leq r \leq p-1$.
3) $(n+1) \mid \binom{2n}{n}$.

Proof:

1) It follows from property (2.6) that $\binom{2n}{n}$ is an even integer, where $n \geq 1$.
2) Since $1 \leq r \leq p-1$, $(p, r) = 1$. So, again by property (2.6), $p \mid \binom{p}{r}$; that is, $\binom{p}{r} \equiv 0 \pmod{p}$.
3) Let $m = 2n$ in (2.7). Since $(2n+1, n) = 1$, it follows that $(n+1) \mid \binom{2n}{n}$. ■

It follows from this corollary that the numbers $\frac{1}{n+1}\binom{2n}{n}$ are integers. The numbers $C_n = \frac{1}{n+1}\binom{2n}{n}$ are the well-known *Catalan numbers*.[*] As can be guessed, the *central binomial coefficient* (CBC) $\binom{2n}{n}$ plays a central role in the study of Catalan numbers. We will encounter both in later chapters.

The property that C_n is an integer follows also from the fact that $\frac{(a+b-1)!}{a!b!}$ is an integer, where $(a, b) = 1$. This result was discovered by the Belgian mathematician Eugene Charles Catalan (1814–1894) in 1874 using elliptic functions.[†]

We will develop an interesting primality test using Hermite's divisibility property (2.6) in Chapter 6.

[†] See the author's *Catalan Numbers with Applications.*

[*] For a detailed discussion of Catalan numbers, see the author's *Catalan Numbers with Applications.*

[†] L. E. Dickson, *History of the Theory of Numbers,* Vol. 1, Chelsea, New York, 1952.

Next, we construct a generating function for the CBC. We will employ both differential and integral calculus in the process.

A Generating Function for the CBC

First, we will derive a recurrence relation for the CBC. To this end, we let $a_n = \binom{2n}{n}$. Then:

$$a_{n+1} = \binom{2n+2}{n+1} = \frac{(2n+2)!}{(n+1)!(n+1)!}$$

$$= \frac{2(2n+1)}{n+1} \cdot \frac{(2n)!}{n!n!} = \frac{(4n+2)}{n+1} a_n$$

Thus a_n satisfies the recurrence relation $(n+1)a_{n+1} = (4n+2)a_n$, where $a_0 = 1$.

To derive a generating function for the CBC, we let $f(x) = \sum_{n=0}^{\infty} a_n x^n$. Then the recurrence relation yields

$$\sum_{n=0}^{\infty} (n+1)a_{n+1}x^n = 4\sum_{n=0}^{\infty} na_n x^n + 2\sum_{n=0}^{\infty} a_n x^n$$

That is, $f'(x) = 4xf'(x) + 2f(x)$, where $f'(x)$ denotes the derivative of $f(x)$ with respect to x. Then

$$\frac{f'(x)}{f(x)} = \frac{2}{1-4x}$$

Integrating this with respect to x, we get

$$\ln f(x) = -\frac{1}{2}\ln(1-4x) + C$$

where C is a constant. Since $f(0) = a_0 = 1$, this implies $C = 0$. Thus $\ln f(x) = -\frac{1}{2}\ln(1-4x)$; that is, $f(x) = \frac{1}{\sqrt{1-4x}}$. Thus

$$\frac{1}{\sqrt{1-4x}} = \sum_{n=0}^{\infty} \binom{2n}{n} x^n$$

is the desired generating function. (We will use this in Chapter 17.)

Least Common Multiple of Binomial Coefficients

The next theorem, proposed as a problem in 1977 by P. L. Montgomery of Huntsville, Alabama, provides a neat mechanism for computing the least common multiple (LCM) of all binomial coefficients, where $0 \le r \le n$. We omit the proof in the interest of brevity.

Theorem 2.4

$$(n+1)\left[\binom{n}{0}, \binom{n}{1}, \binom{n}{3}, \cdots, \binom{n}{n}\right] = [1, 2, 3, \cdots, (n+1)]$$

where $[a, b]$ denotes the *least common multiple* (LCM) of a and b. ■

For example,

$$7 \cdot \left[\binom{6}{0}, \binom{6}{1}, \binom{6}{2}, \binom{6}{3}, \binom{6}{4}, \binom{6}{5}, \binom{6}{6}\right] = 7 \cdot [1, 6, 15, 20, 15, 6, 1]$$

$$= 7 \cdot 60 = 420 = [1, 2, 3, 4, 5, 6, 7]$$

A Palindromic Sum

Next, we evaluate an interesting palindromic sum, proposed in 1948 by E. D. Schell of the United States Air Force. It turns out that this sum is the tetrahedral number T_n. The solution, due to P. Anselone of the College of Puget Sound, Washington, employs the familiar summation formulas.

Example 2.2 Evaluate the sum $n \cdot 1 + (n-1) \cdot 2 + \cdots + 2(n-1) + 1 \cdot n$.

Solution Let S denote the given sum. Then:

$$S = \sum_{k=1}^{n} (n-k+1)k = (n+1)\sum_{k=1}^{n} k - \sum_{k=1}^{n} k^2$$

$$= \frac{n(n+1)^2}{2} - \frac{n(n+1)(2n+1)}{6} = \frac{n(n+1)(n+2)}{6}$$

$$= \binom{n+2}{3} = T_n$$ ■

Infinite Series With Reciprocals of Binomial Coefficients

Infinite series involving the reciprocals of binomial coefficients have been investigated over the years. In 1991, C. Moen of the US Naval Academy, Annapolis, Maryland, using gamma and beta functions, established the following results, where $r \geq 2$:

$$\sum_{n=r}^{\infty} \frac{1}{\binom{n}{r}} = \frac{r}{r-1} \qquad \sum_{r=2}^{\infty} \sum_{n=r+2}^{\infty} \frac{1}{\binom{n}{r}} = \frac{3}{2}$$

$$\sum_{r=2}^{\infty} \sum_{n=r+1}^{\infty} \frac{1}{r\binom{n}{r}} = 1 \qquad \sum_{r=2}^{\infty} \sum_{n=r}^{\infty} \frac{1}{r^2\binom{n}{r}} = 1$$

Iterated Binomial Coefficients

Finally, the idea of a binomial coefficient $\binom{n}{r}$ can be extended to expressions $\binom{\binom{a}{b}}{c}$, where the upper number is itself a binomial coefficient. For example, $\binom{\binom{6}{3}}{5} = \binom{20}{5} = 15{,}504$.

Just as the binomial coefficient $\binom{n}{r}$ denotes the number of r-element subsets of an n-element set, the *iterated binomial coefficient* $\binom{\binom{a}{b}}{c}$ has a natural combinatorial interpretation. It denotes the number of subsets of size c that can be formed from the family of b-element subsets of a set of size a.

For example, there are 15,504 subsets of size 5 of the family of 3-element subsets of a set of size 6.

As a special case, the iterated binomial coefficient $\binom{\binom{m}{2}}{n}$ has an interesting graph-theoretic interpretation. Before introducing it, we will present some graph-theoretic terminology.

Some Graph-Theoretic Terminology

An *undirected graph* or simply a *graph* G is a nonempty set of points, called *vertices* or *nodes*, together with line segments or arcs, called *edges*, joining them. Let V denote the set of vertices and E the set of edges. The graph G can be considered the ordered pair $(V, E) : G = (V, E)$. An edge between vertices v and w is denoted by $v - w$ or $\{v, w\}$. A vertex v is *adjacent* to a vertex w if there is an edge between them.

For example, the graph in Figure 2.2 has four vertices – A, B, C, and D – and seven edges. It has edges with the same vertices; such edges are *parallel edges*. The graph in Figure 2.3 has an edge at Q that starts from and returns at the same vertex; such an edge is a *loop*.

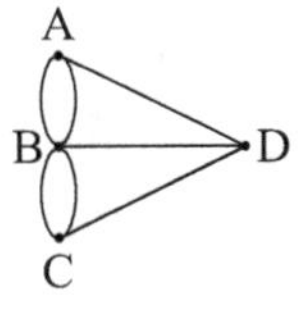

Figure 2.2

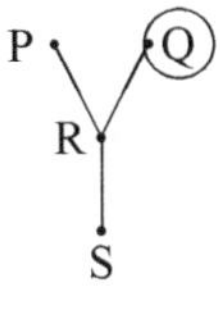

Figure 2.3

A loop-free graph that contains no parallel edges is a *simple graph*. A simple graph with m vertices that contains exactly one edge between every two distinct vertices is a *complete graph*, denoted by K_m. Figure 2.4 shows the complete graphs K_4 and K_5.

Let $H = (V', E')$ be a graph such that $V' \subseteq V$ and $E' \subseteq E$. Then H is a *subgraph* of G. For example, the graph in Figure 2.5 is a subgraph of the graph in Figure 2.2.

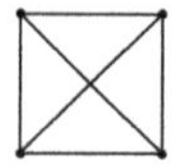

K_4

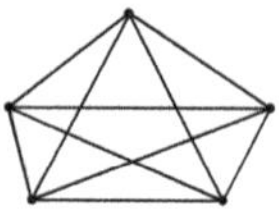

K_5

Figure 2.4

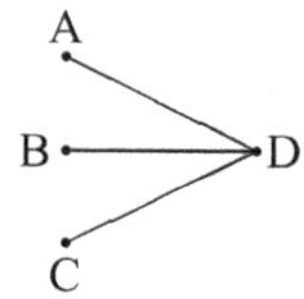

Figure 2.5

We are now ready to give a graph-theoretic interpretation of the iterated binomial coefficient.

A Graph-Theoretic Interpretation

The complete graph K_n has $\binom{n}{2}$ edges. So it has $\binom{\binom{n}{2}}{e}$ subgraphs with e edges.

For example, K_4 has $\binom{\binom{4}{2}}{3} = \binom{6}{3} = 20$ subgraphs consisting of exactly three edges. They are displayed in Figure 2.6.

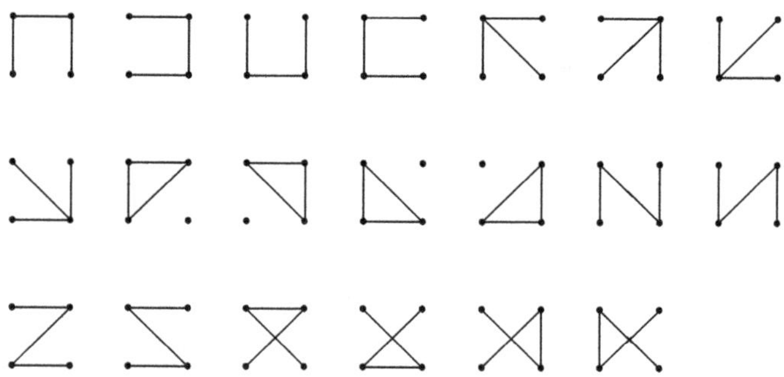

Figure 2.6 Subgraphs of K_4 with Exactly Three Edges

Since an r-element set has 2^r subsets, it follows that K_n has a total of $2^{\binom{n}{2}}$ subgraphs with n vertices. For example, K_4 has $2^{\binom{4}{2}} = 2^6 = 64$ subgraphs, each containing four vertices; this includes the graph with no edges and K_4.

Interated Binomial Coefficients and Labeled Simple Graphs

Interated binomial coefficients occur in the study of labeled simple graphs also. A *labeled graph* is a graph in which every vertex has a label v. For example, the graphs in Figure 2.5 are labeled, but the ones in Figure 2.6 are not.

The graphs G_1 and G_2 in Figure 2.7 are basically the same; they have the same structure; that is, they are *isomorphic*. But G_1 and G_3 are not isomorphic, although they have the same number of vertices and edges; this is so, since $a - c$ is an edge in G_1, but *not* in G_3.

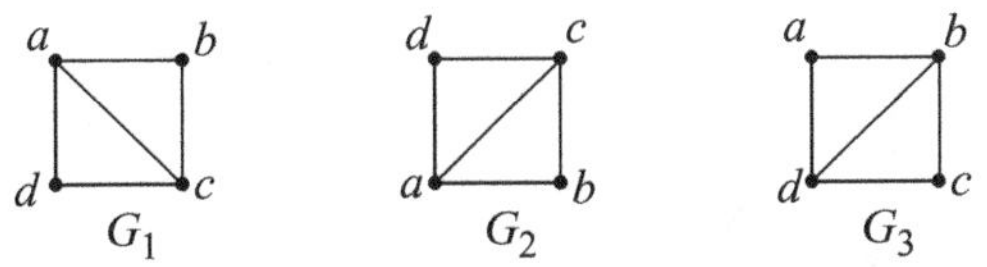

Figure 2.7

How many non-isomorphic labeled simple graphs G with n vertices are there? To answer this, by the handshake problem in Example 1.4, there are at most $\binom{n}{2}$ edges in G, since every edge $u - v$ is an unordered pair of vertices u and v. Then the e edges in G can be chosen in $\binom{\binom{n}{2}}{e}$ ways, where $0 \le e \le \binom{n}{2}$. So, by the addition principle, the total number of non-isomorphic labeled simple graphs G with n vertices is given by

$$\sum_{e=0}^{\binom{n}{2}} \binom{\binom{n}{2}}{e} = 2^{\binom{n}{2}}$$

For example, there are $\binom{\binom{3}{2}}{2} = 8$ such graphs with three vertices; they are given in Figure 2.8.

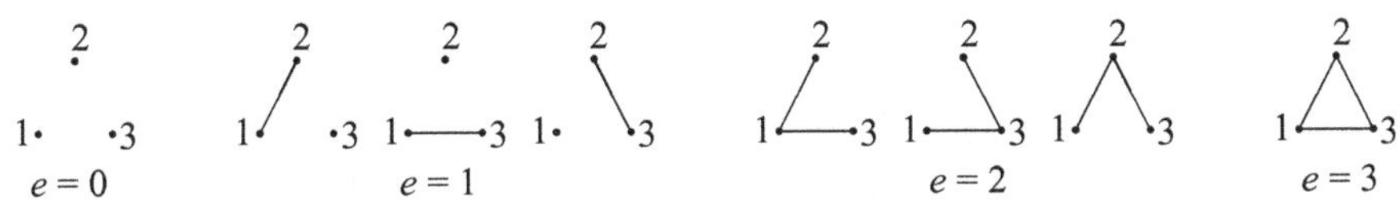

Figure 2.8 Non-isomorphic Labeled Simple Graphs with Three Vertices

In 1980, S. W. Golomb of the University of Southern California, Los Angeles, established a number of identities, inequalities, divisibility properties, and reduction formulas. For example:

1) $\binom{\binom{n}{1}}{2} = 1\binom{n}{2}$ 2) $\binom{\binom{n}{2}}{2} = 3\binom{n+1}{4}$

3) $\binom{\binom{n}{3}}{2} = 6\binom{n+2}{6} + 3\binom{n+1}{6} + \binom{n}{6}$ 4) $\binom{\binom{n}{4}}{2} = 10\binom{n+3}{8} + 15\binom{n+2}{8} + 10\binom{n+1}{8}$ ■

3

The Binomial Theorem

Binomial coefficients and Pascal's identity play a significant role in the development of the binomial theorem, as we will see shortly. The theorem can be employed to develop a vast array of combinatorial identities.

The binomial expansion for $n = 2$ can be found in Euclid's work of around 300 B.C.. There is strong evidence that the prominent Indian mathematician–astronomer Aryabhatta (*c*. 476–550) was familiar with the binomial theorem for $n = 2$ and $n = 3$. The latter case was also known to the Indian mathematician and astronomer Brahmagupta (598?–670?). The binomial theorem for positive integral exponents was known to the Persian poet and mathematician Omar Khayyám (1048–1131).

The English mathematician and physicist Isaac Newton (1642–1727) showed successfully how to expand $(1 + x)^n$ directly without resorting to the expansion of $(1 + x)^{n-1}$. He achieved this by demonstrating that each coefficient can be computed from the previous one using *Cardano's rule of succession*: $\binom{n}{r} = \frac{n-r+1}{r}\binom{n}{r-1}$, where $r \geq 1$. This formula is named after the Italian mathematician Girolano Cardano (1501–1576), a doctor by profession and generally considered one of the most eccentric of mathematicians.

Theorem 3.1 (The Binomial Theorem) Let x and y be arbitrary real numbers, and n an arbitrary nonnegative integer. Then $(x + y)^n = \sum_{r=0}^{n} \binom{n}{r} x^{n-r} y^r$.

Proof: We can invoke induction to prove the theorem, using Pascal's identity and a lot of algebra. So we will resort to a combinatorial argument; it is short and elegant.

Notice that $(x+y)^n = \underbrace{(x+y)(x+y)\cdots(x+y)}_{n \text{ factors}}$. Clearly, every term in the expansion is of the form $Cx^{n-r}y^r$, where $0 \le r \le n$ and C is a constant to be determined. An x can be chosen from any $n-r$ factors on the RHS† and y from the remaining r factors. So the $n-r$ x's can be chosen in $\binom{n}{n-r}$ ways and the r y's in $\binom{r}{r} = 1$ way. Thus, by the multiplication principle, $C = \binom{n}{n-r} \cdot 1 = \binom{n}{r}$. Since this is true for every r, it follows by the addition principle that $(x+y)^n = \sum_{r=0}^{n} \binom{n}{r} x^{n-r} y^r$, as desired. ■

Four Consecutive Binomial Coefficients

Next, we derive an identity involving four consecutive binomial coefficients a, b, c, and d.

Let $a = \binom{n}{r-2}$, $b = \binom{n}{r-1}$, $c = \binom{n}{r}$, and $d = \binom{n}{r+1}$. Then, by Pascal's identity, $a+b = \binom{n+1}{r-1}$, $b+c = \binom{n+1}{r}$, and $c+d = \binom{n+1}{r+1}$. So

$$\begin{aligned}\frac{a}{a+b} + \frac{c}{c+d} &= \frac{\binom{n}{r-2}}{\binom{n+1}{r-1}} + \frac{\binom{n}{r}}{\binom{n+1}{r+1}} = \frac{r-1}{n+1} + \frac{r+1}{n+1} \\ &= \frac{2\binom{n}{r-1}}{\binom{n+1}{r}} = \frac{2b}{b+c}\end{aligned}$$

In particular, let $r = 7$. Then $4a(c^2 - bd) = 3c(b^2 - ac)$.

Next, we would like to ask the following: For what value of r, is the binomial coefficient $\binom{n}{r}$ a maximum?

Largest Binomial Coefficient

The largest binomial coefficients in the expansion of $(x+y)^4$, $(x+y)^5$, $(x+y)^6$, and $(x+y)^7$ reveal a clear pattern:

$$6 = \binom{4}{\lfloor 4/2 \rfloor} \qquad 10 = \binom{4}{\lfloor 5/2 \rfloor} \qquad 20 = \binom{6}{\lfloor 6/2 \rfloor} \qquad 35 = \binom{7}{\lfloor 7/2 \rfloor}$$

Accordingly, we conjecture that the binomial coefficient $\binom{n}{r}$ is the largest when $r = \lfloor n/2 \rfloor$.

The next theorem confirms this.

Theorem 3.2 The largest binomial coefficient is $\binom{n}{\lfloor n/2 \rfloor}$, where $n \ge 0$.

†RHS and LHS are abbreviations of right-hand side and left-hand side, respectively.

Proof: The proof consists of two parts:

First, we will prove that $\binom{n}{r-1} < \binom{n}{r}$ if and only if $r < \frac{n+1}{2}$.

To this end, suppose that $\binom{n}{r-1} < \binom{n}{r}$. Then $\frac{n!}{(r-1)!(n-r+1)!} < \frac{n!}{r!(n-r)!}$. This yields $\frac{1}{n-r+1} < \frac{1}{r}$; that is, $n - r + 1 > r$, so $r < \frac{n+1}{2}$.

Conversely, let $r < \frac{n+1}{2}$. By retracing these steps, it can be shown that $\binom{n}{r-1} < \binom{n}{r}$.

It follows by part 1 that $\binom{n}{0} < \binom{n}{1} < \cdots < \binom{n}{r-1} < \binom{n}{r}$, where $r < \frac{n+1}{2}$. Thus, $\binom{n}{r}$ is maximum when r is the largest integer $\leq \frac{n+1}{2}$.

Let n be odd, say, $n = 2k+1$. Then $\frac{n+1}{2} = k+1$. So r must be $k = \frac{n-1}{2} = \lfloor n/2 \rfloor$.

On the other hand, let n be even, say, $n = 2k$. Then $\frac{n+1}{2} = \frac{2k+1}{2} = k + 1/2$. So r must be $k = n/2 = \lfloor n/2 \rfloor$.

Thus $\binom{n}{r}$ is maximum when $r = \lfloor n/2 \rfloor$. ■

In particular, the CBC is the largest binomial coefficient in the expansion of $(x+y)^{2n}$.

Middle Coefficient in the Expansion of $(1 + x)^m$

We now turn to the middle coefficient in the expansion of $(1 + x)^m$. It is also the same as the largest binomial coefficient. In Theorem 3.2, we found it to be $\binom{m}{\lfloor m/2 \rfloor}$.

In particular, let $m = 2n$. Then the middle coefficient is $\binom{m}{\lfloor m/2 \rfloor} = \binom{2n}{n}$.

On the other hand, let $m = 2n - 1$. Then there are two middle coefficients and they are equal; each is given by $\binom{m}{\lfloor m/2 \rfloor} = \binom{2n-1}{n-1}$.

The binomial theorem can be used to develop an array of *combinatorial identities*, as the next three corollaries demonstrate.

Corollary 3.1

$$\sum_{r=0}^{n} \binom{n}{r} = 2^n \tag{3.1}$$

In words, the sum of the binomial coefficients is 2^n.

Proof: The proof follows by letting $x = 1 = y$ in the binomial theorem. ■

For example,

$$\sum_{r=0}^{4} \binom{4}{r} = \binom{4}{0} + \binom{4}{1} + \binom{4}{2} + \binom{4}{3} + \binom{4}{4} = 1 + 4 + 6 + 4 + 1 = 2^4.$$

An alternate argument: Every r-member subcommittee of an n-member committee can be represented by a binary word of r ones and $n - r$ zeros; a 1 indicates that the corresponding member belongs to the subcommittee and a

0 indicates that he/she does not belong to the subcommittee. Since the 1s can be in any r of the n positions, there are exactly $\binom{n}{r}$ such binary words. So the total number of subcommittees equals $\sum_{r=0}^{n} \binom{n}{r}$.

Since each bit in the string has two choices, it follows by the multiplication principle that there are exactly 2^n n-bit words. Thus $\sum_{r=0}^{n} \binom{n}{r} = 2^n$.

It now follows from Corollary 3.1 that an n-member committee has a total of $\sum_{r=0}^{n} \binom{n}{r} = 2^n$ subcommittees.

Gray Codes

The subcommittees of an n-member committee S_n can be listed systematically using a simple coding scheme, called a *Gray code*, where $n \geq 1$. It is named after the physicist Frank Gray, of AT&T Bell Telephone Laboratories, who developed it in the 1940s. Gray codes are often used to minimize the effect of errors in the conversion of analog signals to digital ones.

The essence of the scheme lies in representing each subcommittee by an n-bit word. Consequently, the 2^n subsets of S_n can be represented by the n-bit words 000...0, ... , 111...1, where a 1 indicates that the corresponding member belongs to the subcommittee and a 0 indicates otherwise.

We now demonstrate the construction of a Gray code for $n = 1, 2$, and 3.

Case 1 Let $n = 1$. Then $S_1 = \{a\}$ has two subcommittees: $\emptyset, \{a\}$. They can be represented by the binary words 0 and 1, respectively.

Case 2 Let $n = 2$. Then $S_2 = \{a, b\}$ has four subcommittees: $\emptyset, \{a\}, \{a, b\}, \{b\}$. They can be represented by the binary words 00, 10, 11, and 01, respectively.

These four binary words can be constructed from the words 0 and 1 in Case 1 in an elegant way: Append the 0 with a 0 or 1, and the 1 with a 1 or 0. This yields 00, 10, 11, and 01, as expected. Such an ordered list is a Gray code for $n = 2$. (Notice that 00, 01, 11, 10 is also a Gray code.)

These binary words satisfy an interesting property: Every two adjacent words differ in exactly one position. The number of positions in which two words differ is called the *Hamming distance* between them, after Richard W. Hamming (1915–1998). The Hamming distance between every two adjacent words in a Gray code is one. Figure 3.1 shows the cyclic nature of the code; every word can be obtained from its predecessor by simply changing exactly one bit.

These binary words display another property also. To see this, first we list the words 0 and 1 from Case 1 in a column (see Column 1 in Table 3.1). Then, list them again, and concatenate it with the list in the reverse order;

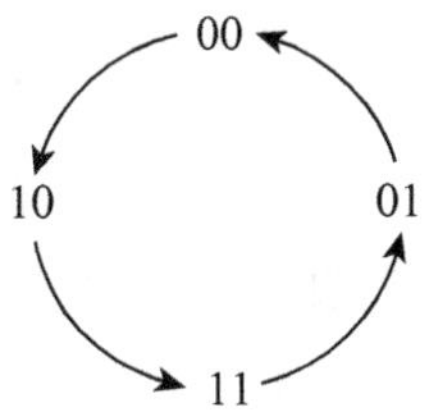

Figure 3.1

Table 3.1

Column 1	Column 2	Column 3
0	0	00
1	1	10
	1	11
	0	01

see Column 2. Finally, append a 0 to each word in the upper half and a 1 to each in the lower half; see Column 3. This algorithm produces the same Gray code in Column 3.

Case 3 Clearly, the recursive algorithm illustrated for $n = 2$ can be applied to Gray codes of any length n. Table 3.2 shows the steps for $n = 3$.

Table 3.2

Gray code for $n = 2$	Repeat the code and list it in reverse order	Append the upper half with a 0	Append the lower half with a 1
00	00	000	000
10	10	100	100
11	11	110	110
01	01	010	010
	01	01	011
	11	11	111
	10	10	101
	00	00	001

Gray Codes and Venn Diagrams

Gray codes for $n = 3$, for example, can also be obtained using a *Venn diagram*, which is a pictorial representation of relationships among sets. It is named after the English logician, John Venn (1834–1923), who used it in his study of symbolic logic.

The three circles in Figure 3.2 represent three arbitrary sets A, B, and C, and they divide the universe U into eight non-overlapping regions. The first bit in each binary word indicates whether or not the corresponding region is part of set A; the other two bits are interpreted similarly. Consequently, the above Gray code can be recovered by threading through the various regions, beginning with 000; see Figure 3.2.

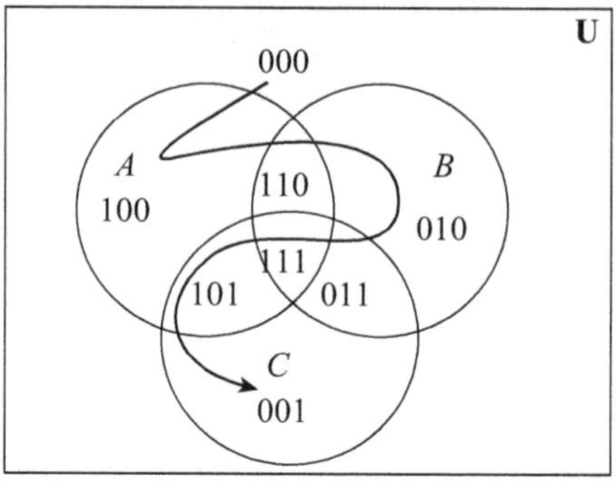

Figure 3.2

Notice, for example, that the set $S_3 = \{a, b, c\}$ has exactly eight subsets; $\emptyset$, $\{a\}$, $\{a, b\}$, $\{b\}$, $\{b, c\}$, $\{a, b, c\}$, $\{a, c\}$, $\{c\}$; each can be obtained from the previous subset by changing exactly one element in it.

Parity of the CBC Revisited

Corollary 3.1 can be used to re-confirm the even parity of the CBC. To see this, we have:

$$\sum_{r=0}^{2n} \binom{2n}{r} = 2^{2n}$$

$$\sum_{r=0}^{n-1} \binom{2n}{r} + \binom{2n}{n} + \sum_{r=n+1}^{2n} \binom{2n}{r} = 2^{2n}$$

$$2\sum_{r=0}^{n-1} \binom{2n}{r} + \binom{2n}{n} = 2^{2n}$$

This re-confirms the even parity of the CBC, where $n \geq 1$.

Submatrices of a Matrix

The next example, yet another application of Corollary 3.1, was proposed in 1943 by the mathematics historian H. Eves of Syracuse University. The obvious solution, due to A. Wayne of Flushing, New York, employs the multiplication and addition principles.

Example 3.1 Find the total number of nonempty submatrices of an $m \times n$ matrix.

Solution Since r rows can be chosen in $\binom{m}{r}$ ways, the total number of combinations of rows from m rows equals $\sum_{r=1}^{m} \binom{m}{r} = 2^m - 1$, by Corollary 3.1. Similarly, the total number of columns we can choose is $2^n - 1$. Thus there are $(2^m - 1)(2^n - 1)$ ways of choosing rows and columns; that is, there are $(2^m - 1)(2^n - 1)$ submatrices of an $m \times n$ matrix. ∎

Next, we present another interesting byproduct of Corollary 3.1.

Corollary 3.2

$$\sum_{r=0}^{n} \binom{n}{r} = 2\sum_{r=0}^{n-1} \binom{n-1}{r} \tag{3.2}$$ ∎

The binomial theorem yields yet another byproduct, as the next corollary shows.

Corollary 3.3

$$\sum_{r=0}^{\lfloor n/2 \rfloor} \binom{n}{2r} = \sum_{r=0}^{\lfloor \frac{n-1}{2} \rfloor} \binom{n}{2r+1} = 2^{n-1}$$

That is, the sum of the 'even' binomial coefficients equals that of the 'odd' binomial coefficients.

Proof: By the binomial theorem, we have $\sum_{r=0}^{n} \binom{n}{r} x^{n-r} y^r = (x+y)^n$. Now, let $x = 1$ and $y = -1$. Then this equation yields $\sum_{r=0}^{n} \binom{n}{r}(-1)^r = 0$; that is,

$$\sum_{r \text{ odd}} \binom{n}{r} = \sum_{r \text{ even}} \binom{n}{r} \tag{3.3}$$

But, by Corollary 3.1,

$$\sum_{r \text{ odd}} \binom{n}{r} + \sum_{r \text{ even}} \binom{n}{r} = 2^n$$

This, coupled with equation (3.3), yields the desired result. ∎

Divisors of Binomial Coefficients

Corollary 3.3 has an interesting byproduct, as the next example shows. It was proposed as a problem in 1970 by N. S. Mendelsohn of the University of Manitoba,

Canada. The solution presented here is based on the one given in the following year by students at St. Olaf College.

Example 3.2 Find the gcd of the binomial coefficients $\binom{2n}{1}, \binom{2n}{3}, \binom{2n}{5}, \cdots,$ and $\binom{2n}{2n-1}$.

Solution By Corollary 3.3,

$$\binom{2n}{1} + \binom{2n}{3} + \binom{2n}{5} + \cdots + \binom{2n}{2n-1} = 2^{2n-1}$$

so the gcd of the binomial coefficients must be of the form 2^k, where k is odd. Let $n = 2^m q$, where q is odd. Then $\binom{2n}{1} = 2^{m+1}q$; so a common divisor of the given binomial coefficients cannot be greater than 2^{m+1}.

Now, we show that 2^{m+1} divides all of them. To this end, we consider the binomial coefficient $\binom{2^{m+1}q}{k}$. It can be rewritten as $\binom{2^{m+1}q}{k} = \frac{2^{m+1}q}{k}\binom{2^{m+1}q-1}{k-1}$. Since the binomial coefficients are integers by Theorem 2.3 and k is odd, $Q = \frac{q}{k}\binom{2^{m+1}q-1}{k-1}$ is an integer for $k = 1, 3, \cdots, 2n-1$. So $\binom{2^{m+1}q}{k} = 2^{m+1}Q$. Thus, the desired gcd is 2^{m+1}, where $m = \lg(n/q)$.[†] ■

For instance, consider the binomial coefficients

$$\binom{12}{1} = 12 = \binom{12}{11}, \binom{12}{3} = 220 = \binom{12}{9}, \text{ and } \binom{12}{5} = 792 = \binom{12}{7}$$

Clearly, the gcd of 12, 220, and 792 is $4 = 2^2 = 2^{1+1} = 2^{\lg(6/3)+1}$.

The next corollary also follows from the binomial theorem.

Corollary 3.4 Let x be any real number. Then

$$(1+x)^n = \sum_{r=0}^{n} \binom{n}{r} x^r \tag{3.4}$$ ■

In particular, this corollary implies that

$$\sum_{r=0}^{n} 2^r \binom{n}{r} = 3^n \tag{3.5}$$

A Square Sum

The next result was developed in 1963 by L. Moser of the University of Alberta, Alberta, Canada. The solution, given in the following year by H. W. Gould of West Virginia University, is an application of Corollary 3.4, and involves a new family of numbers.

[†] $\lg x = \log_2 x$.

Example 3.3 Prove that $\sum_{r=1}^{2n-1}\binom{4n-2}{2r}2^{r-1}$ is a square.

Proof: By Corollary 3.4,

$$(1+\sqrt{x})^m=\sum_{r=0}^{m}\binom{m}{r}x^{r/2} \quad \text{and} \quad (1-\sqrt{x})^m=\sum_{r=0}^{m}(-1)^r\binom{m}{r}x^{r/2}$$

Adding,

$$(1+\sqrt{x})^m+(1-\sqrt{x})^m=2\sum_{k=0}^{\lfloor m/2\rfloor}\binom{m}{2k}x^k$$

Now, let $m=4n-2$ and $x=2$. Then:

$$\sum_{k=1}^{2n-1}\binom{4n-2}{2k}2^{k-1}=\frac{(1+\sqrt{2})^{4n-2}+(1-\sqrt{2})^{4n-2}-2}{4}$$

$$=\left[\frac{(1+\sqrt{2})^{2n-1}+(1-\sqrt{2})^{2n-1}}{2}\right]^2$$

as desired. ■

For example, $\sum_{r=1}^{5}\binom{10}{2r}2^{r-1}=41^2$, a square.

The integers $Q_n=\frac{(1+\sqrt{2})^n+(1-\sqrt{2})^n}{2}$ are the *Pell-Lucas numbers* $1, 3, 7, 17, 41, \cdots$.

Calculus, and the binomial theorem and its corollaries, can be used to develop a number of additional combinatorial identities, as the next corollary demonstrates.

Corollary 3.5

$$1\binom{n}{1}+2\binom{n}{2}+3\binom{n}{3}+\cdots+\binom{n}{n}=n2^{n-1} \tag{3.6}$$

$$1\binom{n}{1}+3\binom{n}{3}+\cdots=2\binom{n}{2}+4\binom{n}{4}+\cdots=n2^{n-2} \tag{3.7}$$

Proof: By Corollary 3.4, we have

$$(1+x)^n=\sum_{r=0}^{n}\binom{n}{r}x^r \tag{3.8}$$

Differentiating both sides with respect to x, this yields

$$n(1+x)^{n-1}=\sum_{r=1}^{n}r\binom{n}{r}x^{r-1} \tag{3.9}$$

Substituting $x=1$ in this equation, we get identity (3.6).

To establish identity (3.7), substitute $x = -1$ in equation (3.9): $0 = \sum_{r=1}^{n} r\binom{n}{r}(-1)^{r-1}$. That is,

$$\sum_{r \text{ even}}^{n} r\binom{n}{r} = \sum_{r \text{ odd}}^{n} r\binom{n}{r}$$

This, coupled with identity (3.6), yields identity (3.7):

$$1\binom{n}{1} + 3\binom{n}{3} + \cdots = 2\binom{n}{2} + 4\binom{n}{4} + \cdots = n2^{n-2}$$ ■

Identity (3.6) was developed in 1952 by Moser. It can also be derived by letting S denote the sum, considering the same sum in the reverse order, and then by computing the sum $2S$.

Interestingly, identity (3.7) follows from identity (2.4) also:

$$\begin{aligned}\sum_{r=1}^{n} r\binom{n}{r} &= \sum_{r=1}^{n} n\binom{n-1}{r-1}\\ &= n\sum_{r=1}^{n}\binom{n-1}{r-1} = n\sum_{k=0}^{n-1}\binom{n-1}{k}\\ &= n2^{n-1}, \quad \text{by Corollary 3.1}\end{aligned}$$

Binomial Coefficients and Factorials

Interestingly, binomial coefficients can be used to compute factorials, as Euler did. For example,

$$\sum_{r=0}^{4}(-1)^{4-r}\binom{4}{r}r^4 = \binom{4}{0}0^4 - \binom{4}{1}1^4 + \binom{4}{2}2^4 - \binom{4}{3}3^4 + \binom{4}{4}4^4 = 24 = 4!$$

whereas $\sum_{r=0}^{5}(-1)^{5-r}\binom{5}{r}r^5 = 0$.

More generally, we have the following result. Its proof uses induction twice and is a bit long, so we omit the proof for the sake of brevity.

Theorem 3.3 (*Euler's formula*)

$$\sum_{i=0}^{n}(-1)^{n-i}\binom{n}{i}i^r = \delta_{nr}n! \tag{3.10}$$

where $0 \le r \le n$ and δ_{nr} denotes the *Kronecker delta*, defined by

$$\delta_{nr} = \begin{cases} 1 & \text{if } n = r \\ 0 & \text{otherwise} \end{cases}$$ ■

We close this chapter with a brief introduction to a new class of binomial coefficients, called Gaussian binomial coefficients (GBCs), studied by Gauss. They have interesting applications to abstract algebra, combinatorics, linear algebra, and finite geometry.

Gaussian Binomial Coefficients

Analogous to the ordinary binomial coefficient $\binom{n}{r} = \frac{n(n-1)\cdots(n-r+1)}{1\cdot 2\cdot 3\cdots r}$, the *Gaussian binomial coefficient* $\begin{bmatrix} n \\ r \end{bmatrix}$ is defined by

$$\begin{bmatrix} n \\ r \end{bmatrix} = \frac{q^n - 1}{q - 1} \cdot \frac{q^{n-1} - 1}{q^2 - 1} \cdots \frac{q^{n-r+1} - 1}{q^r - 1} \tag{3.11}$$

where $\begin{bmatrix} n \\ 0 \end{bmatrix} = 1$, q is a dummy variable, and $0 < r \le n$.

For example, $\begin{bmatrix} 3 \\ 1 \end{bmatrix} = \frac{q^3-1}{q-1} = 1+q+q^2$, $\begin{bmatrix} 3 \\ 2 \end{bmatrix} = \frac{q^3-1}{q-1} \cdot \frac{q^2-1}{q^2-1} = 1+q+q^2$, and

$\begin{bmatrix} 3 \\ 3 \end{bmatrix} = \frac{q^3-1}{q-1} \cdot \frac{q^2-1}{q^2-1} \cdot \frac{q-1}{q^3-1} = 1.$

The GBCs satisfy a number of interesting properties; they have strikingly similar counterparts to ordinary binomial coefficients:

$$\begin{bmatrix} n \\ r \end{bmatrix} = \begin{bmatrix} n \\ n-r \end{bmatrix} \tag{3.12}$$

$$\begin{bmatrix} n \\ r \end{bmatrix} = \begin{bmatrix} n-1 \\ r \end{bmatrix} + \begin{bmatrix} n-1 \\ r-1 \end{bmatrix} q^{n-r} \tag{3.13}$$

$$\begin{bmatrix} n \\ r \end{bmatrix} = \begin{bmatrix} n-1 \\ r \end{bmatrix} q^r + \begin{bmatrix} n-1 \\ r-1 \end{bmatrix} \tag{3.14}$$

These properties follow algebraically from the definition.

For example,

$$\begin{bmatrix} 3 \\ 1 \end{bmatrix} = \frac{q^3 - 1}{q - 1} = 1 + q + q^2 = \begin{bmatrix} 3 \\ 2 \end{bmatrix}$$

and

$$\begin{bmatrix} 4 \\ 2 \end{bmatrix} + \begin{bmatrix} 4 \\ 1 \end{bmatrix} q^3 = \frac{q^4-1}{q-1} \cdot \frac{q^3-1}{q^2-1} + \frac{q^4-1}{q-1} \cdot q^3$$
$$= \frac{q^4-1}{(q-1)(q^2-1)} [(q^3-1) + q^3(q^2-1)]$$
$$= \frac{q^5-1}{q-1} \cdot \frac{q^4-1}{q^2-1} = \begin{bmatrix} 5 \\ 2 \end{bmatrix}$$

It follows from equation (3.13) by induction that every GBC is a polynomial in q with positive integer coefficients. In fact,

$$\begin{bmatrix} n \\ r \end{bmatrix} = \sum_{i=0}^{r(n-r)} A_i q^i \tag{3.15}$$

where A_i is a positive integer. In particular, $\begin{bmatrix} n \\ 1 \end{bmatrix} = 1 + q + q^2 + \cdots + q^{n-1}$.

For example,

$$\begin{bmatrix} 5 \\ 2 \end{bmatrix} = (q^4 + q^3 + q^2 + q + 1)(q^2 + 1) = 1 + q + 2q^2 + 2q^3 + 2q^4 + q^5 + q^6$$

Notice that the sum of the coefficients equals $1+1+2+2+2+1+1 = 10 = \binom{5}{2}$.

Next, we evaluate $\lim\limits_{q \to 1} \begin{bmatrix} n \\ r \end{bmatrix}$ using L'Hospital's rule to see an interesting relationship between the two families of binomial coefficients. To this end, assume that q is a real variable:

$$\lim_{q\to 1} \begin{bmatrix} n \\ r \end{bmatrix} = \lim_{q\to 1} \frac{q^n-1}{q-1} \cdot \lim_{q\to 1} \frac{q^{n-1}-1}{q^2-1} \cdots \lim_{q\to 1} \frac{q^{n-r+1}-1}{q^r-1}$$
$$= \lim_{q\to 1} \frac{nq^{n-1}}{1} \cdot \lim_{q\to 1} \frac{(n-1)q^{n-2}}{2q} \cdots \lim_{q\to 1} \frac{(n-r+1)q^{n-r}}{rq^{r-1}}$$
$$= \frac{n(n-1)\cdots(n-r+1)}{r!}$$
$$= \binom{n}{r} \tag{3.16}$$

Thus, every ordinary binomial coefficient is the limit of the corresponding GBC as $q \to 1$.

Notice that Pascal's identity follows from (3.13). It follows from equation (3.15) that

$$\binom{n}{r} = \lim_{q\to 1} \begin{bmatrix} n \\ r \end{bmatrix} = \sum_{i=0}^{r(n-r)} A_i$$

and from equations (3.13) and (3.14) that $\binom{n}{r} = \frac{n-r+1}{r}\binom{n}{r-1}$.

Because of the significance of the next property, we state it as a theorem. It follows from (3.13) by induction.

Theorem 3.4

$$(1+x)(1+qx)(1+q^2x)\cdots(1+q^{n-1}x) = \sum_{r=0}^{n} \begin{bmatrix} n \\ r \end{bmatrix} q^{r(r-1)/2}x^r \qquad \blacksquare$$

For example,

$$\begin{aligned}(1+x)(1+qx)(1+q^2x) &= 1 + (1+q+q^2)x + (1+q+q^2)qx^2 + q^3x^3 \\ &= \begin{bmatrix} 3 \\ 0 \end{bmatrix} + \begin{bmatrix} 3 \\ 1 \end{bmatrix} x + \begin{bmatrix} 3 \\ 2 \end{bmatrix} qx^2 + \begin{bmatrix} 3 \\ 3 \end{bmatrix} q^3x^3\end{aligned}$$

Interestingly, Theorem 3.4, coupled with property (3.16), yields the binomial theorem.

We now add several interesting applications of the GBCs to combinatorics, vector spaces over finite fields F_q, group theory, projective geometry, and the theory of partitions of an integer, where q is a prime power:

1) The coefficients A_i in equation (3.15) have a delightful geometric interpretation, as G. Polya of Stanford University found around 1970. To see this, first we define a *lattice point*; it is a point (x, y) on the cartesian plane with integer coordinates x and y.

   Suppose we would like to walk from the origin to the lattice point $(n-r, r)$; we can walk one unit at a time in the easterly or northerly direction. There are $\binom{n}{r} = \sum_{i=0}^{r(n-r)} A_i$ such lattice paths.

   For example, there are $\binom{4}{2} = 6$ such lattice paths from the origin to (2,2); see Figure 3.3. They form areas of 0, 1, 2, 3, and 4 square units with the x-axis and the vertical line through (2,2). Interestingly, there is one path contributing area 0, one path contributing area 1, two paths contributing area 2, and so on; the number of paths contributing area i equals the coefficient of q^i in $\begin{bmatrix} 4 \\ 2 \end{bmatrix} = 1 + q + 2q^2 + q^3 + q^4$.

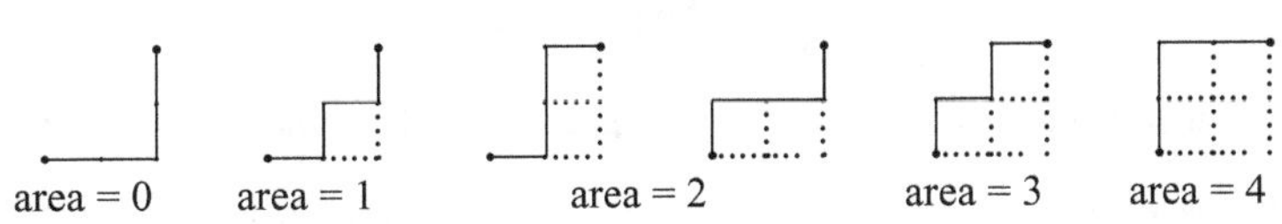

Figure 3.3

More generally, A_i is the number of paths from the origin to the point $(n-r, r)$, contributing an area of i square units, where $0 \le i \le r(n-r)$.

2) The coefficient A_i of q^i in the polynomial (3.15) is also the number of partitions $a_1 + a_2 + \cdots + a_k$ of the integer i, where $a_1 \ge a_2 \ge \cdots \ge a_k$.[†] Such partitions can be represented by a row of a_1 cells, followed by a row of a_2 cells, and so on, enclosed in an $r \times (n-r)$ rectangle. Such a diagram is called a *Ferrers diagram*, after the English mathematician, lawyer, and priest Norman Macleod Ferrers (1829–1903).

   For example, the partitions

$$0 = 0 \qquad 1 = 1 \qquad 2 = 2 = 1 + 1 \qquad 3 = 3 = 2 + 1$$

$$4 = 3 + 1 = 2 + 2 \qquad 5 = 3 + 2 \qquad 6 = 3 + 3$$

   of the integers 0 through 6 can be represented by the Ferrers diagrams in Figure 3.4.

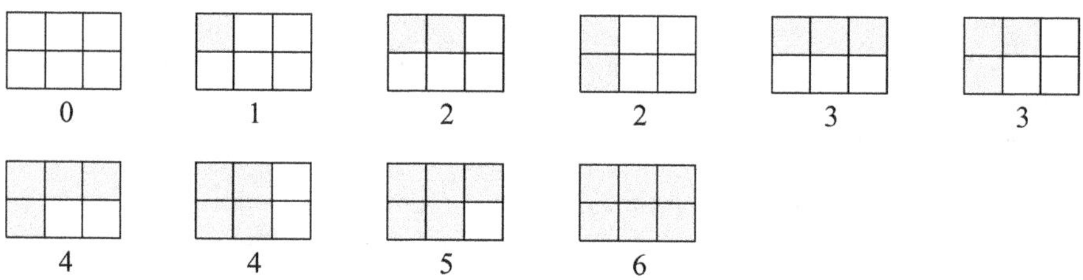

Figure 3.4 Ferrers Diagrams

3) The GBC $\begin{bmatrix} n \\ r \end{bmatrix}$ is the number of r-dimensional subspaces of the n-dimensional vector space $V_n(q) = F_q^n = \{(a_1, a_2, \cdots, a_n) | a_i \in F_q\}$,[*] where F_q denotes a finite field with q elements.

   This can be established by counting the number N of linearly independent r-tuples $(v_1, v_2, \cdots, v_r)$ in $V_n(q)$ in two different ways. Since v_1 can be any non-zero element in $V_n(q)$, it can be chosen in $q^n - 1$ ways; since v_1 and v_2 must be linearly independent, v_2 can be chosen in $(q^n - 1) - (q - 1) = q^n - q$ ways; likewise, v_3 can be chosen in $q^n - q^2$ ways. In general, v_i can be chosen in $q^n - q^{i-1}$ ways, where $1 \le i \le r$. Thus, by the multiplication principle,

$$\begin{aligned} N &= (q^n - 1)(q^n - q) \cdots (q^n - q^{r-1}) \\ &= q^{r(r-1)/2}(q^n - 1)(q^{n-1} - 1) \cdots (q^{n-r+1} - 1) \qquad (3.17) \end{aligned}$$

[†]See J. H. van Lint and R. M. Wilson, *A Course in Combinatorics*, Cambridge University Press, New York, 1992.

[*]See R. P. Stanley, *Enumerative Combinatorics*, Vol. 1, Cambridge University Press, New York, 1997; P. J. Cameron, *Combinatorics: Topics, Techniques, Algorithms*, Cambridge University Press, New York, 1994.

On the other hand, let W be an r-dimensional subspace of $V_n(q)$. Let $(w_1, w_2, \cdots, w_r)$ be an r-tuple of linearly independent vectors in W. Then, as above, there are $q^{r(r-1)/2}(q^r - 1)(q^{r-1} - 1)\cdots(q - 1)$ such r-tuples. Suppose there are G r-dimensional subspaces W of $V_n(q)$. Then the total number of linearly independent r-tuples in $V_n(q)$ is given by

$$N = q^{r(r-1)/2}(q^r - 1)(q^{r-1} - 1)\cdots(q - 1)G \tag{3.18}$$

It follows from equations (3.17) and (3.18) that $G = \begin{bmatrix} n \\ r \end{bmatrix}$, as desired. ■

For example, $V_3(q)$ contains $\begin{bmatrix} 3 \\ 1 \end{bmatrix} = 1 + q + q^2$ one-dimensional subspaces and $\begin{bmatrix} 3 \\ 2 \end{bmatrix} = 1 + q + q^2$ two-dimensional subspaces.

4) Since the number of linearly independent r-tuples in $V_n(q)$ equals $(q^n - 1)(q^n - q)\cdots(q^n - q^{r-1})$, by property (3), it follows that the number of $n \times n$ matrices over F_q equals $(q^n - 1)(q^n - q)\cdots(q^n - q^{n-1})$.
5) Since the $n \times n$ non-singular matrices over F_q form a group, it follows that the order of this group is $(q^n - 1)(q^n - q)\cdots(q^n - q^{n-1})$.
6) The n-dimensional *projective space* over F_q is an $(n + 1)$-dimensional vector space over F_q. It has $\begin{bmatrix} n+1 \\ 1 \end{bmatrix} = \frac{q^{n+1}-1}{q-1}$ points, and $\begin{bmatrix} n+1 \\ r+1 \end{bmatrix}$ r-flats, each containing $\begin{bmatrix} r+1 \\ 1 \end{bmatrix}$ points.[†]

[†]See P. J. Cameron, *Combinatorics: Topics, Techniques, Algorithms*, Cambridge University Press, New York, 1994.

4

Binomial Congruences

This chapter presents a number of interesting congruences involving binomial coefficients. We will study several primality tests for a positive integer, investigate binomial coefficients modulo a prime, and characterize odd binomial coefficients using binary numbers.

We begin our discussion with the elementary binomial congruence established in Corollary 2.2.

Theorem 4.1 Let p be a prime and $0 < k < p$. Then $\binom{p}{k} \equiv 0 \pmod{p}$.

Proof: By identity (2.1), $k\binom{p}{k} = p\binom{p-1}{k-1}$. Since $k < p, (p, k) = 1$. So $p|\binom{p}{k}$; that is, $\binom{p}{k} \equiv 0 \pmod{p}$, as desired. (Both the division algorithm and identity (2.2) also can be used to prove the congruence.) ■

Notice that Theorem 1.10 follows by PMI on a, using the binomial theorem and Theorem 4.1.

Theorem 4.1 plays a pivotal role in the proof of the following result. It shows how to establish Corollary 1.1 by invoking Theorem 4.1, but without Fermat's little theorem.

Theorem 4.2 Let p be a prime, and a and b any integers. Then $(a + b)^p \equiv a^p + b^p \pmod{p}$.

Proof: By the binomial theorem,

$$\begin{aligned}(a+b)^p &= a^p + \sum_{r=1}^{p-1} \binom{p}{r} a^{p-k} b^k + b^p \\ &\equiv a^p + 0 + b^p \equiv a^p + b^p \pmod{p}\end{aligned}$$ ■

There is a plethora of identities involving binomial coefficients. One such identity is the well-known *Lagrange's identity*: $\sum_{r=0}^{n} \binom{n}{r}^2 = \binom{2n}{n}$. (We will prove it in the next chapter.) For example,

$$\sum_{r=0}^{5} \binom{5}{r}^2 = 1^2 + 5^2 + 10^2 + 10^2 + 5^2 + 1^2 = 252 = \binom{10}{5}$$

Lagrange's identity, coupled with Theorem 4.1, provides an interesting result, which was proposed as a problem in 1959 by J. M. Gandhi of Belgaum, India. The proof given in the following example is based on one given in the same year by D. Zeitlin of Remington Rand Univac.

Example 4.1 If p is a prime, then $\binom{2p}{p} \equiv 2 \pmod{p}$.

Proof: Using Lagrange's identity and Theorem 4.1,

$$\begin{aligned}\binom{2p}{p} &= \sum_{r=0}^{p} \binom{p}{r}^2 = 2 + \sum_{r=1}^{p-1} \binom{p}{r}^2 \\ &\equiv 2 + 0 \equiv 2 \pmod{p}\end{aligned}$$

■

An Alternate Proof

We now present an alternate proof of Gandhi's congruence by T. M. Little of the University of California. His proof uses Corollary 3.4, and Theorems 1.10 and 4.1:

$$\begin{aligned}(1+1)^{2p} &= \sum_{r=0}^{2p} \binom{2p}{r} \\ 4^p &= 2 + \sum_{r=1}^{2p-1} \binom{2p}{r} \\ 4 &\equiv 2 + \binom{2p}{p} \pmod{p}\end{aligned}$$

This yields the desired result. ■

Moser extended Gandhi's congruence to modulo $2p$: $\binom{2p}{p} \equiv 2 \pmod{2p}$. In fact, he generalized it even further: $\frac{(pq)!}{(p!)^q} \equiv q! \pmod{pq}$, where p and q are primes.

For example, let $p = 5$ and $q = 3$. Then $\binom{2p}{p} = \binom{10}{5} = 252 \equiv 2 \pmod{10}$ and $\frac{(pq)!}{(p!)^q} = \frac{15!}{(5!)^3} = 756$, $756 \equiv 6 \equiv 3! \pmod{15}$.

Gandhi's congruence can be generalized as the next theorem shows. An application of Wilson's theorem, it appeared as an independent solution in 1959 by J. H. Hodges of Cornell Aeronautical Laboratory.

Theorem 4.3 Let m be an arbitrary positive integer and p any prime. Then

$$\binom{mp}{p} \equiv m \pmod{p} \tag{4.1}$$

Proof: Using Wilson's theorem, we have:

$$\begin{aligned}\binom{mp}{p} &= \frac{(mp)!}{p!(mp-p)!}\\ &= \frac{m(mp-1)\cdots(mp-p+1)}{(p-1)!} \equiv \frac{m(p-1)!(-1)^{p-1}}{(p-1)!} \pmod{p}\\ &\equiv m(-1)^{p-1} \equiv m \pmod{p}\end{aligned}$$

whether $p = 2$ or p is odd. This concludes the proof. ■

For example, when $m = 4$ and $p = 7$, $\binom{28}{7} = 1,184,040 \equiv 4 \pmod{7}$.

Four Additional Congruences

In 1819, the English mathematician Charles Babbage (1791–1871), known as the father of the *analytic engine*, reported in the *Edinburgh Philosophical Journal*[†] that if p is a prime, then $\binom{2p-1}{p-1} \equiv 1 \pmod{p^2}$; and $\binom{p+k}{p} \equiv 1 \pmod{p}$ if and only if p is a prime.

For example, let $p = 11$. Then $\binom{2p-1}{p-1} = \binom{21}{10} \equiv 352,716 \equiv 1 \pmod{11^2}$. Now let $k = 6$. Then $\binom{p+k}{p} = \binom{17}{11} \equiv 12,376 \equiv 1 \pmod{11}$.

About four decades later, J. Wolstenholme[*] proved that $\binom{2p-1}{p-1} \equiv 1 \pmod{p^3}$, if p is a prime > 3. For example, when $p = 17$, $\binom{2p-1}{p-1} = \binom{33}{16} \equiv 1,166,803,110 \equiv 1 \pmod{17^3}$.

In 1874, Catalan generalized this result slightly differently: If p is a prime, then $\binom{mp-1}{p-1} \equiv 1 \pmod{p}$. For instance, when $m = 3$ and $p = 13$, $\binom{3p-1}{p-1} = \binom{38}{12} \equiv 2,707,475,148 \equiv 1 \pmod{13}$.

Next, we present an interesting congruence, developed in 1951 by C. D. Olds of San Jose State College. It generalizes Catalan's congruence further. The proof, given in the following year by E. Trost of Techicum Winterthur, Zürich,

[†]Vol. 1, 1819, p. 40.

[*]*Quarterly Journal of Mathematics* 5 (1862), 35–39.

Switzerland, uses the following two facts,[‡] which we shall state as lemmas without proofs.

Lemma 4.1 If p is a prime, then $x^{p-1} \equiv (x-1)(x-2)\cdots(x-p+1) \pmod{p}$. ■

Lemma 4.2 Let p be an odd prime and s_r the sum of the products of the numbers $1, 2, \cdots, p-1$, taken r at a time, where $1 \leq r \leq p-2$. Then $s_r \equiv 0 \pmod{p}$. ■

Theorem 4.4 Let p be a prime > 3 and m an arbitrary positive integer. Then

$$\binom{mp-1}{p-1} \equiv 1 \pmod{p^3}$$

Proof: Let x be any integer. Then

$$\begin{aligned} x^{p-1} &\equiv (x-1)(x-2)\cdots(x-p+1) \pmod{p} \\ &\equiv x^{p-1} - s_1 x^{p-2} + s_2 x^{p-3} - \cdots - s_{p-2}x + s_{p-1} \pmod{p} \end{aligned}$$

We have $s_{p-1} = (p-1)!$. When $x = p$ and $p > 3$, by Lemma 4.2, this yields

$$\begin{aligned} (p-1)! &\equiv s_{p-3}p^2 - s_{p-2}p + s_{p-1} \pmod{p^3} \\ s_{p-2} &\equiv ps_{p-3} \equiv 0 \pmod{p^2} \end{aligned}$$

Now, let $x = mp$. Then:

$$\begin{aligned} \prod_{i=1}^{p-1}(mp-i) &\equiv s_{p-3}m^2p^2 - s_{p-2}mp + s_{p-1} \pmod{p^3} \\ &\equiv s_{p-1} \equiv (p-1)! \pmod{p^3} \end{aligned}$$

$$\frac{(mp-1)(mp-2)\cdots(mp-p+1)}{(p-1)!} \equiv 1 \pmod{p^3}$$

$$\binom{mp-1}{p-1} \equiv 1 \pmod{p^3}$$

as desired. ■

For example, let $m = 3$ and $p = 13$. Then $\binom{mp-1}{p-1} = \binom{38}{12} = 2{,}707{,}475{,}148 \equiv 1 \pmod{p^3}$.

The numbers s_i in Theorem 4.4 are the *Stirling numbers of the first kind*; see Chapter 20 for a discussion of such numbers.

[‡]G. H. Hardy and E. M. Wright, *An Introduction to the Theory of Numbers*, 5th edition, Oxford, 1979.

The following corollary follows from this theorem.

Corollary 4.1 Let p be a prime > 3, and m and n arbitrary positive integers. Then

$$\binom{mp-1}{p-1} \equiv \binom{np-1}{p-1} \pmod{p^3} \qquad \blacksquare$$

Interestingly enough, Theorem 4.1 can now be generalized, as the next theorem shows. Its proof uses Theorem 4.2 and Lagrange's theorem.[†]

Theorem 4.5 Let p be a prime and n a positive integer. Then $\binom{p^n}{r} \equiv 0 \pmod{p}$, where $0 < r < p^n$.

Proof: By the binomial theorem,

$$(x+1)^{p^n} = x^{p^n} + \sum_{r=1}^{p^n-1} \binom{p^n}{r} x^{p^n-r} + 1$$

$$x^{p^n} + 1 \equiv x^{p^n} + \sum_{r=1}^{p^n-1} \binom{p^n}{r} x^{p^n-r} + 1$$

$$\sum_{r=1}^{p^n-1} \binom{p^n}{r} x^{p^n-r} \equiv 0 \pmod{p}$$

This congruence in x has p incongruent solutions modulo p, namely, $0, 1, 2, \cdots, p-1$, so it can be written as a polynomial congruence in x with degree less than p. By Lagrange's theorem, a polynomial in x with degree m has at most m incongruent solutions modulo p; so the congruence has fewer than p incongruent solutions modulo p. This is possible only if the congruence is identically zero: that is, only if each coefficient is congruent to zero modulo p. Thus $\binom{p^n}{r} \equiv 0 \pmod{p}$, as desired. $\blacksquare$

For example, let $p = 3$, $n = 2$, and $1 \le r \le p^n - 1$. We have $\binom{p^n}{r} = \binom{9}{r}$. Then $\binom{9}{r} \equiv 0 \pmod{3}$, for every r, where $1 \le r \le 8$.

In 1972, W. A. Broomhead of Newcastle-upon-Tyne, England, proved Theorem 4.5 using the floor function.

Theorem 4.5 can be extended even further, as the next theorem shows. It appeared in 1959 as a solution to the Gandhi problem in Example 4.1 by N. G. Gunderson of the University of Rochester, Rochester, New York. The proof is a bit complicated, so we omit it.

[†] See the author's *Elementary Number Theory with Applications*.

Theorem 4.6 Let p be a prime > 3, and m and n arbitrary positive integers. Then

$$\binom{mp^n}{p} \equiv mp^{n-1} \pmod{p^{2n+1}}$$ ■

For example, let $p = 5$ and $m = 3 = n$. Then $mp^{n-1} = 75$, $p^{2n+1} = 78125$, and $\binom{mp^n}{p} = \binom{375}{5} = 60, 105, 468, 825$. Notice that $60, 105, 468, 825 \equiv 75 \pmod{78125}$, as expected.

Suppose we let $n = 1$ in Theorem 4.6. Then $\binom{mp}{p} \equiv m \pmod{p^3}$. So

$$\binom{mp}{p} \equiv m \pmod{p^2} \tag{4.2}$$

This result, true for every prime, appeared as a problem in 1966 by D. Singer of the University of California, Berkeley.

We now present a short, delightful proof of congruence (4.2), given in 1967 by P. R. Chernoff of Harvard University; it is an application of Corollary 3.4.

Chernoff's Proof

We have:

$$(1+x)^{mp} = [(1+x)^p]^m = \left[\sum_{r=0}^{p} \binom{p}{r} x^r\right]^m$$

Equating the coefficients of x^p from both sides, we get

$$\binom{mp}{p} = \sum_{r_1+\cdots+r_m=p} \binom{p}{r_1} \cdots \binom{p}{r_m}$$

Since $p|\binom{p}{r}$ for $0 < r < p$, every term in this sum is divisible by p^2, except when $r_i = p$; when $r_i = p$, $r_j = 0$ for $j \neq i$. Then each binomial coefficient equals 1, so RHS $= m$. Thus $\binom{mp}{p} = m \pmod{p^2}$. ■

For example, let $m = 6$ and $p = 5$. Then $\binom{mp}{p} = \binom{30}{5} = 142, 506 \equiv 6 \pmod{5^2}$, as expected.

Binomial Coefficients and Mersenne Numbers

The next theorem establishes a close link between odd binomial coefficients and Mersenne numbers. It was initially proposed in 1946 by D. H. Browne of Buffalo, New York. Solutions by E. P. Starke of Rutgers University, P. Bateman of

Philadelphia, and M. S. Knebelman of Washington State College, appeared a few months later. It was rediscovered in 1981 by R. M. Dacić of Instituit Mathematique, Belgrade, Serbia.[†]

Theorem 4.7 Let n be a fixed positive integer. The binomial coefficient $\binom{n}{r}$ is odd for all r if and only if n is a Mersenne number, where $0 \leq r \leq n$. ■

For example, $n = 2^3 - 1 = 7$ is a Mersenne number. It can be verified that every binomial coefficient $\binom{7}{r}$ is odd.

Theorem 4.7 also provides an interesting bridge between odd binomial coefficients $\binom{n}{r}$ and the binary expansion of n. To see this, let n be a Mersenne number $2^m - 1$. Then the binary expansion of n consists of m ones: $n = 2^m - 1 = \underbrace{11 \cdots 11}_{m \text{ ones}}$. Thus $\binom{n}{r}$ is odd for all r if and only if the binary expansion of n consists of all ones. Accordingly, Theorem 4.7 can be restated, as the following corollary shows.

Corollary 4.2 Let n be a fixed integer. Then every binomial coefficient $\binom{n}{r}$ is odd for all r if and only if the binary expansion of n consists of all ones, where $0 \leq r \leq n$. ■

For instance, $7 = 111_{\text{two}}$ and the binary expansion of 7 consists of 1s. Accordingly, every binomial coefficient $\binom{7}{r}$ is odd.

Number of Odd Binomial Coefficients

In Theorem 4.7, we learned that every binomial coefficient $\binom{n}{r}$ is odd if and only if n is a Mersenne number. Suppose n is not a Mersenne number and $n \geq 1$. Since $\binom{n}{0} = 1 = \binom{n}{n}$, there are at least two odd binomial coefficients in row n. When $n = 2$, there are exactly two odd binomial coefficients, and when $n = 6$, there are four such numbers.

More generally, the next theorem shows that the number of odd binomial coefficients is always a power of 2. It appeared as a problem in the 1956 William Lowell Putnam Mathematical Competition. Its proof uses *Cardano's rule* and the fact that $x \equiv y \pmod 2$ if and only if $hx \equiv ky \pmod 2$, where both h and k are odd integers.

Theorem 4.8 The number of odd binomial coefficients $\binom{n}{r}$ is a power of 2, where $0 \leq r \leq n$.

Proof: (by contradiction:) Clearly, when $n = 0$, 1, or 2, the number of odd binomial coefficients is a power of 2. So, assume that it is not a power of 2

[†]For a proof, see the author's *Elementary Number Theory with Applications*.

when $n > 2$. Let n be the least integer ≥ 3 such that the number of odd binomial coefficients $\binom{n}{r}$ is not a power of 2.

Case 1 Let $n = 2k$, where $k \geq 2$. Suppose r is odd. Then $n - r$ is odd and hence $\binom{n}{r} = \frac{n-r+1}{r}\binom{n}{r-1}$ is even. Thus r must be even for $\binom{n}{r}$ to be odd. Let $r = 2q$. Then:

$$\begin{aligned}
\binom{n}{r} &= \binom{2k}{2q} \\
&\equiv \frac{2 \cdot 4 \cdot 6 \cdots (2k)}{[2 \cdot 4 \cdot 6 \cdots (2q)][2 \cdot 4 \cdot 6 \cdots (2k-2q)]} \pmod{2} \\
&\equiv \frac{2^k\, k!}{(2^q\, q!)\left[2^{k-q}\,(k-q)!\right]} \pmod{2} \\
&\equiv \frac{k!}{q!(k-q)!} \equiv \binom{k}{q} \pmod{2}
\end{aligned}$$

Thus, $\binom{n}{r} = \binom{2k}{2q}$ is odd if and only if $\binom{k}{q}$ is odd. Consequently, the number of odd binomial coefficients $\binom{n}{r}$ is the same as the number of odd binomial coefficients $\binom{k}{q}$, where $k < n$. This contradicts our choice of n.

Case 2 Let $n = 2k + 1$, where $k \geq 1$. As in Case 1, it can be seen that r must be odd; let $r = 2q + 1$. Since $\binom{n}{r} = \binom{n}{n-r}$ and $r \not\equiv n - r \pmod{2}$, the number of odd binomial coefficients equals twice the number of binomial coefficients with r odd.

As before, $\binom{n}{r} = \binom{2k+1}{2q+1} = \binom{k}{q} \pmod{2}$, so the number of odd binomial coefficients $\binom{n}{r}$ equals twice the number of odd binomial coefficients $\binom{k}{q}$, where $k < n$. Once again, this is a contradiction.

Thus, the number of odd binomial coefficients is a power of 2. ■

For example, there are exactly eight odd binomial coefficients $\binom{11}{r}$:

$$\binom{11}{0} = \binom{11}{11},\ \binom{11}{1} = \binom{11}{10},\ \binom{11}{2} = \binom{11}{9},\ \binom{11}{3} = \binom{11}{8}$$

Piza's Congruence

The next congruence, proposed as a problem in 1948 by P. A. Piza of San Juan, Puerto Rico, provides an interesting link between binomial coefficients and the floor function. The proof is based on the one provided in 1950 by J. Barlaz of Rutgers University.

Theorem 4.9 (Piza, 1948) Let p be a prime. Then $\binom{n}{p} \equiv \lfloor n/p \rfloor \pmod{p}$.

Proof: Consider the p consecutive integers $n, n-1, \cdots, n-p+1$. They are congruent to the integers $1, 2, \cdots, p$ modulo p in some order. Besides, one of them, say, N, is divisible by p, so $\lfloor n/p \rfloor = N/p$. After removing N from the list, the remaining $p-1$ integers are congruent to $1, 2, \cdots, (p-1)$ modulo p in some order; thus:

$$\begin{aligned}
\frac{n(n-1)\cdots(n-p+1)}{N} &\equiv 1 \cdot 2 \cdot 3 \cdots (p-1) \pmod{p} \\
&\equiv (p-1)! \pmod{p} \\
\frac{n(n-1)\cdots(n-p+1)}{p} &\equiv \frac{N(p-1)!}{p} \pmod{p} \\
\frac{n(n-1)\cdots(n-p+1)}{p!} &\equiv \frac{N}{p} \pmod{p} \\
\binom{n}{p} &\equiv \frac{N}{p} \pmod{p} \\
&\equiv \lfloor n/p \rfloor \pmod{p}
\end{aligned}$$

as desired. (*Note*: Since $p|N$, $\frac{N}{p}$ is a positive integer.) ■

For example, $\binom{12}{7} = 792 \equiv 1 \equiv \lfloor 12/7 \rfloor \pmod{7}$ and $\binom{11}{5} = 462 \equiv 2 \equiv \lfloor 11/5 \rfloor \pmod{5}$.

This theorem has an interesting byproduct.

Corollary 4.3 (Piza, 1948) If $p^s | \lfloor n/p \rfloor$, then $p^s | \binom{n}{p}$. ■

For example, let $n = 110$ and $p = 3$. Then $\lfloor 110/3 \rfloor = 3^2 \cdot 4$, so $3^2 | \lfloor 110/3 \rfloor$. Notice that $3^2 | \binom{110}{3}$, where $\binom{110}{3} = 215820$. In fact, $2^2 | \lfloor 110/3 \rfloor$ and $2^2 | \binom{110}{3}$ also.

The next result, established by the French mathematician Francois-Edouard-Anatole Lucas (1842–1891)[†] around 1877, provides a delightful algorithm for computing the least residue of the binomial coefficient $\binom{m}{n}$ modulo a prime p. It takes advantage of the binary expansions of m and n in base p. The proof, based on the one given in 1947 by N. J. Fine of the University of Pennsylvania, uses both the binomial theorem and Corollary 1.1.

Theorem 4.10 (Lucas' Theorem) Let p be a prime, $m = m_0 + m_1 p + \cdots + m_k p^k$, and $n = n_0 + n_1 p + \cdots + n_k p^k$, where $0 \le m_i, n_i < p$. Then $\binom{m}{n} \equiv \binom{m_0}{n_0}\binom{m_1}{n_1}\cdots\binom{m_k}{n_k} \pmod{p}$.

[†]See L. E. Dickson, *History of the Theory of Numbers*, Vol. 1, Chelsea, New York, p. 271.

Proof: By the binomial theorem, we have:

$$\sum_{n=0}^{m}\binom{m}{n}x^n = (1+x)^m = \prod_{i=0}^{k}\left[(1+x)^{p^i}\right]^{m_i}$$

$$\equiv \prod_{i=0}^{k}\left[(1+x^{p^i})\right]^{m_i} \equiv \prod_{i=0}^{k}\left[\sum_{s_i=0}^{m_r}\binom{m_i}{n_i}x^{s_ip_i}\right] \pmod{p}$$

$$\equiv \sum_{n=0}^{m}\left[\sum\prod_{i=0}^{k}\binom{m_i}{s_i}\right]x^n \pmod{p}$$

where the inner sum is taken over all $(k+1)$-tuples $(s_0, s_1, \cdots, s_k)$ such that $\sum_{i=0}^{k} s_ip^i = n$. But $0 \le s_i \le m_i < p$, so there is at most one such $(k+1)$-tuple, given by $s_i = n_i$ if every $n \le m_i$, where $0 \le i \le k$; otherwise, the sum is zero. Since $\binom{m_i}{n_i} = 0$ for $m_i < n_i$, by equating the coefficients of x^n from both sides, we get $\binom{m}{n} \equiv \prod_{i=0}^{k}\binom{m_i}{n_i} \pmod{p}$, as desired. ■

For example, let $m = 29$, $n = 16$, and $p = 3$. Then:

$$\binom{m}{n} = \binom{29}{16} = 67,863,195 \equiv 0 \pmod{3}$$

$$29 = 1\cdot 3^3 + 0\cdot 3^2 + 0\cdot 3 + 2 = 1002_{\text{three}}$$

$$16 = 0\cdot 3^3 + 1\cdot 3^2 + 2\cdot 3 + 1 = 0121_{\text{three}}$$

$$\binom{1}{0}\binom{0}{1}\binom{0}{2}\binom{2}{1} = 1\cdot 0\cdot 0\cdot 2 = 0 \equiv \binom{29}{16} \pmod{3}$$

as expected.

Interestingly, Piza's theorem follows from Lucas' theorem, as shown below.

Corollary 4.4 (Piza's Theorem) Let p be a prime. Then $\binom{n}{p} \equiv \lfloor n/p \rfloor \pmod{p}$.

Proof: Let $n = n_0 + n_1p + \cdots + n_kp^k$. Since $p = 0 + 1\cdot p + 0\cdot p^2 + \cdots + 0\cdot p^k$, by Lucas' theorem, $\binom{n}{p} = \binom{n_1}{1} \equiv n_1 \pmod{p}$. But $n_1 \equiv \lfloor n/p \rfloor \pmod{p}$, so $\binom{n}{p} \equiv \lfloor n/p \rfloor \pmod{p}$. ■

Lucas' theorem provides another interesting byproduct, as the next corollary demonstrates. It was originally established in 1947 by N. J. Fine, and was proved independently 12 years later by E. H. Lockwood and P. Gant of Felsted School, Essex, England.

Corollary 4.5 (Fine, 1947) Let m be a fixed nonnegative integer and p a prime. Then the number of binomial coefficients $\binom{m}{n}$ *not* divisible by p is given by $\prod_{i=0}^{k}(m_i + 1)$, where $m = \sum_{i=0}^{k} m_i p^i = (m_k \cdots m_1 m_0)_p$ and $0 \le m_i < p$.

Proof: Let $n = (n_k \cdots n_1 n_0)_p$, where $0 \le n_i < p$. Since $0 \le n_i \le m_i < p$, there are $m_i + 1$ numbers $\binom{m_i}{n_i}$ not divisible by p. So there are $\prod_{i=0}^{k}(m_i + 1)$ products $\prod_{i=0}^{k}\binom{m_i}{n_i}$ not divisible by p. Thus, by Lucas' theorem, there are $\prod_{i=0}^{k}(m_i + 1)$ binomial coefficients $\binom{m}{n}$ such that $\binom{m}{n} \not\equiv 0 \pmod{p}$. ■

For example, since $29 = 2001_{\text{three}}$, there are exactly $(2 + 1)(0 + 1)(0 + 1)(1+1) = 6$ binomial coefficients $\binom{29}{r}$ that are not divisible by 3; they are $\binom{29}{0} = 1 = \binom{29}{29}$, $\binom{29}{1} = 29 = \binom{29}{28}$, and $\binom{29}{2} = 1 = \binom{29}{27}$. The remaining binomial coefficients are congruent to 0 modulo 3.

Number of Odd Binomial Coefficients

Corollary 4.5 can be successfully employed to determine the number of odd binomial coefficients. To this end, let $p = 2$. Then $m_i + 2 = 1$ if and only if $m_i = 1$; that is, each non-zero bit in the binary expansion of m contributes a factor of 2 to the product $\prod_{i=0}^{k}(m_i + 1)$ and vice versa. In other words, suppose there are e ones in the binary expansion of m. Then there are exactly 2^e odd binomial coefficients $\binom{m}{r}$.

Thus, we have the following result, which was also discovered independently by Lockwood and Gant.

Corollary 4.6 Let m be a fixed positive integer. Suppose the binary expansion of m contains e ones. Then there are exactly 2^e odd binomial coefficients $\binom{m}{r}$, where $0 \le r \le m$. ■

For example, since the binary representation of $23 = 11101_{\text{two}}$ contains four 1s, there are $2^4 = 16$ odd binomial coefficients $\binom{23}{r}$, as we saw earlier.

In 1967, Leonard Carlitz (1907–1999) of Duke University studied the number of binomial coefficients divisible by a fixed power of a prime; four years later, F. T. Howard of Wake Forest University, North Carolina, investigated the number of binomial coefficients divisible by a fixed power of 2.

Notice that there are at least two odd binomial coefficients $\binom{m}{r}$, namely, $\binom{m}{0}$ and $\binom{m}{1}$. So, next we determine the parity of the coefficient $\binom{m}{r}$.

Parity of a Binomial Coefficient

By Lucas' theorem, we have $\binom{m}{n} \equiv \binom{m_0}{n_0}\binom{m_1}{n_1}\cdots\binom{m_k}{n_k} \pmod 2$. Clearly, $\binom{m}{n}$ is odd if and only if every $\binom{m_i}{n_i}$ is odd; that is, if and only if each $\binom{m_i}{n_i} \equiv 1 \pmod 2$. In other words, $\binom{m}{n}$ is even if and only if $\binom{m_i}{n_i} \equiv 0 \pmod 2$ for some i.

Let us now consider the four possible cases:

Case 1 Let $m_i = 0 = n_i$. Then $\binom{m_i}{n_i} = 1$.
Case 2 Let $m_i = 0$ and $n_i = 1$. Then $\binom{m_i}{n_i} = 0$.
Case 3 Let $m_i = 1$ and $n_i = 0$. Then $\binom{m_i}{n_i} = 1$.
Case 4 Let $m_i = 1 = n_i$. Then $\binom{m_i}{n_i} = 1$.

Thus $\binom{m_i}{n_i} = 0$ if and only if $m_i = 0$ and $n_i = 1$. In other words, the binomial coefficient $\binom{m}{n}$ is even if and only if the binary expansion of m contains a 0 and the corresponding bit in the expansion of n is 1; that is, if and only if $m_i < n_i$ for some i.

This gives us the next corollary, which was also known to Lockwood and Gant in 1959.

Corollary 4.7 Let $m = (m_k \cdots m_1 m_0)_{\text{two}}$ and $n = (n_k \cdots n_1 n_0)_{\text{two}}$. Then the binomial coefficient $\binom{m}{n}$ is odd if and only if $m_i \geq n_i$ for every i, where $0 \leq i \leq k$. ■

For example, $23 = 10111_{\text{two}}$ and $10 = 01010_{\text{two}}$. Their binary representations contain a pair $\binom{m_3}{n_3} = \binom{0}{1}$, where $m_3 < n_3$, so the binomial coefficient $\binom{23}{10}$ is even: $\binom{23}{10} = 1,144,066$, an even number.

On the other hand, $23 = 10111_{\text{two}}$ and $19 = 10001_{\text{two}}$. Since $m_i \geq n_i$ for every i, $\binom{23}{19}$ is odd: $\binom{23}{19} = 8{,}855$, an odd number.

Lucas' theorem has yet another delightful application. It can successfully be employed to compute the remainder when $\binom{n}{r}$ is divided by p, as the next example illustrates.

Example 4.2 Compute the remainder when $\binom{527}{123}$ is divided by 7.

Solution We have $527 = 1354_{\text{seven}}$ and $121 = 0232_{\text{seven}}$. So, by Lucas' theorem,

$$\binom{527}{123} \equiv \binom{1}{0}\binom{3}{2}\binom{5}{3}\binom{4}{2} \equiv 1 \cdot 3 \cdot 10 \cdot 6 \equiv 5 \pmod 7$$

Thus, when $\binom{527}{121}$ is divided by 7, the remainder is 5. ■

Next, we pursue a primality test.

A Prime Characterization

In 1951, Piza developed another primality test. The test, furnished in the following theorem, is based on the proof provided in the following year by E. P. Starke of Rutgers University.

Theorem 4.11 A positive integer $2n + 1$ is a prime if and only if $\binom{n+r}{2r} \equiv 0 \pmod{2r + 1}$, where $0 < r < n$.

Proof: Let $Q_r = \frac{1}{2r+1}\binom{n+r}{2r}$. Suppose $2n + 1$ is composite. Let $p(< n)$ be a prime factor of $2n + 1$. Then

$$Q_{n-p} = \frac{(2n-p)!}{p!(2n-2p+1)!} = \frac{(2n-p)(2n-p-1)\cdots(2n-2p+2)}{p!}$$

Since none of the factors in the numerator are divisible by p, it follows that Q_{n-p} is not an integer; in other words, $\binom{n+r}{2r} \not\equiv 0 \pmod{2r + 1}$, when $r = n - p$.

On the other hand, let $2n + 1$ be a prime. Then $n - r$ and $n + r + 1$ are relatively prime. Since

$$(n-r)\binom{n+r+1}{2r+1} = (n+r+1)\binom{n+r}{2r+1}$$

it follows that $\binom{n+r}{2r+1}$ is divisible by $n - r$. Therefore, $Q_r = \frac{1}{2r+1}\binom{n+r}{2r} = \frac{1}{n-r}\binom{n+r}{2r+1}$ is an integer for every $n - r = 1, 2, 3, \cdots, (n - 1)$; that is, for $r = 1, 2, 3, \cdots, (n - 1)$.

This completes the proof. ■

For example, let $n = 6$. Then $2n + 1 = 13$ is a prime. It can be verified that $\binom{6+r}{2r} \equiv 0 \pmod{2r + 1}$ for each value of r, where $1 \le r \le 5$. Thus 13 is a prime, as expected.

On the other hand, let $n = 7$. When $r = 1$, $\binom{7+r}{2r} = 28 \not\equiv 0 \pmod 3$; so, by Theorem 4.11, 15 is not a prime, as we already know.

Next, we state the binomial congruence discovered in 1964 by H. Flanders of Purdue University. We omit its proof for convenience.

Theorem 4.12 Let $2^e \le n < 2^{e+1}$ and $0 \le k \le n - 2^e$. Then $\binom{n}{r} \equiv \binom{n-2^e}{r} \pmod 2$; that is, $\binom{n}{r}$ and $\binom{n-2^e}{r}$ have the same parity. ■

For example, let $e = 4$, $n = 25$, and $r = 7$. Then

$$\binom{n}{r} = \binom{25}{7} = 480700 \equiv 36 = \binom{9}{7} \equiv \binom{n-2^e}{r} \pmod 2$$

In 1965, E. S. Langford of North American Aviation, Anaheim, California, generalized Flanders' result, as the next theorem shows. Again, we omit the proof.

Theorem 4.13 Let $p^e \leq n < p^{e+1}$, $0 \leq r \leq n - p^e$, and $s \geq 0$. Then $\binom{n}{r} \equiv \binom{n-p^e}{r} \pmod{p^s}$. ■

For example, let $p = 3, e = 2, n = 17, r = 6$, and $s = 1$. Then $\binom{n}{r} = \binom{17}{6} = 12,376$ and $\binom{n-p^e}{r} = \binom{8}{6} = 28$. Notice that $12,376 \not\equiv 28 \pmod 3$.

Primality Tests

In 1994, E. T. H. Wang of Wilfrid Laurier University, Waterloo, Ontario, Canada, proposed an interesting primality test involving binomial coefficients. The solution presented here is based on the one given by D. Mubayi, then a student at Davidson College, Davidson, North Carolina.

Theorem 4.14 A positive integer $n + 1$ is a prime if and only if $r + 1 | \binom{n}{r}$ for all $0 \leq r \leq n - 1$.

Proof: We have

$$\binom{n+1}{r+1} = \frac{n+1}{r+1}\binom{n}{r}$$

Consequently, if $n + 1$ is a prime, then $(r + 1) | \binom{n}{r}$.

Conversely, suppose $n + 1$ is composite. Let p be the smallest prime factor of $n + 1$, so $p \leq \frac{n+1}{2}$. Consider the product

$$\frac{1}{p-1}\binom{n}{p-1} = \frac{n(n-1)\cdots(n-p+2)}{p!}$$

Since $p|(n + 1)$, the greatest integer $< (n + 1)$ that is divisible by p is $n + 1 - p$. Consequently, the RHS and hence $\frac{1}{p-1}\binom{n}{p-1}$ is not an integer. In other words, $p \nmid \binom{n}{p-1}$; that is, $(r + 1) \nmid \binom{n}{r}$, when $r = p - 1$.

This concludes the proof. ■

For example, let $n = 6$. Then $n + 1 = 7$ is prime and $(r + 1) | \binom{6}{r}$ for each value of r, where $0 \leq r \leq 5$.

In 1996, E. Deutsch of Polytechnic University, Brooklyn, New York, and I. M. Gessel of Brandeis University, Waltham, Massachusetts, presented an interesting characterization of prime numbers using binomial coefficients. The proof given here was given in the following year by H. M. Martson of Princeton, New Jersey.

Theorem 4.15 An integer $n \geq 2$ is prime if and only if $\binom{n-1}{r} \equiv (-1)^r \pmod n$, where $0 \leq r \leq n - 1$.

Proof: Let n be a prime. Then every i has a multiplicative inverse i^{-1} modulo n, where $1 \le i \le n-1$. Therefore:

$$\binom{n-1}{r} = \frac{(n-1)(n-2)\cdots(n-r)}{r!}$$

$$= \prod_{i=1}^{r} \frac{n-i}{i} \equiv \prod_{i=1}^{r} (n-i)i^{-1} \pmod{n}$$

$$= \prod_{i=1}^{r} (-1) \equiv (-1)^r \pmod{n}$$

Conversely, suppose n is composite. Let p be the smallest prime factor of n. Then $\binom{n-1}{p-1} \equiv (-1)^{p-1} \pmod{n}$ as above. But:

$$\binom{n-1}{p} = \binom{n-1}{p-1}\left(\frac{n}{p} - 1\right)$$

$$\equiv (-1)^{p-1}\left(\frac{n}{p} - 1\right) \not\equiv (-1)^p \pmod{n}$$

since $n/p \not\equiv 0 \pmod{n}$. This is a contradiction, so the desired result follows. ■

In 1997, S. Noltie of Ohio University, Lancaster, Ohio, gave an alternate proof of Theorem 4.15, based on Theorem 4.1.

Noltie's Alternate Proof

Let n be a prime. Then, by Theorem 4.1, $\binom{n}{r} \equiv 0 \pmod{n}$, for all $0 < r < n$. Clearly, $\binom{n-1}{0} = 1 \equiv (-1)^0 \pmod{n}$.

When $1 \le r < n$, by Pascal's identity,

$$\binom{n-1}{r} = \binom{n}{r} - \binom{n-1}{r-1}$$

$$= \binom{n-1}{r-1}\left(\frac{n}{r} - 1\right) \equiv -\binom{n-1}{r-1} \pmod{n}$$

Thus, by induction on r, $\binom{n-1}{r} \equiv (-1)^r \pmod{n}$.

Conversely, suppose that $\binom{n-1}{r} \equiv (-1)^r \pmod{n}$ for $0 \le r \le n-1$, and that n is composite. Let p be a prime factor of n. Then $\binom{n-1}{p-1} \equiv (-1)^{p-1} \pmod{n}$ and $\binom{n-1}{p} \equiv (-1)^p \pmod{n}$; so

$$\binom{n-1}{p-1} + \binom{n-1}{p} \equiv (-1)^{p-1} + (-1)^p \equiv 0 \pmod{n}$$

By Pascal's identity, this implies that $\binom{n}{p} \equiv 0 \pmod{n}$.

Let p^e be the largest power of p dividing n. Then $p^{e-1} \mid \binom{n}{p}$, but $p^e \nmid \binom{n}{p}$. Therefore,

$$\binom{n}{p} = \binom{n-1}{p-1} + \binom{n-1}{p} \not\equiv 0 \pmod{n}$$

This is a contradiction, so n must be a prime. ■

The fact that $\binom{p-1}{r} \equiv (-1)^r \pmod{p}$, where p is a prime and $r \leq p-1$, appears in Lucas' *Théorie des Nombres*, published in 1891.[†] The following congruences also appear in his book:

$$\binom{p-2}{r} \equiv (-1)^r(r+1) \pmod{p}$$

$$\binom{p-3}{r} \equiv (-1)^r\binom{r+2}{2} \pmod{p}$$

For example, let $p = 13$ and $r = 4$. Then $\binom{11}{4} = 330 \equiv 5 \equiv (-1)^4 5 \pmod{13}$ and $\binom{10}{4} = 21 \equiv 2 \equiv (-1)^4\binom{6}{2} \pmod{13}$.

[†]See L. E. Dickson, *History of the Theory of Numbers*, Vol. 1, Chelsea, New York, 1952, p. 272.

5

Binomial Coefficients Revisited

This chapter explores the binomial coefficients further and develops some important combinatorial identities. Many of them employ the binomial theorem; we will revisit them in the next two chapters.

One of the well-known identities is *Lagrange's identity*, given in the following theorem.

Theorem 5.1 (*Lagrange's identity*) Let $0 \leq r \leq n$. Then

$$\sum_{r=0}^{n} \binom{n}{r}^2 = \binom{2n}{n} \qquad (5.1) \quad ■$$

The proof follows easily by equating the coefficients of x^n from both sides of the identity $(1+x)^{2n} = (1+x)^n(x+1)^n$; so we omit the proof.

Lagrange's identity is a special case of the identity

$$\sum_{r=0}^{n+k} \binom{n}{r}\binom{n}{r-k} = \binom{2n}{n+k} \qquad (5.2)$$

This can be established by equating the coefficients of x^{n+r} from both sides of the identity $(1+x)^{2n} = (1+x)^n(x+1)^n$.

We will now establish Lagrange's identity combinatorially.

Two Combinatorial Proofs

Recall from Chapter 2 that $\binom{n}{r}$ counts the number of lattice paths from the origin to $(n, n-r)$, where $0 \leq r \leq n$. So $\binom{2n}{n}$ counts the number of lattice paths from the origin to $(2n, n)$.

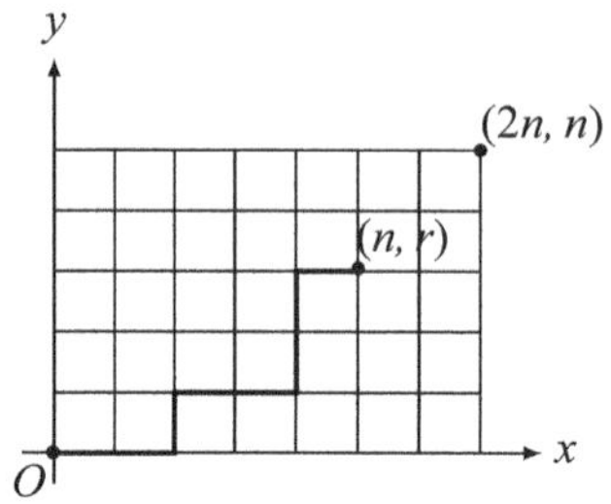

Figure 5.1

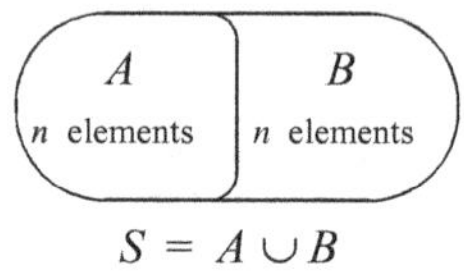

Figure 5.2

We will now count it in a different way; see Figure 5.1. Since the number of lattice paths from (n, r) to $(2n, n)$ = number of paths from O to $(n, n-r) = \binom{n}{r}$, it follows by the multiplication principle that:

$$\begin{pmatrix}\text{Number of lattice}\\ \text{paths from } O \text{ to } (2n, n)\\ \text{passing through } (n, r)\end{pmatrix} = \begin{pmatrix}\text{Number of}\\ \text{lattice paths from}\\ O \text{ to } (n, r)\end{pmatrix}\begin{pmatrix}\text{Number of}\\ \text{lattice paths from}\\ (n, r) \text{ to } (2n, n)\end{pmatrix}$$

$$= \binom{n}{r}\binom{n}{r} = \binom{n}{r}^2$$

So the total number of lattice paths from O to $(2n, n)$ is given by $\sum_{r=0}^{n} \binom{n}{r}^2$. Thus $\sum_{r=0}^{n} \binom{n}{r}^2 = \binom{2n}{n}$, as desired. ■

The second argument is basically the same. Consider a $2n$-member committee S. It has $\binom{2n}{n}$ n-member subcommittees.

Now, partition the committee into two subsets A and B of n elements each; see Figure 5.2. To form a subcommittee of n people from S, we choose r members from A and $n-r$ members from B; they can be chosen in $\binom{n}{r}$ and $\binom{n}{n-r} = \binom{n}{r}$ different ways, respectively. So we can form $\binom{n}{r}\binom{n}{r} = \binom{n}{r}^2$ such n-member subcommittees. Since this is true for every r, we can form a total of $\sum_{r=0}^{n} \binom{n}{r}^2$ n-member subcommittees. Thus, $\sum_{r=0}^{n} \binom{n}{r}^2 = \binom{2n}{n}$, as expected. ■

Vandermonde's Identity

Identities (5.1) and (5.2) are special cases of the well-known Vandermonde's identity, named after the French mathematician and musician Alexandre-Théophile Vandermonde (1735–1796), and presented in the following theorem. Vandermonde published it in 1772 in *Memoires de l'Academie Royal de Sciences*. However, it had already appeared in 1303 in *The Precious Mirror of the*

Four Elements by the Chinese mathematician Chu Shi-Kié. We will prove the identity in two ways: first using the binomial theorem and then using a simple combinatorial argument.

Theorem 5.2 (Vandermonde's identity) Let $m, n, r \geq 0$, where $0 \leq r \leq m, n$. Then

$$\binom{m+n}{r} = \sum_{k=0}^{r} \binom{m}{k}\binom{n}{r-k} \tag{5.3}$$

Proof: The algebraic proof follows by equating the coefficients of x^r from both sides of the equation $(1+x)^{m+n} = (1+x)^m(1+x)^n$. We omit the details in the interest of brevity. ∎

We now give a combinatorial proof of Vandermonde's identity.

A Combinatorial Argument

Suppose there are m apples and n oranges in a basket, and we would like to select r of the $m+n$ fruits. Clearly, this can be done in $\binom{m+n}{r}$ ways.

Here is another way of selecting r fruits from the $m+n$ fruits: pick k fruits from the m apples and the remaining $r-k$ fruits from the n oranges. By the multiplication principle, this can be accomplished in $\binom{m}{k}\binom{n}{r-k}$ ways. This is true for each k, where $0 \leq k \leq r$; so, by the addition principle, the total number of ways r fruits can be selected is given by $\sum_{k=0}^{r} \binom{m}{k}\binom{n}{r-k}$. Thus

$$\binom{m+n}{r} = \sum_{k=0}^{r} \binom{m}{k}\binom{n}{r-k}$$

as desired. ∎

For example, let $m = 7, n = 5$, and $r = 4$. Then:

$$\sum_{k=0}^{4} \binom{7}{k}\binom{5}{4-k} = \binom{7}{0}\binom{5}{4} + \binom{7}{1}\binom{5}{3} + \binom{7}{2}\binom{5}{2} + \binom{7}{3}\binom{5}{1} + \binom{7}{4}\binom{5}{0} = 495 = \binom{7+5}{4}$$

Interesting Byproducts

Vandermonde's identity can be employed to deduce several interesting results. For example, Lagrange's identity follows by letting $m = n = r$ and $k = 0$. By letting

$m = n$, it yields the identity

$$\sum_{r=0}^{n-1} \binom{n}{r}\binom{n}{r+1} = \binom{2n}{n+1}$$

By letting $r = m - t$, Vandermonde's identity yields another identity:

$$\binom{m+n}{m-t} = \sum_{k=0}^{m-t} \binom{m}{m-t-k}\binom{n}{k}$$

That is,

$$\sum_{k=0}^{m} \binom{m}{t+k}\binom{n}{k} = \binom{m+n}{n+t} \tag{5.4}$$

where we have changed the upper limit in the sum to m, although $k \le m - t$.

The binomial coefficient $\binom{m+n}{m} = \binom{m+n}{n}$ has a nice geometric interpretation. It was developed in 1938 in the Italian journal *Periodico di Matematiche* by C. Ciamberlini and A. Margengoni. J. Ginsburg of Yeshiva University, New York reported it in *Scripta Mathematica* in the same year.

Before presenting it, we need the following result from combinatorics. We omit its proof for convenience.

Theorem 5.3 The number of permutations of n items of which n_1 items are of one type, n_2 are of a second type, $\cdots$, and n_k are of a kth type, is $\frac{n!}{n_1!n_2!\cdots n_k!}$, where $n = n_1 + n_2 + \cdots + n_k$. ■

For example, there are $\frac{6!}{3!2!} = 60$ permutations of the letters in the word BANANA. In other words, suppose there are three different books, B, N, and A; they have 1, 2, and 3 identical copies, respectively. Then they can be arranged on a book shelf in $\frac{6!}{2!3!} = 60$ different ways.

We now present a geometric interpretation of the binomial coefficient $\binom{m+n}{m}$.

Lattice Walking

Consider a city divided into avenues and streets, as Figure 5.3 shows. The avenues run west to east and the streets run south to north. Suppose a tourist would like to find the number of paths she can choose from point O to point A. She can travel in the easterly or northerly direction only.

Two possible paths are shown in the figure. The heavy route can be represented by the nine-letter word *EENNENEEN*; it means, travel two blocks east, two blocks north, one block east, one block north, two blocks east, and one block north. The second route, *NEEEENNNE*, can be interpreted similarly.

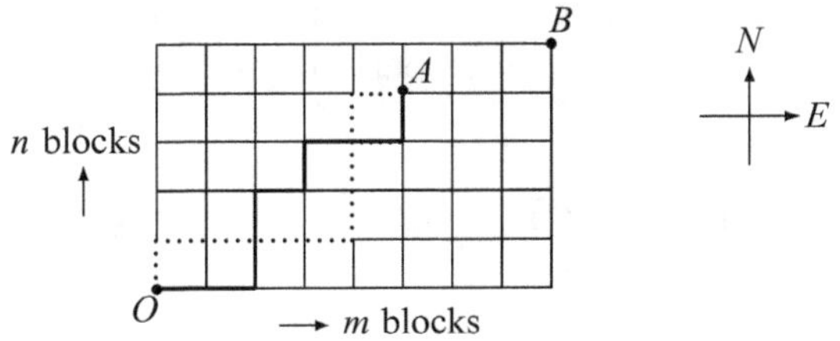

Figure 5.3 A Portion of a City Map

In fact, every route from O to A can be represented by a word of nine letters; five of them are alike (Es) and four are alike (Ns). Therefore:

$$\begin{pmatrix}\text{total number of}\\ \text{paths from } O \text{ to } A\end{pmatrix} = \begin{pmatrix}\text{number of 9-letter words of which five}\\ \text{are } E\text{s and the other four are } N\text{s}\end{pmatrix}$$

$$= \frac{9!}{5!4!} = \binom{9}{5}$$

More generally, suppose point B is m blocks east of O and n blocks north of O; see Figure 5.3. Then there are $\binom{m+n}{m} = \binom{m+n}{n}$ paths from O to B, covering exactly $m+n$ blocks. In particular, if $m = n$, then there are $\binom{2n}{n}$ such paths from O to B.

The following example derives a new combinatorial identity, using the binomial theorem. We will revisit it in Chapter 6.

Example 5.1 Prove that $\sum_{r=0}^{\lfloor n/2 \rfloor} \binom{n}{r}\binom{n-r}{r}2^{n-2r} = (n+1)C_n$, where C_n denotes the nth Catalan number.

Proof: The proof follows by equating the coefficients of x^n from both sides of the equation $(1+x)^{2n} = [(1+x)^2]^n = [1+x(2+x)]^n$. We omit the details in the interest of brevity. ■

For example, let $n = 7$. Then:

$$\sum_{r=0}^{3}\binom{7}{r}\binom{7-r}{r}2^{7-2r} = \binom{7}{0}\binom{7}{0}2^7 + \binom{7}{1}\binom{6}{1}2^5 + \binom{7}{2}\binom{5}{0}2^3$$
$$+ \binom{7}{3}\binom{4}{3}2 = 3,432 = 8C_7$$

For those familiar with Catalan numbers and the theory of trees, we add that there are C_n ordered rooted trees with n edges each; $(n+1)C_n$ counts the total number of vertices on them, and exactly one-half of them are leaves.[†]

[†]See L. W. Shapiro, Problem 19753, *American Mathematical Monthly*, 106 (1999), 777.

The following example develops an identity using Corollary 3.4 and differential calculus. Interestingly, Catalan numbers pop up again. We will revisit this formula also in the next chapter.

Example 5.2 Prove that $\sum_{r=1}^{n} r\binom{n}{r}^2 = n(2n-1)C_{n-1}$.

Proof: By Corollary 3.4, we have $(1+x)^n = \sum_{r=0}^{n} \binom{n}{r}x^r$. Differentiate both sides with respect to x:

$$n(1+x)^{n-1} = \sum_{r=1}^{n} r\binom{n}{r}x^{r-1}$$

Then

$$nx(1+x)^{n-1} = \sum_{r=1}^{n} r\binom{n}{r}x^r$$

Now, replace x with $\frac{1}{x}$:

$$\frac{n}{x}\left(1+\frac{1}{x}\right)^{n-1} = \sum_{r=1}^{n} \frac{r}{x^r}\binom{n}{r}$$

$$n(1+x)^{n-1} = \sum_{r=1}^{n} r\binom{n}{r}x^{n-r}$$

Multiply both sides by $(1+x)^n = \sum_{r=0}^{n} \binom{n}{r}x^r$:

$$n(1+x)^{2n-1} = \left[\sum_{r=1}^{n} r\binom{n}{r}x^{n-r}\right]\left[\sum_{r=0}^{n} \binom{n}{r}x^r\right]$$

Equating the coefficients of x^n from both sides yields the desired result. ■

For example,

$$\sum_{r=1}^{6} r\binom{6}{r}^2 = 1\binom{6}{1}^2 + 2\binom{6}{2}^2 + 3\binom{6}{3}^2 + 4\binom{6}{4}^2 + 5\binom{6}{5}^2 + 6\binom{6}{6}^2$$

$$= 2,772 = 6(2\cdot 6-1)C_5$$

as expected.

The next example provides a delectable convergence of CBCs, differential and integral calculus, and generating functions. We will revisit the resulting identity in the next chapter.

Example 5.3 Prove that

$$\sum_{r=0}^{n} \binom{2r}{r}\binom{2n-2r}{n-r} = 4^n \tag{5.5}$$

Proof: Recall from Chapter 2 that the CBCs are generated by the function $\frac{1}{\sqrt{1-4x}}$:

$$\frac{1}{\sqrt{1-4x}} = \sum_{k=0}^{\infty} \binom{2k}{k} x^n$$

Squaring both sides, we get:

$$\frac{1}{1-4x} = \left[\sum_{r=0}^{\infty} \binom{2r}{r} x^r\right]\left[\sum_{k=0}^{\infty} \binom{2k}{k} x^k\right]$$

That is,

$$\sum_{n=0}^{\infty} 4^n x^n = \sum_{n=0}^{\infty}\left[\sum_{r=0}^{n} \binom{2r}{r}\binom{2n-2r}{n-r}\right] x^n$$

Equating the coefficients of x^n from both sides yields the desired result. ■

For example,

$$\sum_{r=0}^{4} \binom{2r}{r}\binom{8-2r}{4-r} = \binom{0}{0}\binom{8}{4} + \binom{2}{1}\binom{6}{3} + \binom{4}{2}\binom{4}{2} + \binom{6}{3}\binom{2}{1} + \binom{8}{4}\binom{0}{0}$$

$$= 70 + 40 + 36 + 40 + 70 = 4^4, \quad \text{as expected.}$$

Combinations with Repetitions

Just as there are permutations with repeated elements, there are combinations with repeated elements; they are called *multisets*. For brevity, the multiset $\{a, a, a, b, b, c\}$, for example, is denoted by $\{a^3, b^2, c\}$

Suppose three friends go to a local restaurant for beverages – coffee (c) or hot tea (t). The waitress puts all three requests on the same order. How many different orders are possible? Clearly, the order in which the beverages are selected is immaterial; the same beverage can be selected by more than one person; and not every beverage need be selected.

The various 3-combinations of the two-element set $\{c, t\}$ are $\{c^3\}$, $\{c^2, t\}$, $\{c, t^2\}$, and $\{t^3\}$; so there are four possible different beverage orders. Likewise, there are ten 3-combinations with repetitions of the set $\{x, y, z\}$; we ask you to list them.

More generally, the following theorem gives the number of r-combinations with repetitions.

Theorem 5.4 The number of r-combinations with repetitions of a set of n elements is $\binom{n+r-1}{r}$.

Proof: Let $S = \{x_1, x_2, \cdots, x_n\}$. Each r-combination with repeated elements from S elements can be considered a word of r Xs and $n-1$ commas. Each word contains $r+n-1 = n+r-1$ characters; r of them are Xs and $n-1$ are commas: $\underbrace{X\cdots X, X\cdots X, \cdots, X\cdots X}_{r\ X\text{s and } n-1 \text{ commas}}$. Consequently, by Theorem 5.3, the number of such strings, that is, the number of r-combinations, equals $\frac{n+r-1}{r!(n-1)!} = \binom{n+r-1}{r}$. ■

The following example illustrates this theorem.

Example 5.4 There are six choices of a soft drink – Coke Classic, Diet Coke, root beer, Mountain Dew, orange juice, and Sprite – at a local restaurant. Find the number of beverage orders 11 patrons can make.

Solution There are six choices of soft drinks, so $n = 6$. Each beverage order is an 11-combination of the soft drinks, with repetitions. Therefore, by Theorem 5.4, there are $\binom{n+r-1}{r} = \binom{6+11-1}{11} = \binom{16}{11} = 4{,}368$ possible beverage orders. ■

Example 5.4 has a delightful number-theoretic interpretation. To see this, let x_i denote the number of patrons ordering soft drink i, where $1 \le i \le 6$. Then $x_1 + x_2 + x_3 + x_4 + x_5 + x_6 = 11$, where x_i is a nonnegative integer. The number of nonnegative integer solutions of this *linear diophantine equation* (LDE)† is the same as the number of possible beverage orders, so it has 4,368 such solutions. For example, (2,0,3,5,1,1) and (0,1,0,2,9,0) are two of them.

The next theorem, which is a re-statement of Theorem 5.4, generalizes this result.

Theorem 5.5 Let $x_1, x_2, \cdots, x_n$ be nonnegative integer variables and r a nonnegative integer. The LDE $x_1 + x_2 + \cdots + x_n = r$ has $\binom{n+r-1}{r}$ nonnegative integer solutions. ■

For example, the LDE $x_1 + x_2 + x_3 + x_4 = 7$ has $\binom{n+r-1}{r} = \binom{4+7-1}{7} = \binom{10}{7} = 120$ solutions.

Since there are $\binom{n+r-1}{r}$ nonnegative integer solutions to the LDE $x_1 + x_2 + \cdots + x_n = r$, it follows that there are $\binom{n+r-1}{r}$ ordered ways of expressing r as a sum of at most n nonnegative integers.

For example, the LDE $x_1 + x_2 + x_3 + x_4 = 2$ has $\binom{4+2-1}{2} = 10$ solutions: 0011, 0101, 0110, 1010, 1001, 1100, 2000, 0200, 0020, and 0002. Consequently, there are 10 ordered ways of writing 2 as a sum of at most four nonnegative integers: $0+0+1+1, 0+1+0+1, \cdots$, and $0+0+0+2$.

†A *linear diophantine equation* is an equation of the form $x_1 + x_2 + \cdots + x_n = r$, where each x_i is an integer variable.

Next, we present some additional applications of the binomial coefficient $\binom{n+r-1}{r}$.

Triangular and Tetrahedral Numbers Revisited Again

Returning to the triangular numbers t_n and tetrahedral numbers T_n in Chapter 1, it is interesting to note that their formulas do indeed fit into the formula in Theorem 5.4:

$$t_n = \binom{n+1}{2} = \binom{n+2-1}{2}$$

$$T_n = \binom{n+2}{3} = \binom{n+3-1}{3}$$

For Loops

The next two examples illustrate how these two formulas pop up in computer science.

Example 5.5 Find the number of times the assignment statement $x \leftarrow x + 1$ is executed by the following *nested for* loops:†

for $i = 1$ to n do

for $j = 1$ to i do

$x \leftarrow x + 1$

Solution Notice that the variable i in the outer loop controls the value of j: namely, the number of times the inner loop is executed, where $1 \leq j \leq i \leq n$. In the inner loop, the assignment statement is executed for each value of j. Since $1 \leq j \leq i$, there are i legal values for j; in other words, the inner loop is executed i times. But $1 \leq i \leq n$, so the total number of times the assignment statement is executed is given by $\sum_{i=1}^{n} i = \frac{n(n+1)}{2} = t_n$.

Unfortunately, this approach does not reveal the applicability of Theorem 5.4 in this example, so we take a different route.

A Combinatorial Approach

It follows from the above solution that the assignment statement is executed for each pair (i, j), where $1 \leq j \leq i \leq n$. Notice that the statement is *not* executed

†The *assignment statement* $x \leftarrow E(x)$ means, store the value of expression $E(x)$ in location x; the left-hand arrow $\leftarrow$ is the *assignment operator*.

when $j > i$. Thus the number of executions equals the number of positive integer pairs (j, i) with $j \leq i$, where repetitions are allowed and $i \leq n$; in other words, it equals the number of 2-combinations of the set $\{1, 2, \cdots, n\}$. The number of such 2-combinations is $\binom{n+2-1}{2}$, so the assignment statement is executed in $\binom{n+2-1}{2} = \binom{n+1}{2} = t_n$ different ways.

We now redo this example with one more level of *for* loop added.

Example 5.6 Find the number of times the assignment statement $x \leftarrow x+1$ is executed by the following nested *for* loops:

for $i = 1$ to n do

for $j = 1$ to i do

for $k = 1$ to j do

$x \leftarrow x + 1$

Solution Algebraically, the number of times the given assignment statement is executed is given by:

$$\sum_{i=1}^{n}\sum_{j=1}^{i}\sum_{k=1}^{j} k = \sum_{i=1}^{n}\sum_{j=1}^{i} j = \sum_{i=1}^{n} \frac{i(i+1)}{2}$$

$$= \frac{1}{2}\sum_{i=1}^{n}(i^2 + i) = \frac{1}{2}\left[\frac{n(n+1)(n+2)}{6} + \frac{n(n+1)}{2}\right]$$

$$= \frac{n(n+1)(n+2)}{6} = \binom{n+2}{3} = T_n$$

Using a combinatorial argument, the assignment statement is executed for every triple (k, j, i), where repetitions are allowed and $1 \leq k \leq j \leq i \leq n$. The number of such 3-combinations equals $\binom{n+3-1}{3}$, so the assignment statement is executed $\binom{n+3-1}{3} = \binom{n+2}{3} = T_n$ times. ■

Next, we investigate a generalization of these nested *for* loops.

A Generalized Formula

Consider the following sequence of nested *for* loops consisting of r levels:

for $i_1 = 1$ to n do

for $i_2 = 1$ to i_1 do

for $i_3 = 1$ to i_2 do

⋮

for $i_r = 1$ to i_{r-1} do

$x \leftarrow x + 1$

Using the above combinatorial argument, it follows that the number of times the assignment statement is executed by these nested *for* loops equals the number of r-combinations of the set $\{1, 2, 3, \cdots, n\}$ with repetitions; namely, $\binom{n+r-1}{r}$.

An Interesting Byproduct

An interesting combinatorial identity can be derived by applying the formulas in Examples 5.5 and 5.6. The assignment statement inside the following nested *for* loops is executed $\binom{i+1}{2}$ times:

$$\begin{array}{l} \text{for } j = 1 \text{ to } i \text{ do} \\ \quad \text{for } k = 1 \text{ to } j \text{ do} \\ \qquad x \leftarrow x + 1 \end{array}$$

So the total number of times the statement is executed by the given nested *for* loops in Example 5.5 equals $\sum_{i=1}^{n} \binom{i+1}{2}$. Thus

$$\sum_{i=1}^{n} \binom{i+1}{2} = \binom{n+2}{3}$$

In other words,

$$\binom{2}{2} + \binom{3}{2} + \binom{4}{2} + \cdots + \binom{n}{2} = \binom{n+1}{3}$$

Using exactly the same argument, it follows from the general discussion of the above r nested *for* loops that

$$\sum_{k=1}^{n} \binom{k+r-2}{r-1} = \binom{n+r-1}{r}$$

That is,

$$\binom{r-1}{r-1} + \binom{r}{r-1} + \binom{r+1}{r-1} + \cdots + \binom{n+r-2}{r-1} = \binom{n+r-1}{r}$$

In other words,

$$\binom{r}{r} + \binom{r+1}{r} + \binom{r+2}{r} + \cdots + \binom{n}{r} = \binom{n+1}{r+1} \tag{5.6}$$

For example, $\binom{4}{4} + \binom{5}{4} + \binom{6}{4} + \binom{7}{4} + \binom{8}{4} + \binom{9}{4} = 252 = \binom{10}{5}$.

We will revisit formula (5.6) in the next chapter.

Additional Byproducts

Finally, using identity (5.6), we can develop the following familiar summation formulas and more:

$$\sum_{k=1}^{n} k = \frac{n(n+1)}{2} \qquad \sum_{k=1}^{n} k^2 = \frac{n(n+1)(2n+1)}{6} \qquad \sum_{k=1}^{n} k^3 = \left[\frac{n(n+1)}{2}\right]^2$$

For example, since $k^3 = 6\binom{k}{3} + 6\binom{k}{2} + \binom{k}{1}$, by identity (5.6), we have:

$$\begin{aligned}\sum_{k=1}^{n} k^3 &= 6\sum_{k=1}^{n}\binom{k}{3} + 6\sum_{k=1}^{n}\binom{k}{2} + \sum_{k=1}^{n}\binom{k}{1} \\ &= 6\binom{n+1}{4} + 6\binom{n+1}{3} + \binom{n+1}{2} \\ &= 6\left[\binom{n+1}{4} + 6\binom{n+1}{3}\right] + \binom{n+1}{2} = 6\binom{n+2}{4} + \binom{n+1}{2} \\ &= 6 \cdot \frac{(n+2)(n+1)n(n-1)}{24} + \frac{(n+1)n}{2} \\ &= \frac{n(n+1)}{4}\left[(n^2+n-2)+2\right] = \left[\frac{n(n+1)}{2}\right]^2\end{aligned}$$

Similarly, it can be shown that:

$$\sum_{k=1}^{n} k^4 = \frac{n(n+1)(6n^3+9n^2+n-1)}{30} \qquad \sum_{k=1}^{n} k^5 = \frac{n^2(n+1)(2n^3+4n^2+n-1)}{12}$$

Identity (5.6) can be used to evaluate other combinatorially related sums as well. For example:

$$\begin{aligned}\sum_{k=1}^{n} k(k+1)\cdots(k+m-1) &= \sum_{k=1}^{n} \frac{(k+m-1)!}{(k-1)!} = m!\sum_{k=1}^{n}\binom{k+m-1}{k-1} \\ &= m!\sum_{k=1}^{n}\binom{k+m-1}{m} = \binom{n+m}{m+1}m!\end{aligned}$$

In particular, $\sum_{k=1}^{n} k(k+1)(k+2)(k+3) = 24\binom{n+4}{5}$.

6

Pascal's Triangle

In the preceding chapters, we saw many intriguing properties of binomial coefficients. In this chapter, we will revisit a number of them through the lens of *Pascal's triangle*. Pascal wrote *Treatise on the Arithmetic Triangle* about the triangular array in 1653 (published posthumously in 1665). In his book, Pascal shows a relationship between binomial coefficients and polynomials. Although the array is named after Pascal, it had appeared in 1529 on the title page of an arithmetic book by the German astronomer Petrus Apianus (1495–1552) – the first appearance of the triangular array in the Western world.

Similar forms of the triangular array were given by Stifel in his mathematical work *Arithmetica Integra*, published at Nuremburg in 1544; by the German mathematician Johannes Scheubel (1494–1570) in 1545; by the German mathematician Christoff Rudolff (*c*.1500–1545) in the revised version of his 1525 algebra book *Die Coss*, published in 1553; by the Italian mathematician Niccolo Tartaglia (1499?–1557) in 1556; by the Dutch mathematician Simon Stevin (1548–1620); by the Dutch mathematician Albert Girard (1595–1632) in 1629; by the English mathematician William Oughtred (1574–1660) in 1631; and by the English mathematician Henry Briggs (1561–1631) in 1633. Tartaglia used Pascal's triangle to expand $(x + y)^{12}$ and Cardano used it in his study of combinations and polygonal numbers.

In fact, the array made an appearance more than two centuries earlier, in Shi-Kié's *The Precious Mirror of the Four Elements* (see Figure 6.2). According to Shi-Kié, the triangle appeared even earlier, in one of the works by Yang Hui Yang around 1275. The Persian poet, astronomer, and mathematician Omar Khayyàm knew of it about 1100, probably from even earlier Chinese or Indian sources.

However, it is generally accepted that Pascal was the first person to investigate extensively the properties of the triangular array and their applications to the theory of probability. So the array has come to be known as *Pascal's triangle*. According to Eves, "in Pascal's treatise on the triangle appears one of the earliest acceptable statements of the method of mathematical induction."

Cardano presents two forms of Pascal's triangle (see Figures 6.5 and 6.6) and its various applications in his *Opus Novum de Proportionibus*, published in 1570. In his determining of the roots of numbers, Cardano used the familiar form, citing Stifel as the putative discoverer. He gave the various elements of the array through $n = 17$. Later on in the *Opus Novum*, however, he gave the array through $n = 11$ in virtually the same form that made Pascal famous.

The Swiss mathematician James Bernoulli (1654–1705) states in his 1713 book *Ars Conjectandi* (published posthumously) that "the essence of combinations is concealed in it, but those who are intimately acquainted with geometry know also that capital secrets of all mathematics are hidden in it." In fact, Pascal himself wrote, "I leave out many more than I include; it is extraordinary how fertile in properties this is. Everyone can try his hand."

In 1987, P. Hilton of the State University of New York at Binghamton and J. Pedersen of Santa Clara University, Santa Clara, California, wrote: "Pascal's triangle provides a laboratory for students [and professionals alike] to conduct experiments: looking for patterns, formulating from their observations some general statements, and then trying to prove them." So we begin our rewarding investigation of Pascal's triangle.

Pascal's Triangle

The various binomial coefficients $\binom{n}{r}$, where $0 \leq r \leq n$, can be arranged in the form of an equilateral triangular array. Replacing each binomial coefficient with its actual numeric value yields the familiar Pascal's triangle in Figure 6.1.

```
                        1                    ← row 0
                    1       1
                1       2       1
            1       3       3       1
        1       4       6       4       1
    1       5      10      10       5       1
1       6      15      20      15       6       1
1   7      21      35      35      21       7       1
```

Figure 6.1 Pascal's Triangle

It is worth noting that Pascal worked with the rectangular array in Figure 6.5 (see Pascal's own drawing in Figure 6.4), the same as Tartaglia's version, while both the Chinese (Figure 6.2) and Japanese (Figure 6.3) versions mirror the present

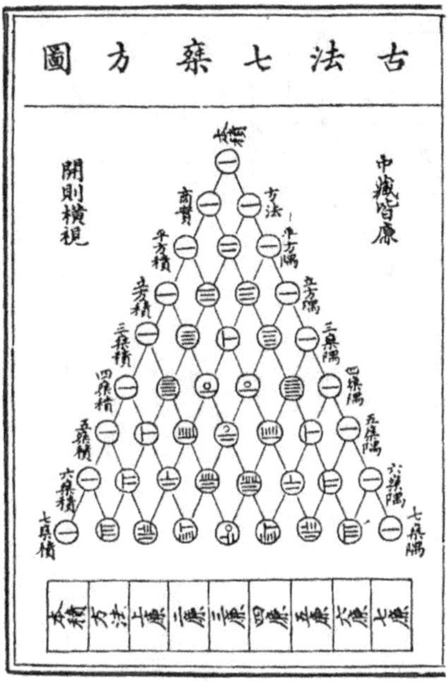

Figure 6.2 Chinese Version of Pascal's Triangle

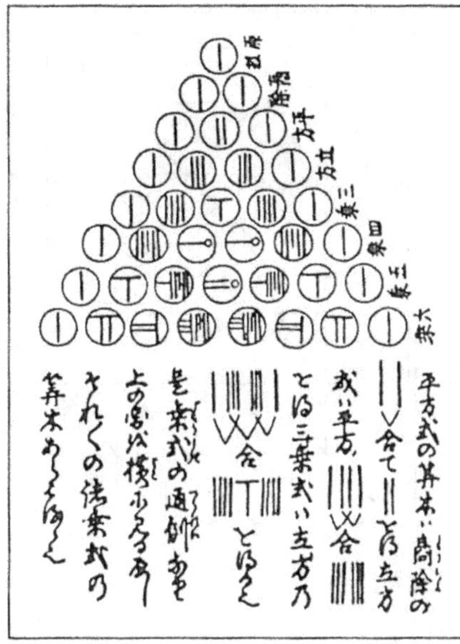

Figure 6.3 Japanese Version of Pascal's Triangle

popular version in Figure 6.1. Notice that the Chinese version lists $\binom{7}{2}$ incorrectly. Another version is given in Figure 6.6, which is the same as Figure 6.1 left-justified.

Figure 6.7 shows Cardano's combinatorial triangle from *Opus Novum* and Figure 6.8 shows the combinatorial triangle from Bernoulli's *Ars Conjectandi*, which is the same as the left-justified array in Figure 6.6.

Since $\binom{n}{r} = \frac{n!}{r!(n-r)!}$, the element $\binom{n}{r}$ lies on row n, and on the rth northeast and the $(n-r)$th southeast diagonals, the same numbers involved in the binomial coefficient; see Figure 6.9.

Pascal's triangle possesses a myriad of delightful properties:

1) Every row begins with and ends in a 1, since $\binom{n}{0} = \binom{n}{n}$.
2) Pascal's triangle is symmetric about the vertical line through the middle. This is so since $\binom{n}{r} = \binom{n}{n-r}$.
3) Every internal entry in each row is the sum of the numbers to its left and to its right in the preceding row; see Figure 6.1. This is true by virtue of Pascal's identity:

$$\binom{n}{r} = \binom{n-1}{r-1} + \binom{n-1}{r}$$

This provides us with a simple algorithm to compute each row, except the first, from the previous row, as Figures 6.10 and 6.11 show.

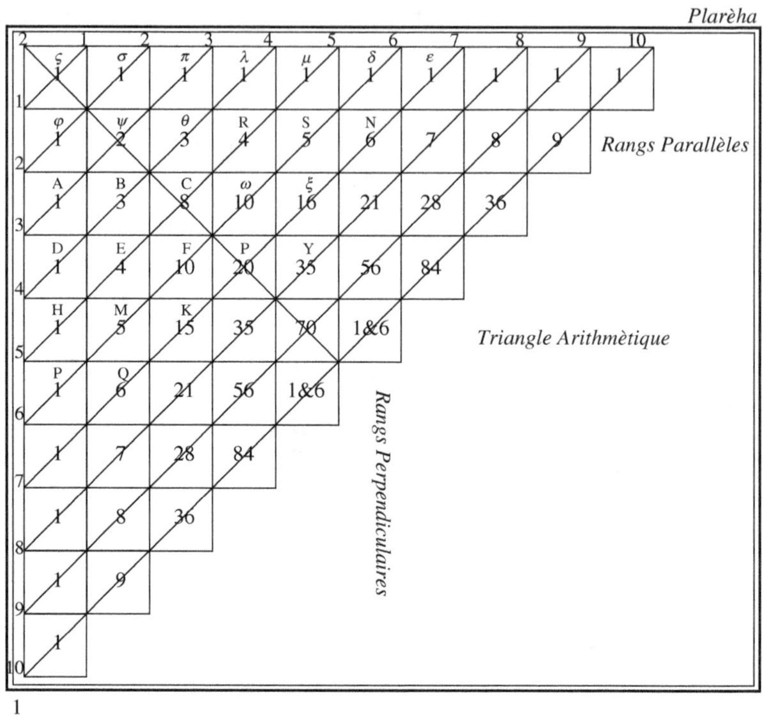

Figure 6.4 Pascal's Own Drawing

1	1	1	1	1	1	1	
1	2	3	4	5	6	7	
1	3	6	10	15	21	28	⋯
1	4	10	20	35	56	84	
1	5	15	35	70	126	210	
1	6	21	56	126	252	462	
			⋮				

Figure 6.5

1					
1	1				
1	2	1			
1	3	3	1		
1	4	6	4	1	
1	5	10	10	5	1
			⋮		

Figure 6.6

4) The sum of the numbers in row n is 2^n: $\sum_{r=0}^{n} \binom{n}{r} = 2^n$. See Corollary 3.1.
5) The cumulative sum of the numbers in rows 0 through n is given by $\sum_{k=0}^{n} 2^k = 2^{n+1} - 1$, a Mersenne number.
6) Numbers along the northeast diagonals reveal interesting number patterns; see Figure 6.12:

    - Natural numbers $\quad 1, 2, 3, 4, 5, \cdots$

1	2	3	4	5	6	7	8	9	10	11
1	1	1	1	1	1	1	1	1	1	1
2	3	4	5	6	7	8	9	10	11	
3	6	10	15	21	28	36	45	55		
4	10	20	35	56	84	120	165			
5	19	35	70	126	210	330				
6	21	56	126	252	462					
7	28	84	210	462						
8	36	120	330							
9	45	165								
10	55									
11										

Figure 6.7 Cardano's Combinatorial Triangle

Tabula

Combinationum, feu Numerorum Figuratorum.

Exponentes Combinationum.

Numeri Rerum Combinandarum.

	I.	II.	III.	IV.	V.	VI.	VII.	VIII.	IX.	X.	XI.	XII.
1.	1	0	0	0	0	0	0	0	0	0	0	0
2.	1	1	0	0	0	0	0	0	0	0	0	0
3.	1	2	1	0	0	0	0	0	0	0	0	0
4.	1	3	3	1	0	0	0	0	0	0	0	0
5.	1	4	6	4	1	0	0	0	0	0	0	0
6.	1	5	10	10	5	1	0	0	0	0	0	0
7.	1	6	15	20	15	6	1	0	0	0	0	0
8.	1	7	21	35	35	21	7	1	0	0	0	0
9.	1	8	28	56	70	56	28	8	1	0	0	0
10.	1	9	36	84	126	126	84	36	9	1	0	0
11.	1	10	45	120	210	252	210	120	45	10	1	0
12.	1	11	55	165	330	462	462	330	165	55	11	1

Figure 6.8 Bernoulli's Combinatorial Triangle

- Triangular numbers 1, 3, 6, 10, 15, $\cdots$
- Tetrahedral numbers 1, 4, 10, 20, 35, $\cdots$

7) Since $t_n = \sum_{i=1}^{n} i$ and $T_n = \sum_{i=1}^{n} t_i$, triangular and tetrahedral numbers can also be computed as sums of the first n consecutive integers and the first n triangular numbers, respectively; see the loops in Figure 6.12.

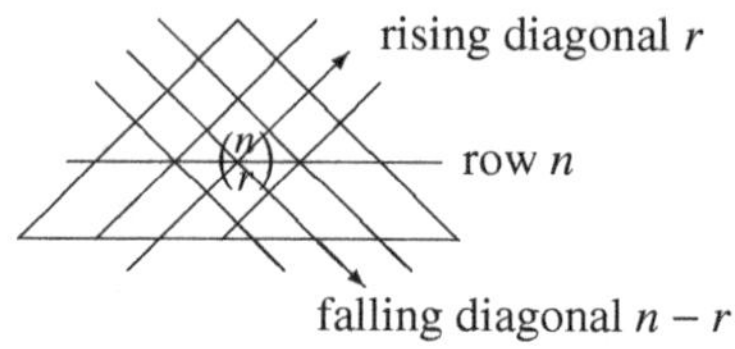

Figure 6.9

Figure 6.10

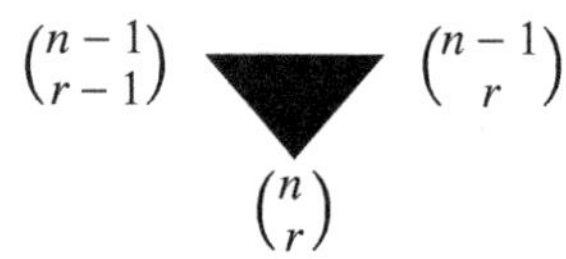

Figure 6.11

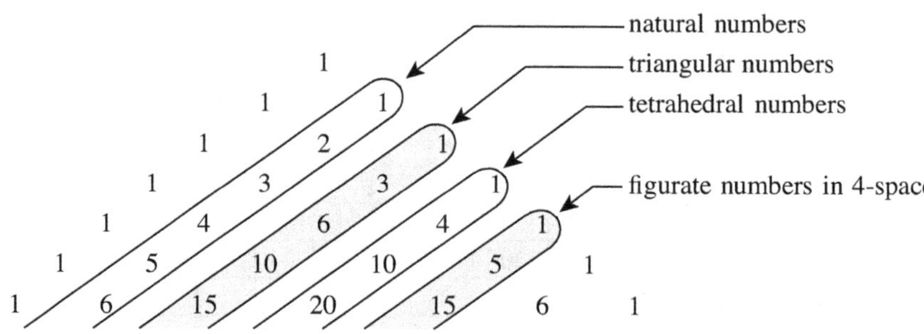

Figure 6.12

8) Since both triangular and tetrahedral numbers can be read from Pascal's triangle, the answers to the two questions in the *Twelve Days of Christmas puzzle* can be read from it:

$$\text{Number of gifts given on the } n\text{th day} = t_n = \binom{n+1}{2}$$

$$\text{Total number of gifts given in } n \text{ days} = T_n = \binom{n+2}{3}$$

9) The sum of any two consecutive elements on diagonal 2 is a square.
For example, $1 + 3 = 2^2$, $3 + 6 = 3^2$, and $6 + 10 = 4^2$.
This follows from the fact that the sum of any two consecutive triangular numbers is a square: $t_n + t_{n-1} = n^2$. This result was discovered by Theon of Smyrna around A.D. 100.

10) The rising diagonal 2 manifests another interesting pattern:

$$3^2 - 1^2 = (3-1)^3$$
$$6^2 - 3^2 = (6-3)^3$$
$$10^2 - 6^2 = (10-6)^3$$

To establish this property, discovered in 1933 by V. F. Ivanoff of San Francisco, California, we have $t_n = \binom{n}{2}$ and $t_{n-1} = \binom{n-1}{2}$. So $t_n - t_{n-1} = n - 1$ and $t_n + t_{n-1} = (n-1)^2$. Therefore,

$$\begin{aligned} t_n^2 - t_{n-1}^2 &= (t_n + t_{n-1})(t_n - t_{n-1}) \\ &= (n-1)^3 = (t_n - t_{n-1})^3 \end{aligned}$$

11) Consider the positive differences of alternate numbers on diagonal 3.
For example, $10 - 1 = 3^2$, $20 - 4 = 4^2$, and $35 - 10 = 5^2$. In each case, the positive difference is always a square. This is always the case, since

$$\begin{aligned} T_n - T_{n-2} &= \binom{n+2}{3} - \binom{n}{3} \\ &= \binom{n+1}{2} + \binom{n+1}{3} - \binom{n}{3} \\ &= \binom{n+1}{2} + \binom{n}{2} \\ &= t_n + t_{n-1} = n^2 \qquad (6.1) \end{aligned}$$

This intriguing property was observed by A. K. Fong when she was a student at Santa Clara University, Santa Clara, California.

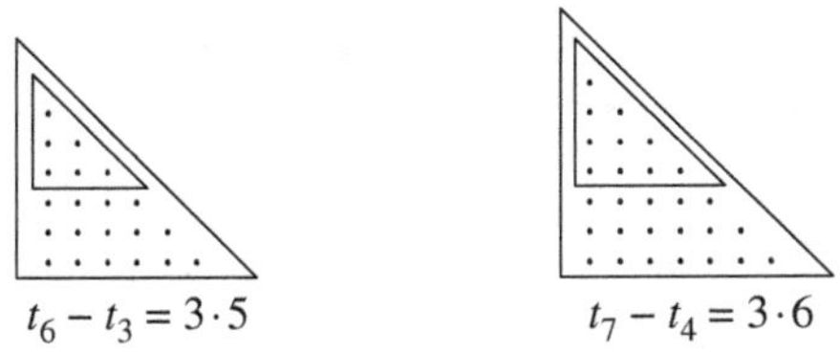

Figure 6.13

Identity (6.1) can be established sightly differently also, using a geometric approach. To this end, it follows from Figure 6.13 that $t_6 - t_3 = 3 \cdot 5$ and

$t_7 - t_4 = 3 \cdot 6$. More generally, $t_{n+1} - t_{n-2} = 3n$, where $n \geq 3$. That is,

$$\frac{(n+1)(n+2)}{2} - \frac{(n-2)(n-1)}{2} = 3n$$

$$\frac{n(n+1)(n+2)}{6} - \frac{(n-2)(n-1)n}{6} = n^2$$

$$T_n - T_{n-2} = n^2, \quad \text{as desired.}$$

12) The largest element in row n is $\binom{n}{\lfloor n/2 \rfloor}$; see Theorem 3.2.
For example, the largest element in row 5 is $\binom{5}{\lfloor 5/2 \rfloor} = \binom{5}{2} = \binom{5}{3} = 10$, and that in row 6 is $\binom{6}{\lfloor 6/2 \rfloor} = \binom{6}{3} = 20$.
13) Row n can be used to compute 11^n, where $n \geq 0$.
For example, $11^0 = 1$, $11^1 = 11$, $11^2 = 121$, $11^3 = 1{,}331$, and $11^4 = 14{,}641$. This remarkable pattern, observed by Newton, holds for all $n \geq 0$, since $(10+1)^n = \sum_{r=0}^{n} \binom{n}{r} 10^{10-r}$. For example, $11^4 = \sum_{r=0}^{4} \binom{4}{r} 10^{4-r} = 1 \cdot 10^4 + 4 \cdot 10^3 + 6 \cdot 10^2 + 4 \cdot 10 + 1 = 14,641$.
14) Recall from Chapter 3 that $\sum_{r=0}^{n} 2^r \binom{n}{r} = 3^n$. In other words, if we multiply each entry $\binom{n}{r}$ in row n of Pascal's triangle by the weight 2^r, and add up the products, then we would get 3^n.
For example, consider row 4:

row 4:	1	4	6	4	1
corresponding weights:	2^0	2^1	2^2	2^3	2^4
multiply:	1	8	24	32	16
sum:	3^4				

15) By Corollary 3.5, we have

$$1\binom{n}{1} + 2\binom{n}{2} + 3\binom{n}{3} + \cdots + n\binom{n}{n} = n2^{n-1}$$

That is, if we multiply the various elements $\binom{n}{r}$ by r and then add up the products, the result would be $n2^{n-1}$.
For example, consider row 4:

row 4:	1	4	6	4	1
multiply by 0, 1, 2, 3, and 4, respectively:	$0 \cdot 1$	$1 \cdot 4$	$2 \cdot 6$	$3 \cdot 4$	$4 \cdot 1$

Now, add the products: $0 \cdot 1 + 1 \cdot 4 + 2 \cdot 6 + 3 \cdot 4 + 4 \cdot 1 = 32 = 4 \cdot 2^3$. We also found that

$$2\binom{n}{2} + 4\binom{n}{4} + 6\binom{n}{6} + \cdots + 2\lfloor n/2 \rfloor \binom{n}{2\lfloor n/2 \rfloor} = n2^{n-2}$$

That is, if we multiply the various elements $\binom{n}{2r}$ by $2r$ and then add the products, the result would be $n2^{n-2}$.

For example, consider the "even" binomial coefficients in row 6, indicated by asterisks:

$$1^* \quad 6 \quad 15^* \quad 20 \quad 15^* \quad 6 \quad 1^* \quad \leftarrow \text{row } 6$$

Multiply them by 0, 2, 4, and 6, respectively; and then add the products. The resulting sum is $0 \cdot 1 + 2 \cdot 15 + 4 \cdot 15 + 6 \cdot 1 = 96 = 6 \cdot 2^4$.

16) The identity $\sum_{r=0}^{n-1} \binom{n}{r}\binom{n}{r+1} = \binom{2n}{n+1}$ from Chapter 5 can be interpreted as follows: The sum of the products of every two consecutive elements in row n equals $\binom{2n}{n+1}$, the element immediately to the right (or to the left) of the CBC in row $2n$.

For example, consider row 6:

$$1 \quad 6 \quad 15 \quad 20 \quad 15 \quad 6 \quad 1 \quad \leftarrow \text{row } 6$$

The sum of the products of every two consecutive elements is $1 \cdot 6 + 6 \cdot 15 + 15 \cdot 20 + 20 \cdot 15 + 15 \cdot 6 + 6 \cdot 1 = 792 = \binom{12}{7}$; it is the element to the right (or left) of the CBC in row 12:

$$1 \quad 12 \quad 66 \quad 220 \quad 495 \quad 792 \quad \boxed{924} \quad 792 \quad 495 \quad 220 \quad 66 \quad 12 \quad 1$$

$$\uparrow$$
$$\text{CBC}$$

17) Recall from Example 5.1 that $\sum_{r=0}^{\lfloor n/2 \rfloor} \binom{n}{r}\binom{n-r}{r}2^{n-2r} = (n+1)C_n$. Thus, if we multiply the product $\binom{n}{r}\binom{n-r}{r}$ by the weight 2^{n-2r} and add up the resulting products for the various values of r, the resulting sum would be $n+1$ times the corresponding Catalan number.

For example,

$$\sum_{r=0}^{3}\binom{7}{r}\binom{7-r}{r}2^{7-2r} = \boxed{1} \cdot 1^* \cdot 2^7 + \boxed{7} \cdot 6^* \cdot 2^5 + \boxed{21} \cdot 10^* \cdot 2^3$$
$$+ \boxed{35} \cdot 4^* \cdot 2^1 = 8 \cdot 429 = 8C_7$$

See the truncated Pascal's triangle in Figure 6.14.

			1		4		6		4*		1			
		1		5		10*		10		5		1		
	1		6*		15		20		15		6		1	
[1*]		[7]		[21]		[35]		35		21		7		1

Figure 6.14

18) The identity $\sum_{r=1}^{n} r\binom{n}{r}^2 = n(2n-1)C_{n-1}$ in Example 5.2 also can be interpreted through the lens of Pascal's triangle. To see this, square the elements $\binom{n}{r}$, where $r \geq 1$; multiply each by the weight r; add the products; the resulting sum is $n(2n-1)C_{n-1}$.
For example, consider row 6, except the first element:

resulting row 6:	6	15	20	15	6	1
square each:	36	225	400	225	36	1
corresponding weights:	1	2	3	4	5	6
multiply:	1	36	1200	900	180	6
sum:	2,772					

19) Identity (5.6) can also be interpreted similarly. The binomial coefficients on the left of the identity are the top $n-r+1$ elements on the northeast diagonal r. Their sum is given by the $(n-r+1)$st element on diagonal $r+1$. For example,

$$\binom{2}{2}+\binom{3}{2}+\binom{4}{2}+\binom{5}{2}=\binom{6}{3}=\binom{2}{0}+\binom{3}{1}+\binom{4}{2}+\binom{5}{3}$$

See the Pascal's triangle in Figure 6.15.

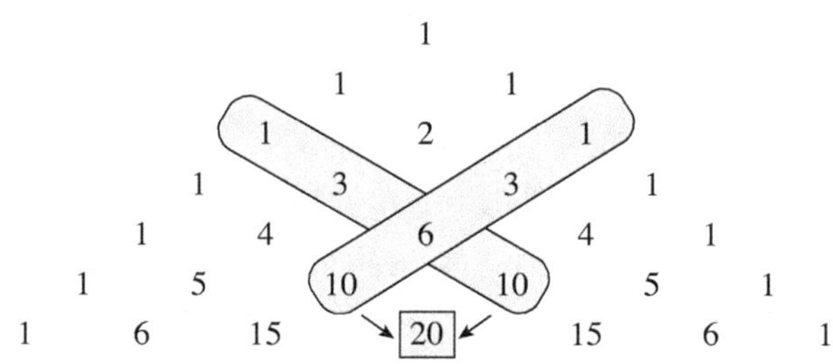

Figure 6.15

20) The difference of the squares of any two consecutive elements on northeast diagonal 2 is a cube:

$$\binom{n}{2}^2-\binom{n-1}{2}^2=\left[\frac{n(n-1)}{2}\right]^2-\left[\frac{(n-1)(n-2)}{2}\right]^2$$
$$=\frac{(n-1)^2}{4}\left[n^2-(n-2)^2\right]=(n-1)^3$$

21) Every entry in row $n \geq 1$ is odd if and only if n is a Mersenne number. For example, every element in row $7 = 2^3 - 1$ is odd; so is every element in row 31.

22) Every element in row n is odd if and only if the binary representation of n consists of 1s.

For example, $15 = 1111_{\text{two}}$. The first eight elements in row 15 are 1, 15, 105, 455, 1365, 3003, 5005, and 6435, and they are all odd. Consequently, every element in the row is odd.

23) The number of odd binomial coefficients in row n is 2^e, where e is the number of non-zero bits in the binary expansion of n.
For example, the binary expansion of $13 = 1101_{\text{two}}$ contains three 1s. Accordingly, row 13 contains $2^3 = 8$ odd numbers: 1, 13, 715, 1287, 1287, 715, 13, and 1; see Table 6.1 also.

Table 6.1

n	Binomial coefficients										n_{two}	e	h	$h+e$
0	1										0	0	0	0
1	1	1									1	1	0	1
2	1	2	1								10	1	1	2
3	1	3	3	1							11	2	1	3
4	1	4	6	4	1						100	1	3	4
5	1	5	10	10	5	1					101	2	3	5
6	1	6	15	20	15	6	1				110	2	4	6
7	1	7	21	35	35	21	7	1			111	3	4	7
8	1	8	28	56	70	56	28	8	1		1000	1	7	8
9	1	9	36	84	126	126	84	36	9	1	1001	2	7	9

24) Let h be the highest power of 2 that divides $n!$:

$$h = \lfloor n/2 \rfloor + \lfloor n/2^2 \rfloor + \lfloor n/2^3 \rfloor + \cdots + \lfloor n/2^i \rfloor + \cdots$$

Then $h + e = n$.
For example, let $n = 7$. Then $h = \lfloor 7/2 \rfloor + \lfloor 7/4 \rfloor = 4$, $e = 3$; and $h + e = 7 = n$; see Table 6.1.

25) If p is a prime, then every internal entry in row p is divisible by p; see Theorem 4.1.
For example, the interior elements 7, 21, 35, 35, 21, and 7 in row 7 are divisible by 7.

26) If p is a prime, then every internal entry in row p^n is divisible by p, where n is an arbitrary positive integer; see Theorem 4.5.
For example, let $p = 3$ and $n = 2$. Row 9 of Pascal's triangle looks like this:

$$1 \quad \underbrace{9 \quad 36 \quad 84 \quad 126 \quad 126 \quad 84 \quad 36 \quad 9}_{\text{all divisible by 9}} \quad 1 \qquad \leftarrow \text{row } 9$$

Clearly, all elements in this row, except the two ends, are divisible by 3. Similarly, every element in row 25, except the two extremes, is divisible by 5.

27) If n is a prime, then every element in row $n-1$ is congruent to $(-1)^r \pmod{n}$, where $0 \leq r \leq n-1$; see Theorem 4.15.
For example, consider row 7. The entries in row 6 are 1, 6, 15, 20, 15, 6, and 1. You may notice that $1 \equiv (-1)^0 \equiv (-1)^6 \pmod{7}$, $6 \equiv (-1)^1 \equiv (-1)^5 \pmod{7}$, and $15 \equiv (-1)^2 \equiv (-1)^4 \pmod{7}$.
28) If $n+1$ is a prime, then $r+1$ divides each element $\binom{n}{r}$, where $0 \leq r \leq n$.
For example, let $n = 10$, so $n+1 = 11$ is a prime. The various elements in row 10 are 1, 10, 45, 120, 210, 252, 210, 120, 45, 10, and 1. Notice that $\binom{10}{r}$ is divisible by $r+1$ for each value of r.
29) By Lagrange's identity, the sum of the squares of the elements in row n equals the CBC in row $2n$.
For example, the sum of the squares of the elements in row 6 equals $1^2 + 6^2 + 15^2 + 20^2 + 15^2 + 6^2 + 1^2 = 924 = \binom{12}{6}$.
30) In Chapter 5, we found that the binomial coefficient $\binom{m+n}{m}$ represents the number of paths consisting of $m+n$ blocks from the origin O to any point on a rectangular grid of easterly avenues and northerly streets. It is the $(m+1)$st element in row $m+n$ in Pascal's triangle.
31) Rows 2, 7, and 14 contain triplets consisting of consecutive elements that form an arithmetic progression. For example, 0–1–2, 7–21–35, and 1001–2002–3003 are three such triplets. In fact, there are infinitely many such triplets in Pascal's triangle; see Chapter 7.
32) The property that

$$\left[\binom{n}{0} - \binom{n}{2} + \binom{n}{4} - \cdots\right]^2 + \left[\binom{n}{1} - \binom{n}{3} + \binom{n}{5} - \cdots\right]^2 = 2^n$$

can be restated in terms of Pascal's triangle: Change the sign of every other "even" binomial coefficient $\binom{n}{2r}$ and that of every other "odd" binomial coefficient $\binom{n}{2r+1}$; add them separately, square each sum, and then add the two squares. The resulting sum is 2^n.
For example, consider row 7, where we have indicated the "even" binomial coefficients with an asterisk:

$$1^* \quad 7 \quad 21^* \quad 35 \quad 35^* \quad 21 \quad 7^* \quad 1$$

Now, change the sign of every other even-numbered binomial coefficient and that of "odd" binomial coefficient:

$$1^* \quad 7 \quad -21^* \quad -35 \quad 35^* \quad 21 \quad -7^* \quad -1$$

Add the elements in the even positions: $1 - 21 + 35 - 7 = 8$
Add the elements in the odd positions: $7 - 35 + 21 - 1 = -8$
The sum of their squares equals $8^2 + (-8)^2 = 128 = 2^7$.

The Pizza Problem Revisited

Recall from Chapter 1 that the maximum number of pieces f_n a pizza can be cut into, with n cuts, is given by:

$$\begin{aligned} f_n &= 1 + \frac{n(n+1)}{2} \\ &= 1 + n + \frac{n(n-1)}{2} = \binom{n}{0} + \binom{n}{1} + \binom{n}{2} \end{aligned}$$

where $n \geq 0$. Accordingly, f_n can be computed directly from Pascal's triangle by adding the first three elements in row n; see Figure 6.16.

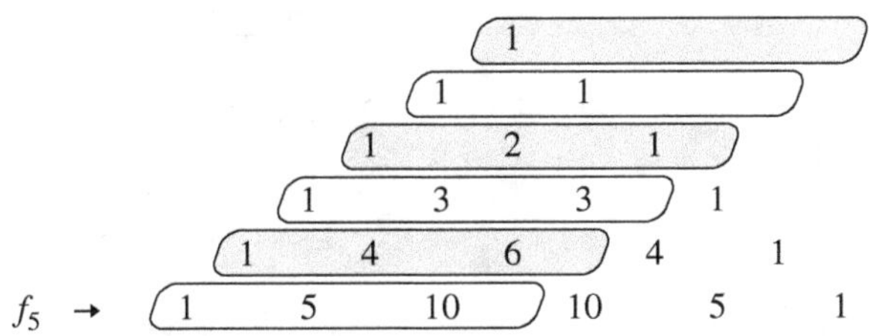

Figure 6.16 Add the Numbers in Each Loop

The Circle Problem Revisited

Again from Chapter 1, the maximum number of non-overlapping regions g_n formed inside a circle by joining n distinct points on it is given by

$$g_n = \binom{n}{0} + \binom{n}{2} + \binom{n}{4} \tag{6.2}$$

which is the sum of the first three even-numbered binomial coefficients in row n. See Figure 6.17.

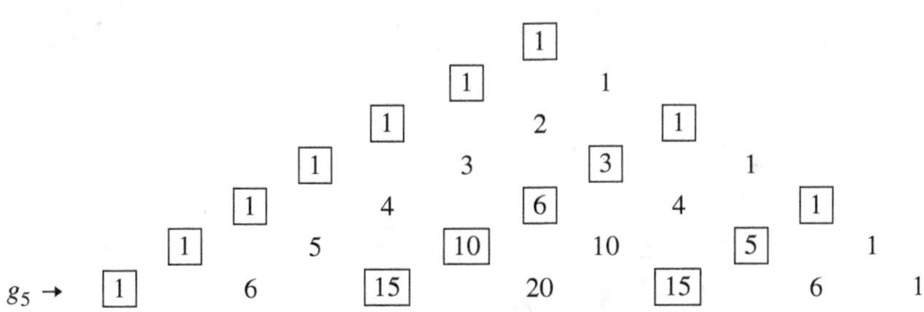

Figure 6.17 Add the Boxed Numbers in Each Row

Using Pascal's identity, however, formula (6.2) can be rewritten as

$$g_n = \binom{n-1}{0} + \binom{n-1}{1} + \binom{n-1}{2} + \binom{n-1}{3} + \binom{n-1}{4} \tag{6.3}$$

which is the sum of the first five numbers in row $n-1$. See the loops in Figure 6.18.

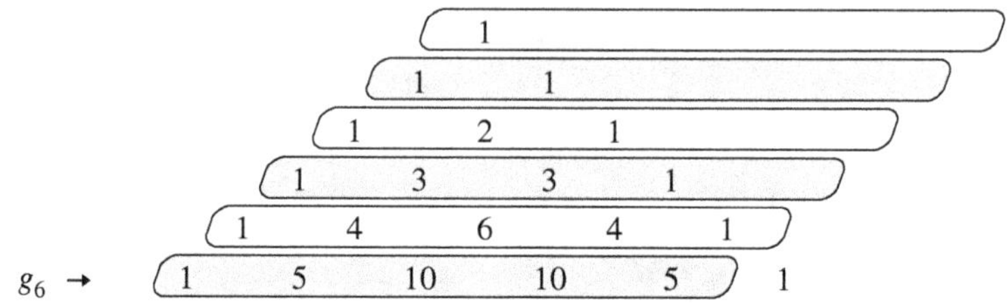

Figure 6.18 Add the Numbers in Each Loop

CBC in Row $2p$

Recall from Example 4.1 that if p is a prime, then $\binom{2p}{p} \equiv 2 \pmod p$; that is, the CBC in row $2p$ leaves the remainder 2 when divided by p. For example, $\binom{22}{11} = 705{,}432 \equiv 2 \pmod{11}$.

More on CBCs

Identity (5.6) describes a remarkable property of the CBCs in Pascal's triangle:

$$\binom{0}{0}\binom{2n}{n} + \binom{2}{1}\binom{2n-2}{n-1} + \binom{4}{2}\binom{2n-4}{n-2} + \cdots + \binom{2n}{n}\binom{0}{0} = 4^n \tag{6.4}$$

For example, let $n = 3$. Then

$$\binom{0}{0}\binom{6}{3}+\binom{2}{1}\binom{4}{2}+\binom{4}{2}\binom{2}{1}+\binom{6}{3}\binom{0}{0}=1\cdot 20+2\cdot 6+6\cdot 2+20\cdot 1=4^3$$

See Figure 6.19.

The next four properties were discovered in 1987 by Hilton and Pedersen. They can be established algebraically, where $n \geq 0$:

- (*The Little Hockey Stick and Puck Theorem*)

$$\binom{n}{0} + \binom{n+2}{1} + \binom{n+4}{2} = \binom{n+5}{2} - \binom{n+5}{0}$$

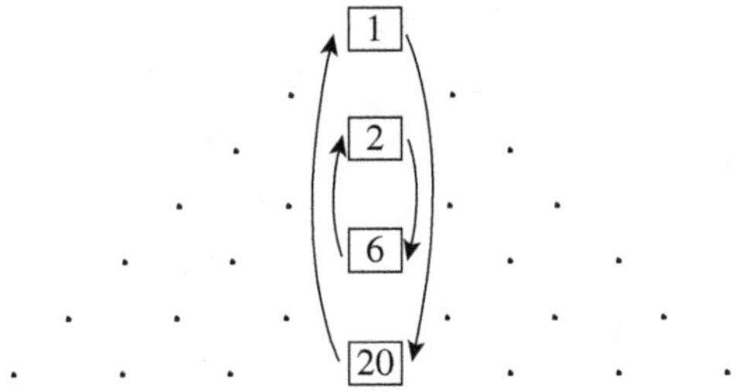

Figure 6.19

- (*The Big Hockey Stick and Puck Theorem*)

$$\binom{n}{0} + \binom{n+2}{1} + \binom{n+4}{2} + \binom{n+6}{3} = \binom{n+7}{3} - \binom{n+6}{1}$$

These two results can be displayed pictorially, as in Figure 6.20, where, for convenience, we have replaced the numbers with dots.

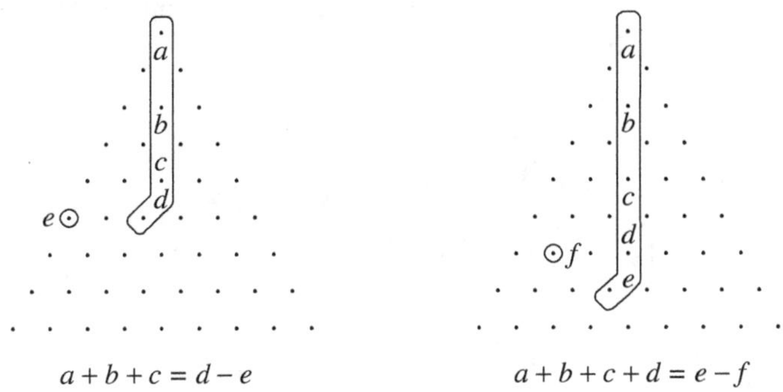

Figure 6.20

The next two properties refer to the left-justified Pascal's triangle in Figure 6.7, where, again for convenience, we have replaced the numbers with dots and labels:

$$\frac{ad}{cb} = \frac{eg}{cf} \quad \text{(see Figure 6.21)} \qquad \frac{ad}{cb} = \frac{eg}{cf} \quad \text{(see Figure 6.22)}$$

Hidden Hexagon Squares

Pascal's triangle still contains a hidden treasure, a magic hexagon. The nearest six elements around $\binom{n}{r}$ follow a hexagonal pattern; see Figure 6.23: The product of

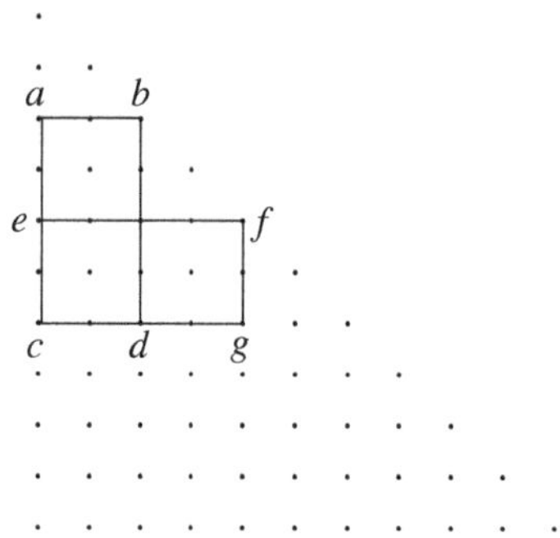

Figure 6.21

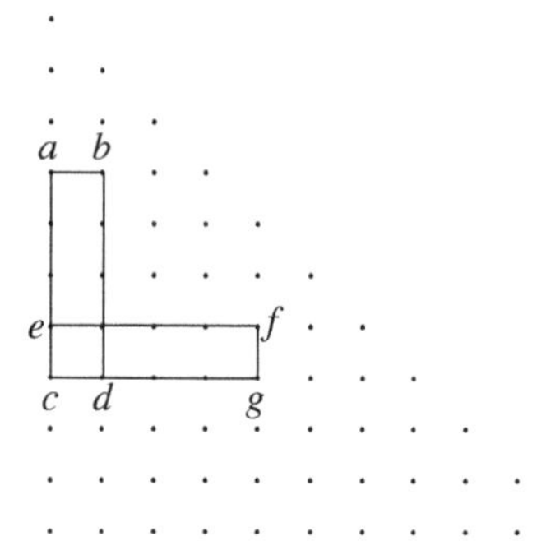

Figure 6.22

the elements at every other vertex equals that of the three remaining elements:

$$\binom{n-1}{r-1}\binom{n}{r+1}\binom{n+1}{r} = \binom{n-1}{r}\binom{n+1}{r+1}\binom{n}{r-1} \tag{6.5}$$

This intriguing property, called the *Hoggatt–Hansell identity*, was discovered in 1971 by V. E. Hoggatt, Jr. of San Jose State College and W. Hansell. In the following year, Gould called it a *Star of David* property; see Figure 6.24.

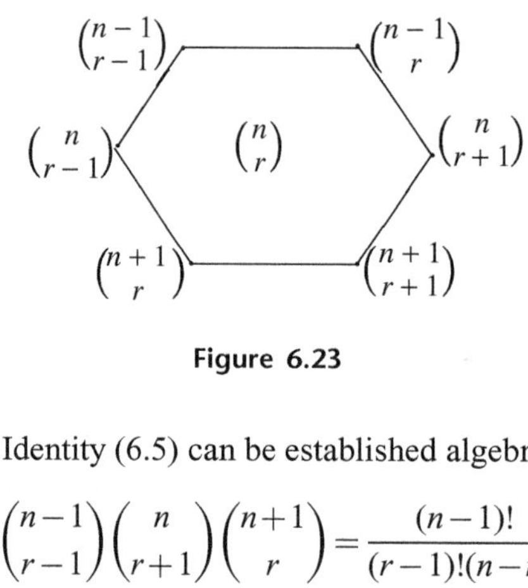

Figure 6.23

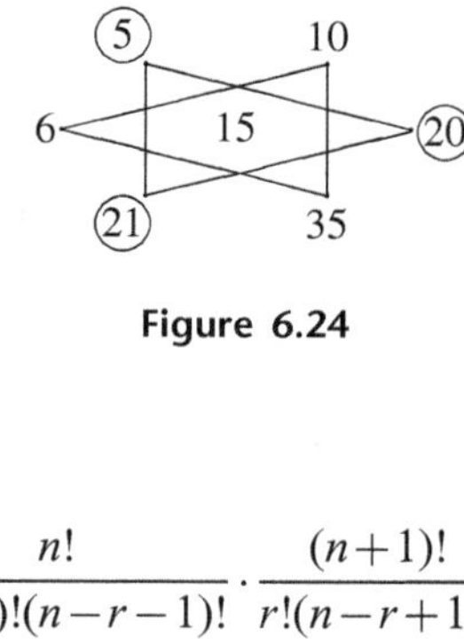

Figure 6.24

Identity (6.5) can be established algebraically:

$$\begin{aligned}\binom{n-1}{r-1}\binom{n}{r+1}\binom{n+1}{r} &= \frac{(n-1)!}{(r-1)!(n-r)!}\cdot\frac{n!}{(r+1)!(n-r-1)!}\cdot\frac{(n+1)!}{r!(n-r+1)!}\\ &= \frac{(n-1)!}{r!(n-r-1)!}\cdot\frac{(n+1)!}{(r+1)!(n-r)!}\cdot\frac{n!}{(r-1)!(n-r+1)!}\\ &= \binom{n-1}{r}\binom{n+1}{r+1}\binom{n}{r-1}\end{aligned}$$

For example, consider the hexagon in Figure 6.24. Clearly, $5\cdot 20\cdot 21 = 10\cdot 35\cdot 6$.

Identity (6.5) implies that the product of the elements at the vertices of the hexagon is a square. For example, using Figure 6.24, notice that $(5 \cdot 20 \cdot 21)(10 \cdot 35 \cdot 6) = 2100^2$.

In 1971, Hoggatt and G. L. Alexanderson of Santa Clara University extended this delightful property to multinomial coefficients.

Gould's Identity

In 1972, Gould generalized the Hoggatt–Hansell identity as follows, where a is any integer:

$$\binom{n-a}{r-a}\binom{n}{r+a}\binom{n+a}{r} = \binom{n-a}{r}\binom{n+a}{r+a}\binom{n}{r-a} \tag{6.6}$$

This identity also can be confirmed algebraically.

For example, let $n = 8, r = 3$, and $a = 2$. Then

$$\binom{6}{1}\binom{8}{5}\binom{10}{3} = 6 \cdot 56 \cdot 120 = 20 \cdot 252 \cdot 8 = \binom{6}{3}\binom{10}{5}\binom{8}{1}; \text{ see Figure 6.25.}$$

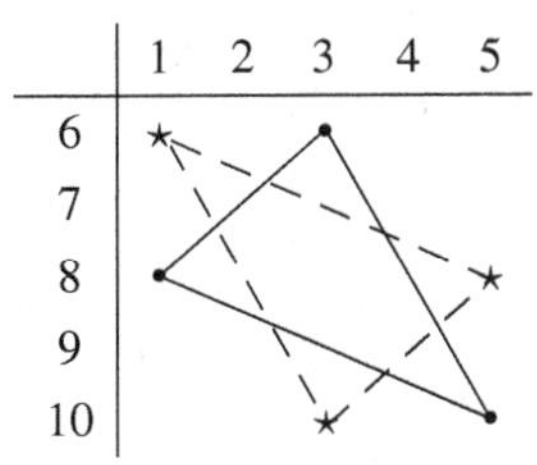

Figure 6.25

Gould's Hexagon Property

As one discovery often leads to another, the Hoggatt–Hansell identity prompted Gould to conjecture an equally intriguing property about the gcd of the alternate elements at the vertices of the hexagon in Figure 6.23:

$$\left(\binom{n-1}{r-1}, \binom{n}{r+1}, \binom{n+1}{r}\right) = \left(\binom{n-1}{r}, \binom{n+1}{r+1}, \binom{n}{r-1}\right) \tag{6.7}$$

This *Star of David property* was established in 1972 by A. P. Hillman of the University of New Mexico, Albuquerque, New Mexico, and Hoggatt.

For example, using Figure 6.24, notice that (5,20,21) = (10,35,6).

The curious-minded might note that Gould's property (6.7) does *not* hold for their lcm. For instance, again using Figure 6.24, $[5, 20, 21] = 420 \neq 210 = [10, 35, 6]$.

A Hidden Treasure

Table 6.2 shows the values of $\left(\binom{n-1}{r}, \binom{n+1}{r+1}, \binom{n}{r-1}\right)$, where $0 \le n \le 16$ and $0 \le r \le \lfloor n/2 \rfloor$. (By virtue of symmetry, the other half is omitted.) In each case, the value is identical with $\left(\binom{n-1}{r-1}, \binom{n}{r+1}, \binom{n+1}{r}\right)$.

Table 6.2

n \ r	0	1	2	3	4	5	6	7	8
0	1								
1	1								
2	1	1							
3	1	1							
4	1	1	1						
5	1	1	1						
6	1	1	1	5					
7	1	1	1	1					
8	1	1	1	7	7				
9	1	1	1	2	14				
10	1	1	1	3	6	42			
11	1	1	1	5	3	6			
12	1	1	1	11	11	33	66		
13	1	1	1	1	11	11	33		
14	1	1	1	13	13	143	143	429	
15	1	1	1	7	91	91	143	143	
16	1	1	1	5	7	91	13	143	715

This table manifests a fascinating property, first observed by Gould. The elements on the main diagonal, except the first element, are alternately Catalan numbers or one-half them:

1 [1] 5 [7] 42 [66] 429 [715] 4862 [8398] ...

Thus

$$\left(\binom{2n-1}{n}, \binom{2n+1}{n+1}, \binom{2n}{n-1}\right) = \begin{cases} C_n & \text{if } 2 \nmid n \\ \frac{1}{2}C_n & \text{otherwise} \end{cases}$$

where C_n denotes the nth Catalan number and $n \ge 1$.

Returning to the binomial coefficients in Gould's identity (6.7), unfortunately the property does not hold for any hexagon. For example, let $n = 8, r = 3$, and $a = 2$. Then $\left(\binom{6}{1}, \binom{8}{5}, \binom{10}{3}\right) = (6, 56, 120) = 2$, whereas $\left(\binom{6}{3}, \binom{10}{5}, \binom{8}{1}\right) = (20, 252, 8) = 4$; see Figure 6.25.

We will revisit the Hoggatt–Hansell identity and Gould's hexagon property in our study of Fibonomial coefficients in Chapter 28.

Usiskin's Square Patterns

In 1973, Z. Usiskin of the University of Chicago, inspired by the Hoggatt–Hansell identity (6.5), discovered a number of similar square patterns. To begin with, we replace, for convenience, each binomial coefficient on the left of the identity by an O and each on the right by an X. This yields Figure 6.26, where the dots represent adjacent elements in Pascal's triangle.

	.		O		X		.
.		X		.		O		.
	.		O		X		.

Figure 6.26

	O		X		O		X	
X				.				O
	O		X		O		X	

Figure 6.27

Now, consider two such adjacent hexagons with a common vertex, as in Figure 6.27. Then the product of the elements on the hexagons is a square (count the common element twice); so is the product of all elements with the common element deleted.

For example, consider the adjacent hexagons with the common vertex 20 in Figure 6.28. Notice that the product of the remaining elements on them is a square: $(5 \cdot 10 \cdot 35 \cdot 21 \cdot 6)(10 \cdot 5 \cdot 6 \cdot 21 \cdot 35) = 220,500^2$.

	5		10		10		5	
6				20				6
	21		35		35		21	

Figure 6.28

	O		.	.	.		X	
X		X		.	.	O		O
	O		.	.	.		X	

Figure 6.29

The square pattern in Figure 6.29 follows from the identity

$$\binom{n-1}{r+a}\binom{n}{r}\binom{n}{r+1}\binom{n+1}{r+a+1} = \binom{n-1}{r}\binom{n}{r+a}\binom{n}{r+a+1}\binom{n+1}{r+1}$$

which can be easily verified algebraically; $a = 4$ in Figure 6.29.

Figure 6.30 shows some aesthetically pleasing square patterns discovered by Usiskin.

All these square patterns can be confirmed algebraically. For example, the fifth pattern in Figure 6.30 follows from the identity

$$\binom{n-k}{r}\binom{n+m}{r+k}\binom{n}{r-m} = \binom{n-k}{r-m}\binom{n+m}{r}\binom{n}{r+k}$$

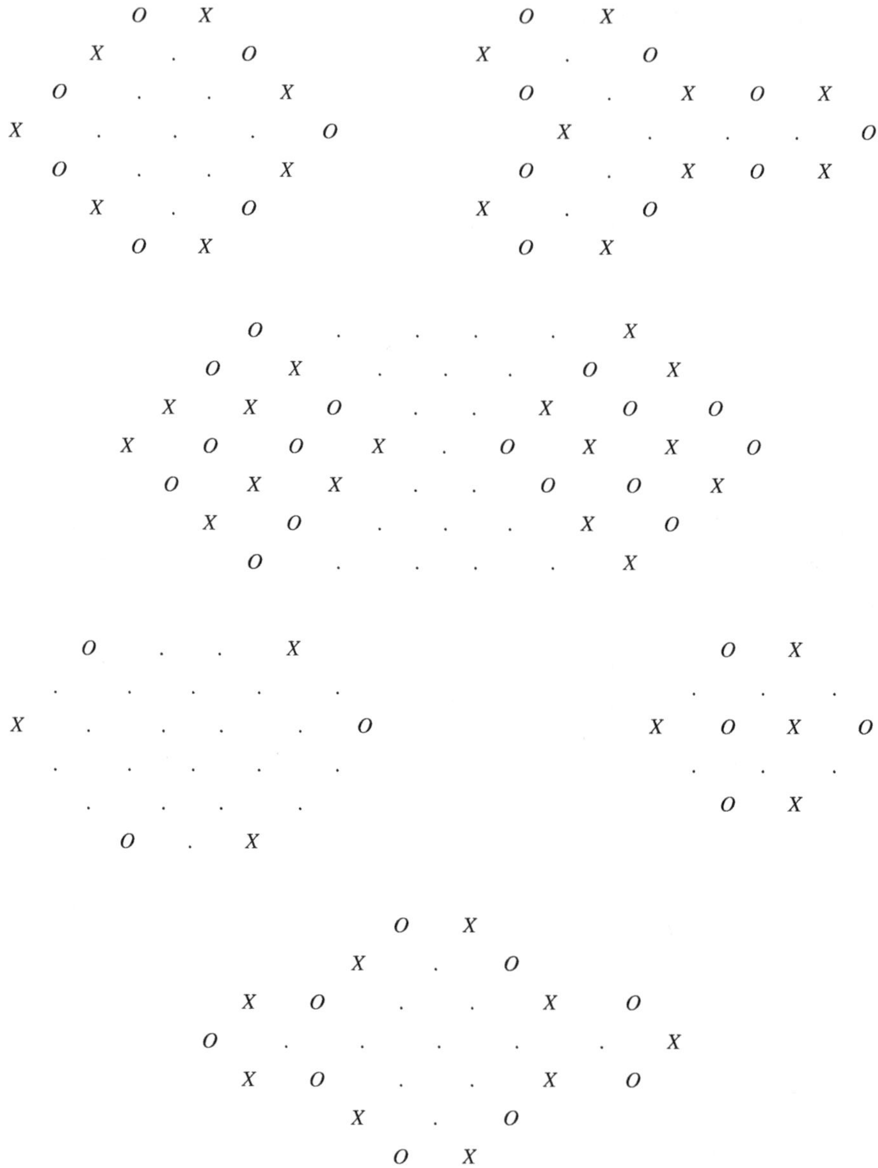

Figure 6.30

Usiskin's Cubic Pattern

Figure 6.31 shows an interesting cubic pattern by Usiskin: The product of the Xs equals the product of the Os, which in turn equals the product of the #s; so the product of all the elements is a cube.

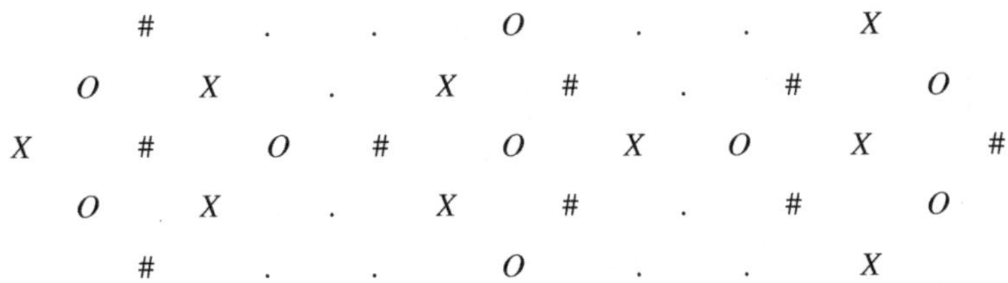

Figure 6.31

The cubic pattern holds since

$$\binom{n}{r}\binom{n-2}{r-1}\binom{n+2}{r+1}\binom{n}{r+2}\binom{n}{r-2}\binom{n-1}{r-4}\binom{n-1}{r+3}\binom{n+1}{r-3}\binom{n+1}{r+4}$$
$$=\binom{n-2}{r-4}\binom{n-1}{r}\binom{n-1}{r+2}\binom{n}{r-3}\binom{n}{r-1}\binom{n}{r+4}\binom{n+1}{r+1}\binom{n+1}{r+3}\binom{n+2}{r-2}$$
$$=\binom{n-2}{r+2}\binom{n-1}{r-3}\binom{n-1}{r-1}\binom{n}{r-4}\binom{n}{r+1}\binom{n}{r+3}\binom{n+1}{r-2}\binom{n+1}{r}\binom{n+2}{r+4}$$

Finally, the polygon in Figure 6.32 has the property that the product of the Xs equals that of the Os, and the product of all of them is a sixth power. Interestingly, the polygon has six interior points.

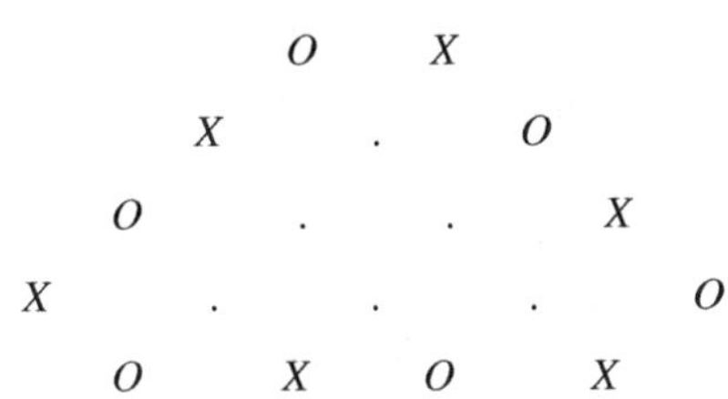

Figure 6.32

Lattice-Walking in Pascal City

Suppose there is a traveler at the apex of the network of roads in Pascal City, in Figure 6.33. At each intersection, he can turn left or right. We would like to find the number of paths he can take to reach the rth intersection in row n, where $0 \leq r \leq n$.

Let $f(n, r)$ denote the number of paths from the apex to the rth intersection in row n. Clearly, $f(0, 0) = f(1, 0) = f(1, 1) = 1$. Also, $f(n, r) = f(n-1, r-1) + f(n-1, r)$. Thus f satisfies exactly the same recursive definition as $\binom{n}{r}$. Consequently, $f(n, r) = \binom{n}{r}$. In other words, the number of paths to each intersection equals the corresponding binomial coefficient.

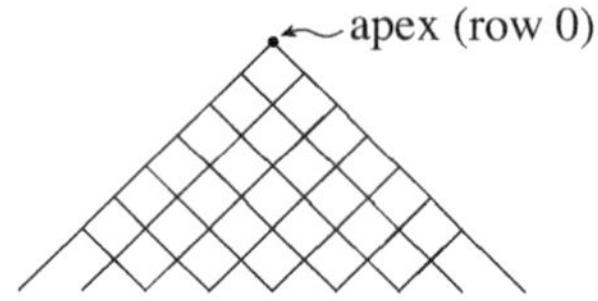

Figure 6.33 Pascal City

We will encounter a closely related combinatorial problem in the next chapter.

Abracadabra

In 1966, G. Polya (1887–1985) of Stanford University, Palo Alto, California, studied a seemingly unrelated problem. Find the number of ways the word ABRACADABRA can be read from the rhombic array in Figure 6.34.

```
          A
        B   B
      R   R   R
    A   A   A   A
  C   C   C   C   C
A   A   A   A   A   A
  D   D   D   D   D
    A   A   A   A
      B   B   B
        R   R
          A
```

Figure 6.34

This problem can be translated into a lattice-walking problem in Pascal City, as in Figure 6.35.

Every path from the apex S (start) to E (end) consists of ten blocks: five blocks in the southeast direction and the rest in the northwest direction. So every path can be represented by a 10-letter word, of which five are alike and the other five are alike. Thus the total number of paths equals $\frac{10!}{5!5!} = 252$. In other words, we can read the magic word in the array in 252 different ways.

Next, we would like to see how this problem is related to Pascal's triangle. To this end, we place a 1 at the apex and then construct the array in Figure 6.36 using Pascal's identity. Each entry in the figure gives the number of distinct paths from the apex to the corresponding location. Thus there are $\binom{10}{5} = 252$ paths from the apex to the bottommost point, and therefore there are 252 ways to read the word.

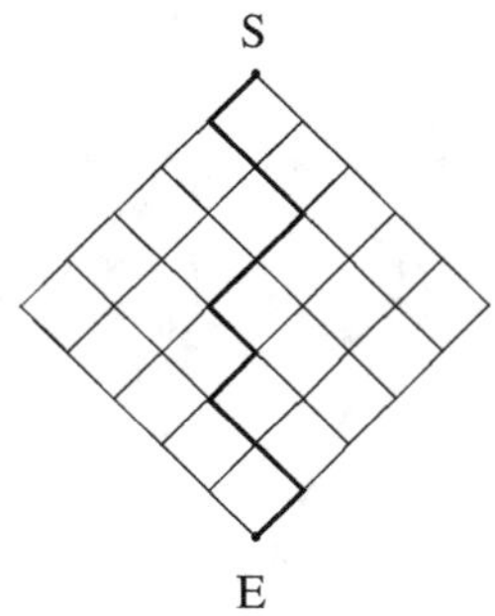

Figure 6.35

1
1 1
1 2 1
1 3 3 1
1 4 6 4 1
1 5 10 10 5 1
6 15 20 15 6
21 35 35 21
56 70 56
126 126
252

Figure 6.36

Lagrange's Identity Revisited

We now give a geometric interpretation to Lagrange's identity. To this end, we would like to find the number of paths N from the apex 1 to the CBC $\binom{2n}{n}$. It follows from Figure 6.37 that each path from 1 to $\binom{2n}{n}$ is a path from 1 to $\binom{n}{r}$ followed by a path from $\binom{n}{r}$ to $\binom{2n}{n}$. Since this is true for each value of r, by symmetry, we have:

$$N = \sum_{r=0}^{n} \left[\text{number of paths from 1 to } \binom{n}{r}\right]\left[\text{number of paths from } \binom{n}{r} \text{ to } \binom{2n}{n}\right]$$

$$= \sum_{r=0}^{n} \binom{n}{r}\binom{n}{r} = \sum_{r=0}^{n} \binom{n}{r}^2$$

But there are exactly $\binom{2n}{n}$ paths from the apex to $\binom{2n}{n}$. Thus $\sum_{r=0}^{n} \binom{n}{r}^2 = \binom{2n}{n}$.

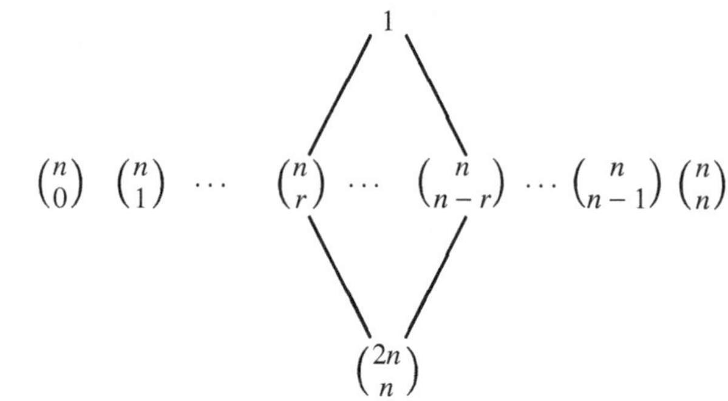

Figure 6.37

A Pascal Card Game

Next, we present a simple card game based on Pascal's triangle, but first we introduce a concept that is closely related to *casting out nines*: namely, the *digital root* $\rho(N)$ of a positive integer N. It is the remainder when N is divided by 9, with one exception; it is 9 if the remainder is 0. Thus

$$\rho(N) = \begin{cases} N \bmod 9 & \text{if } N \bmod 9 \neq 0 \\ 9 & \text{otherwise} \end{cases}$$

For example, since $1976 \equiv 5 \pmod 9$, $\rho(1976) = 5$, whereas $\rho(2079) = 9$.

Now for the puzzle. Place in a row, say, any six playing cards labeled $A(= 1)$ through 10. Use them to build a triangular pyramid with exactly one apex card: Place a card Z between and above any two adjacent cards X and Y in a such way that the label on Z equals $\rho(X + Y)$. Continue this procedure until the apex card is selected.

```
                    5
                 2     3
              9     2     1
           3     6     5     5
        1     2     4     1     4
     [8]   [2]   [9]   [4]   [6]   [7]
```

Figure 6.38

To illustrate this, suppose you select cards 8, 2, 9, 4, 6, and 7. Figure 6.38 shows these labels and the remaining ones in the pyramid. So, the apex label is 5.

Fortunately, we do not have to build the pyramid to determine the apex label, but can let Pascal's triangle do the job for us. Use the elements in row 5 of Pascal's triangle as weights; multiply each label with the corresponding weight, compute the digital root of the sum of the products of the labels with the corresponding weights: $\rho(8 \cdot 1 + 2 \cdot 5 + 9 \cdot 10 + 4 \cdot 10 + 6 \cdot 5 + 7 \cdot 1) = 5$. This yields the desired apex label.

Thus, for the puzzle, if you tell me the original labels and mask the remaining labels, then I can tell you exactly what the apex label is.

Powers of 101 and 1001

Just as Pascal's triangle can be used to compute the various powers of 11, it can also be used to compute the various powers of 101 and 1001, and in general, any power of $1\underbrace{00\cdots00}_{\text{all zeros}}1$.

For example, consider the various powers of 101 in Table 6.3. Each two-digit group in 101^n represents the binomial coefficient $\binom{n}{r}$, where $0 \le r \le n$. Everything works nicely until 101^8.

Table 6.3

101^0	=								01
101^1	=							1	01
101^2	=						1	02	01
101^3	=					1	03	03	01
101^4	=				1	04	06	04	01
101^5	=			1	05	10	10	05	01
101^6	=		1	06	15	20	15	06	01
101^7	=	1	07	21	35	35	21	07	01

But $\binom{9}{4} = \binom{9}{5} = 126$ contains three digits, and row 9 in Pascal's triangle is:

1 09 36 84 ①26 ①26 84 36 09 01

carry carry

To keep the two-digit format, we carry the 1 in the second 126 to the left to yield 127 and then carry the 1 in 127 to its left to yield 85. There are no more three-digit numbers. Thus 101^9 is

1 09 36 85 27 26 84 36 09 01

The various powers of 1001^n in Table 6.4 also reveal a strikingly similar pattern, where we have grouped the digits in threes. According to J. Ginzburg of then Yeshiva College, New York, this list appears in a 16th-century manuscript of an algebra textbook, published in 1902 by Maximilian Curtze.

Table 6.4

1001^0	=								001
1001^1	=							1	001
1001^2	=						1	002	001
1001^3	=					1	003	003	001
1001^4	=				1	004	006	004	001
1001^5	=			1	005	010	100	005	001
1001^6	=		1	006	015	020	015	006	001
1001^7	=	1	007	021	035	035	021	007	001

In general, let $x = 1\underbrace{00\cdots00}_{m \text{ zeros}}1$. Then row n in Pascal's triangle can be used to compute x^n using $(m+1)$-digit blocks. This is so since, by the binomial theorem,

$$x^n = \left(10^{m+1}+1\right)^n = \sum_{r=0}^{n} \binom{n}{r} 10^{(m+1)(n-r)}$$

$$= \binom{n}{0}10^{(m+1)n} + \binom{n}{1}10^{(m+1)(n-1)} + \cdots + \binom{n}{n-1}10^{m+1} + \binom{n}{n}10^0$$

This expansion clearly indicates that we need to use $(m+1)$-digit blocks.

Clearly, the same technique can be applied to compute x^n, where $x = a\underbrace{00\cdots00}_{m \text{ zeros}}b$, where $1 \le a, b \le 9$, and $n \ge 0$. It is obtained by adding the terms a^n, $\binom{n}{1}a^{n-1}b$, $\binom{n}{2}a^{n-2}b^2, \cdots, \binom{n}{n-1}ab^{n-1}$, and b^n in the binomial expansion of $(a \cdot 10^{m+1} + b)^n$, using $(m+1)$-digit blocks.

Mann–Shanks Primality Test

Next, we turn to a primality test developed in 1972 by H. B. Mann and D. Shanks of the University of Arizona, Tucson, Arizona. It involves another displaced Pascal's triangle. Gould also studied the same array in the same year.

Shift row n of Pascal's triangle to the right by $2n$ spaces. Figure 6.39 shows the resulting displaced triangular array. The $n+1$ binomial coefficients $\binom{n}{r}$ occur in columns $2n$ through $3n$ in the new array.

Theorem 6.1 (*Mann–Shanks Primality Test*) A positive integer k is prime if and only if every entry in column k is divisible by the corresponding row number; that is, k is prime if and only if $n | \binom{n}{k-2n}$, where $\lfloor k/3 \rfloor \le n \le \lfloor k/2 \rfloor$.

For example, the entries in column 17 are 6, 35, and 8; they occur in rows 6, 7, and 8, respectively. Since $6|6$, $7|35$, and $8|8$, each entry in column 17 is divisible by the corresponding row label; so 17 is a prime by the test.

n \ k	0	1	2	3	4	5	6	7	8	9	10	11	12	13	14	15	16	17	18	19
0	1																			
1			1	1																
2					1	2	1													
3							1	3	3	1										
4									1	4	6	4	1							
5											1	5	10	10	5	1				
6													1	6	15	20	15	6	1	
7															1	7	21	35	35	21
8																	1	8	28	56
9																			1	9

Figure 6.39 A Displaced Pascal's Triangle

On the other hand, consider column 14. Since $5 \nmid 1$, 15 is not a prime, as expected.

This primality test can be established using Hermite's divisibility property, as Gould did.

Stifel's Version

Finally, Figure 6.40 shows Stifel's version of the figurate triangle, published in 1544. Although it can be extracted from Shi-Kié's, it is very unlikely that Stifel ever knew of the latter.

1				
2				
3	3			
4	6			
5	10	10		
6	15	20		
7	21 —	35 ↓	35	
8	28	[56]	70	
9	36	84	126	126
10	45	120	210	252

Figure 6.40 Stifel's Version of the Figurate Triangle

This array has an important characteristic: Each column, after the first, begins two spaces lower than the preceding column, and starts with the number immediately to its left. Each subsequent number in the column is the sum of the entry immediately above and the entry immediately to the left of the latter, as Figure 6.40 demonstrates: $56 = 35 + 21$.

Suppose we know the expansion of $(x+y)^{n-1}$. Since $(x+y)^n = (x+y)(x+y)^{n-1}$, we can employ Stifel's version and the expansion of $(x+y)^{n-1}$ to find that of $(x+y)^n$. So, it would seem that Stifel had some intuitive knowledge of Pascal's identity.

Next we present a delightful relationship linking the well-known *Tower of Brahma* (TOB) puzzle,* Mersenne numbers M_n, Pascal's triangle, ternary words, and graph theory. But before turning to this, we first state the puzzle and then present some additional graph-theoretic terminology.

The Tower of Brahma Puzzle

According to legend, at the beginning of creation, God stacked 64 golden disks of varying sizes on one of three diamond pegs on a brass platform in the Temple of Brahma at Benares,† India (see Figure 6.41). The priests on duty were asked to transfer them from peg 0 to peg 2, using peg 1 as an auxiliary peg, subject to the following conditions:

- Only one disk can be moved at a time.
- No disk can be placed on top of a smaller one.

The priests were told that the world would come to an end when the job was done.

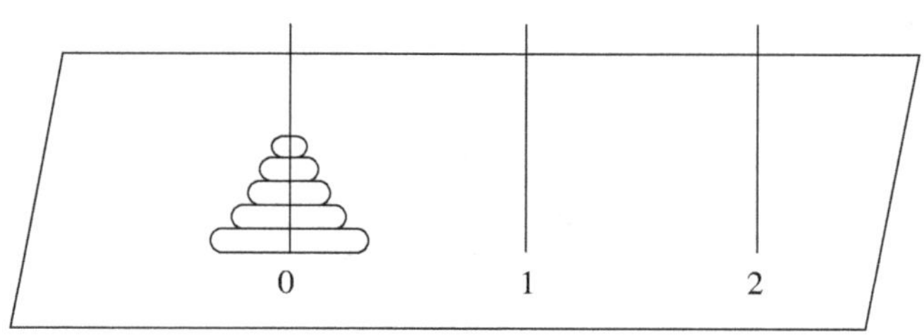

Figure 6.41 Tower of Brahma

Suppose there are n disks in the puzzle; we denote them 0 through $n-1$ in increasing order of size. Let b_n denote the number of moves needed to solve the puzzle. Clearly, $b_1 = 1$. Suppose $n \geq 2$. Then $b_n = b_{n-1} + 1 + b_{n-1} = 2b_{n-1} + 1$ (see Figure 6.42). Using iteration, it follows that $b_n = 2^n - 1 = M_n$ moves are needed to solve the puzzle.

*In 1883, Lucas marketed a puzzle based on the TOB problem, called the *Tower of Hanoi* (TOH); hence the puzzle is also called the TOH problem.

†Benares is now known as Varanasi.

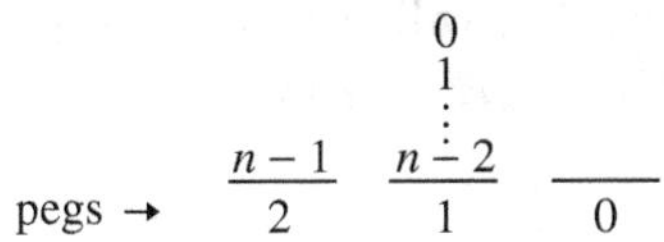

Figure 6.42

Additional Graph-theoretic Terminology

A *path* from vertex v_0 to vertex v_n in a graph is a sequence $v_0 - e_1 - v_1 - e_2 - \cdots - e_n - v_n$ of vertices v_i and edges e_i, where each edge e_i is incident with the vertices v_{i-1} and v_i and $1 \le i \le n$. The *length* of the path is n. A graph is *connected* if there is a path between every two distinct vertices of the graph; otherwise, it is *disconnected.*

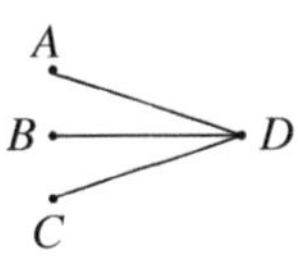

Figure 6.43

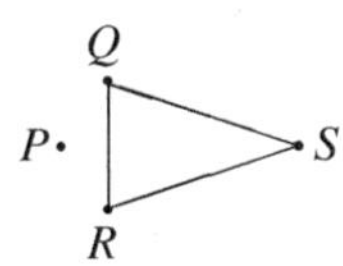

Figure 6.44

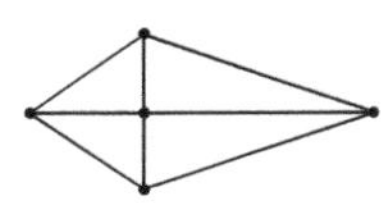

Figure 6.45

For example, in Figure 6.43 A–D–C is a path of length 2. The graph in Figure 6.43 is connected, whereas the one in Figure 6.44 is not.

A path from v_0 to v_n is *simple* if it contains no repeated vertices, except that the endpoints could be the same. The path is *closed* if $v_0 = v_n$; otherwise, it is *open.* A simple closed path is a *cycle.* If a connected graph contains a cycle that includes every vertex, the graph is *Hamiltonian*, after the Irish mathematician William Rowan Hamilton (1805–1865).

For example, Q–R–S–Q is a cycle in the graph in Figure 6.44. The graph in Figure 6.45 is Hamiltonian, but the one in Figure 6.44 is not.

We are now ready to present the relationship linking TOB, Mersenne numbers, Pascal's triangle, ternary words, and graph theory.

Pascal's Triangle and Hanoi Graphs

Recall from Theorem 4.7 that row k in Pascal's triangle consists of odd integers if and only if either $k = 0$ or k is a Mersenne number M_n, where $n \ge 1$. We now construct a graph H_n with vertices at the odd binomial coefficients in rows 0 through M_n; two vertices in H_n are adjacent if the corresponding binomial coefficients are adjacent, either horizontally or diagonally. In 1986, X. M. Lu called H_n the *Hanoi graph*; so we will keep the same practice. Figure 6.46 shows the fractal-like Hanoi graphs H_1, H_2, H_3, and H_4.

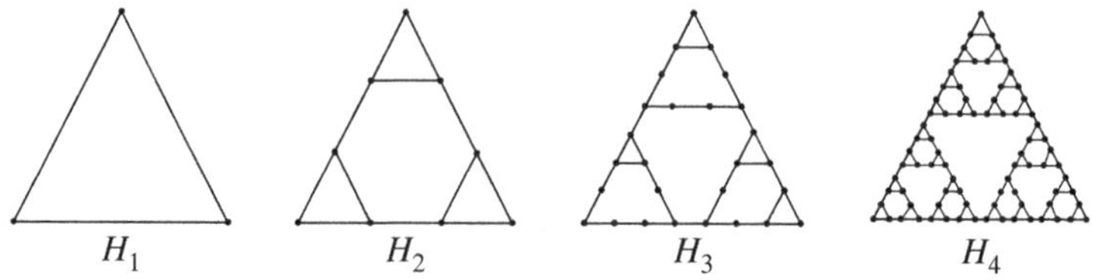

Figure 6.46 The Hanoi Graphs

The Hanoi graph H_n has 3^n vertices. Suppose it has e_n edges. Then $e_1 = 3$ and $e_n = 3e_{n-1} + 3$, where $n \geq 2$. Using iteration, it follows that $e_n = \frac{3}{2}(3^n - 1)$. Thus H_n has $\frac{3}{2}(3^n - 1)$ edges.

We are now ready to exhibit the relationship between H_n and the TOB puzzle.

The Hanoi Graph H_n and the TOB Puzzle

Every legal configuration of the disks can be represented by a *ternary* word $w_{n-1}w_{n-2}\cdots w_0$, where $w_i = j$ if disk i is on peg j and $0 \leq w_i \leq 2$. By the multiplication principle, there are 3^n ternary n-tuples; so there are 3^n legal configurations of the disks in the TOB puzzle with n disks.

For example, consider the configuration for three disks in Figure 6.47: disk 2 at peg 0, disk 0 at peg 1, and disk 1 at peg 2. Correspondingly, we have the ternary word $w_2w_1w_0 = 021$. Clearly, this process is reversible. For example, Figure 6.48 shows the configuration corresponding to the ternary word $w_2w_1w_0 = 121$.

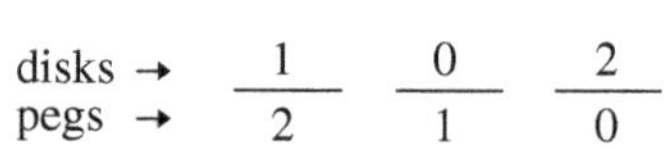

Figure 6.47

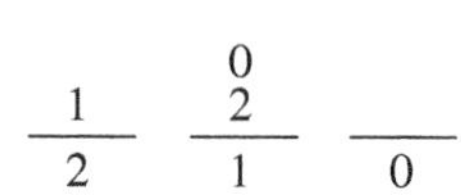

Figure 6.48

Figure 6.49 shows the eight configurations of the disks in the solution of the TOB puzzle with three disks. The corresponding ternary strings are 000, 002, 012, 011, 211, 210, 220, and 222, respectively.

These ternary strings can be used to label the vertices of H_3, as in Figure 6.50. Two vertices in H_3 are adjacent only if the Hamming distance between the corresponding words is one; that is, only if one configuration can be obtained from the other by a single legal move.

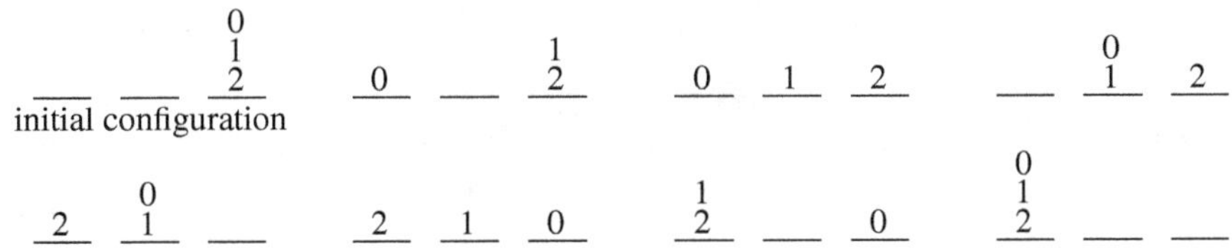

Figure 6.49

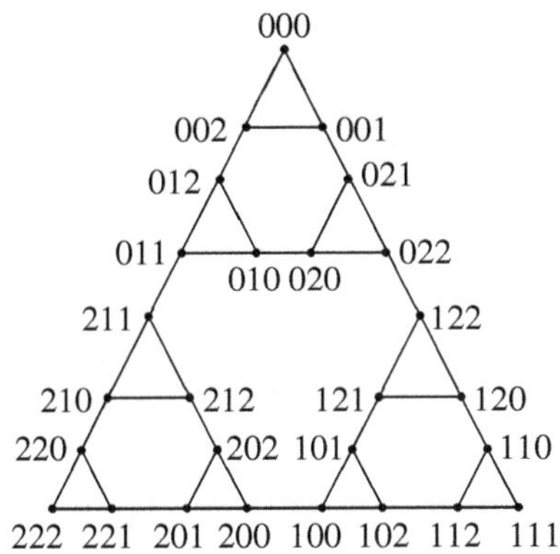

Figure 6.50 The Hanoi Graph H_3

Finally, we note that every Hanoi graph H_n is Hamiltonian. It would be a simple exercise, for example, to find a Hamiltonian cycle for H_3. (For those familiar with vector spaces, we add that the 3^n ternary words $00 \cdots 00$ through $22 \cdots 22$ form an n-dimensional vector space over the finite field Z_3 under scalar multiplication and addition modulo 3.)

7

Pascal's Triangle Revisited

For centuries, Pascal's triangle has been a rich, fertile ground for experimenting, exploring, and conjecturing about binomial coefficients. This chapter continues our investigation of additional properties of Pascal's triangle.

Triples of Consecutive Binomial Coefficients in AP

Consider the Pascal's triangle in Figure 6.1. The numbers $7 = \binom{7}{1}, 21 = \binom{7}{2}$, and $35 = \binom{7}{3}$ in row 7 are in arithmetic progression (AP) with common difference 14; by symmetry, 35, 21, and 7 are also in AP with common difference -14.

This innocent-looking observation tempts us to ask if Pascal's triangle contains other similar consecutive binomial coefficients; if yes, which rows contain such triples? If row n contains such a triple, where does the AP begin? That is, for what values of r, do the binomial coefficients $\binom{n}{r}, \binom{n}{r+1}$, and $\binom{n}{r+2}$ form an AP?

Interestingly, in 1946, T. Motzkin of Jerusalem established that Pascal's triangle contains infinitely many triples of consecutive binomial coefficients that form an AP, and that they occur in rows $n = k^2 - 2$, where $r = \frac{k^2 \pm k - 4}{2}$ and $k \geq 3$; that is, they occur in row n, where $n + 2$ is a square. Ten years later, A. Struyk of Clifton High School, Clifton, New Jersey, also investigated such triples of binomial coefficients.

If $n = k^2 - 2$, then a triple with the desired property begins with the binomial coefficient $\binom{n}{\frac{k^2-k-4}{2}}$. Since Pascal's triangle is symmetric with the vertical line through the middle, the triple beginning with $\binom{n}{\frac{k^2+k-4}{2}}$ also displays the same property.

For example, row $14 = 4^2 - 2$ contains two triples that are in AP. One begins with $r = \frac{k^2-k-4}{2} = \frac{4^2-4-4}{2} = 4$, and the other with $r = \frac{k^2+k-4}{2} = \frac{4^2+4-4}{2} = 8$.

The corresponding triples are $\binom{14}{4} = 1001$, $\binom{14}{5}$, and $\binom{14}{6} = 3003$; and $\binom{14}{8} = 3003$, $\binom{14}{9} = 2002$, and $\binom{14}{10} = 1001$ in the reverse order.

A triple with the desired property does indeed begin with the binomial coefficient $\binom{n}{\frac{k^2+k-4}{2}}$, where $n = k^2 - 2$. To establish this, let r_j denote the value of r in row n such that the triple beginning with $\binom{n}{r}$ forms an AP. The various elements of the sequence $\{r_j\}$ are $-1, 1, 4, 8, 13, 19, 26, \cdots$, where $j \geq 1$. The differences between consecutive elements in this sequence are consecutive positive integers, beginning with 2:

$$\begin{array}{ccccccccccccccc} -1 & & 1 & & 4 & & 8 & & 13 & & 19 & & 26 & & \cdots \\ & 2 & & 3 & & 4 & & 5 & & 6 & & 7 & & ? & \end{array}$$

Consequently, r_j can be defined recursively by $r_j = r_{j-1} + j$, where $r_1 = -1$ and $j \geq 2$.

Using iteration, we have:

$$\begin{aligned} r_j &= r_{j-1} + j \\ &= r_{j-2} + (j-1) + j \\ &= r_{j-3} + (j-2) + (j-1) + j \\ &\vdots \\ &= r_1 + 2 + 3 + \cdots + j \\ &= -1 + \left[\frac{j(j+1)}{2} - 1\right] \\ &= \frac{j(j+1)}{2} - 2 = \frac{j^2 + j - 4}{2} \end{aligned}$$

This can be confirmed using induction.

Thus the binomial coefficients $\binom{n}{r_j}$, $\binom{n}{r_j+1}$, and $\binom{n}{r_j+2}$ form an AP, where $n = k^2 - 2$ and $r_j = \frac{j^2+j-4}{2}$. By symmetry, their counterparts also form an AP.

For example, let $n = 23 = 5^2 - 2$. Then, $j = 4$, so $r_j = r_4 = \frac{4^2+4-4}{2} = 8$. Clearly, the binomial coefficients $\binom{23}{8} = 490{,}314$, $\binom{23}{9} = 817{,}190$, and $\binom{23}{10} = 1{,}144{,}066$ form an AP with common difference 326,876. The next three are $\binom{34}{13}$, $\binom{34}{14}$, and $\binom{34}{15}$, with common ratio 463,991,880.

Suppose row $n = k^2 - 2$ contains a triple with the desired property. Then the common difference of the AP from the first triple is given by:

$$\begin{aligned} \binom{n}{r+1} - \binom{n}{r} &= \frac{n!}{(r+1)!(n-r-1)!} - \frac{n!}{r!(n-r)!} \\ &= \frac{n!}{r!(r+1)(n-r-1)!} - \frac{n!}{r!(n-r-1)!(n-r)} \end{aligned}$$

$$= \frac{n!}{r!(n-r-1)!}\left(\frac{1}{r+1} - \frac{1}{n-r}\right)$$

$$= \frac{(n-2r-1)n!}{(r+1)!(n-r)!} = \frac{n-2r-1}{r+1}\binom{n}{r}$$

$$= \frac{2(k+1)}{k^2-k-2}\binom{n}{\frac{k^2-k-4}{2}} = \frac{2}{k-2}\binom{n}{\frac{k^2-k-4}{2}}$$

For example, when $n = 7 = 3^2 - 2$, the common difference of the first triple is $\frac{2}{1}\binom{7}{1} = 14$; and when $n = 34 = 6^2 - 2$, the common difference is $\frac{2}{4}\binom{34}{13} = 463{,}991{,}880$, as expected.

It follows that the common difference resulting from the second triple in row n is $-\frac{2}{k-2}\binom{n}{\frac{k^2-k-4}{2}}$.

Quadruplets of Consecutive Binomial Coefficients in AP

Next, we examine if it is possible for four consecutive binomial coefficients $\binom{n}{r}$, $\binom{n}{r+1}$, $\binom{n}{r+2}$, and $\binom{n}{r+3}$ to be in AP; and if yes, we would like to identify them. This problem appeared in the 1972 William Lowell Putnam Mathematical Competition.

For the first three coefficients to be in AP, we must have $2\binom{n}{r+1} = \binom{n}{r} + \binom{n}{r+2}$. That is,

$$2 = \frac{r+1}{n-r} + \frac{n-r-1}{r+2} \tag{7.1}$$

Likewise, the last three coefficients yield

$$2 = \frac{r+2}{n-r-1} + \frac{n-r}{r+3} \tag{7.2}$$

which is obtained by changing r to $r+1$ in equation (7.1).

From equations (7.1) and (7.2), we have:

$$\frac{r+1}{n-r} + \frac{n-r-1}{r+2} = \frac{r+2}{n-r-1} + \frac{n-r}{r+3}$$

$$\frac{r+1}{n-r} - \frac{r+2}{n-r-1} = \frac{n-r}{r+3} - \frac{n-r-1}{r+2}$$

$$\frac{n-1}{(n-r)(n-r-1)} = \frac{n-2r-3}{(r+2)(r+3)}$$

$$(n-1)(r+2)(r+3) = (n-r)(n-r-1)(n-2r-3)$$

This implies that $n = 1, r, r+1$, or $2r+3$. Clearly, $n \neq 1, r$, or $r+1$. So, $n = 2r+3$. The corresponding binomial coefficients are $\binom{2r+3}{r}$, $\binom{2r+3}{r+1}$, $\binom{2r+3}{r+2}$, and $\binom{2r+3}{r+3}$. But $\binom{2r+3}{r} = \binom{2r+3}{r+3}$ and $\binom{2r+3}{r+1} = \binom{2r+3}{r+2}$. Since the middle terms are equal, the four terms do not form an increasing or decreasing AP. Thus, there are *no* four consecutive binomial coefficients in AP.

An Alternate Approach

Equations (7.1) and (7.2) show that both r and $r+1$ are solutions of the quadratic equation

$$2 = \frac{x+1}{n-x} + \frac{n-x-1}{x+2} \tag{7.3}$$

Replacing x with $n-x-2$ in equation (7.3) yields the same equation; so equation (7.3) has four solutions: $r, r+1, n-r-3$, and $n-r-2$. But equation (7.3) has only two solutions; so $r = n-r-3$ and hence $n = 2r+3$. The rest of the solution goes as above.

CBC and the Northeast Diagonal *n*

Consider the elements $\binom{2n-r}{n}$ on the northeast diagonal n beginning at the CBC $\binom{2n}{n}$, where $r \geq 0$. Suppose we multiply it by 2^r. The weighted sum $\sum_{r=0}^{n} 2^r \binom{2n-r}{n}$ satisfies an interesting property.

For example, $\sum_{r=0}^{3} 2^r \binom{6-r}{n} = 2^0\binom{6}{3} + 2^1\binom{5}{3} + 2^2\binom{4}{3} + 2^3\binom{3}{3} = 2^{2\cdot 3}$; see Figure 7.1.

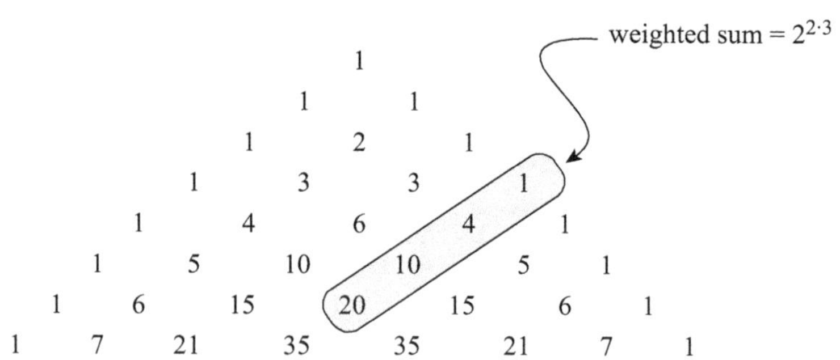

Figure 7.1

More generally, we have the following result:

$$\sum_{r=0}^{n} 2^r \binom{2n-r}{n} = 2^{2n} \tag{7.4}$$

Before we establish this formula, first notice by Corollary 3.1 that $\sum_{r=0}^{2n+1} \binom{2n+1}{r} = 2^{2n+1}$. But

$$\sum_{r=0}^{2n+1} \binom{2n+1}{r} = \sum_{r=0}^{n} \binom{2n+1}{r} + \sum_{r=n+1}^{2n+1} \binom{2n+1}{r}$$

$$2^{2n+1} = \sum_{r=0}^{n} \binom{2n+1}{r} + \sum_{s=0}^{n} \binom{2n+1}{s}$$

$$2^{2n+1} = 2\sum_{r=0}^{n} \binom{2n+1}{r}$$

$$2^{2n} = \sum_{r=0}^{n} \binom{2n+1}{r}$$

For example, $\sum_{r=0}^{3} \binom{7}{r} = \binom{7}{0} + \binom{7}{1} + \binom{7}{2} + \binom{7}{3} = 1 + 7 + 21 + 35 = 64 = 2^{2\cdot 3}$.

To establish the formula, notice that the LHS is really the coefficient of x^n in the polynomial $\sum_{r=0}^{n} 2^r(1+x)^{2n-r}$. We will now evaluate this as a geometric sum, expand the resulting sum as a power series, and then look for the coefficient of x^n again:

$$\begin{aligned}
\sum_{r=0}^{n} 2^r(1+x)^{2n-r} &= (1+x)^{2n} \sum_{r=0}^{n} \left(\frac{2}{1+x}\right)^r \\
&= (1+x)^{2n} \cdot \frac{\frac{2^{n+1}}{(1+x)^{n+1}} - 1}{\frac{2}{1+x} - 1} \\
&= \frac{2^{n+1}(1+x)^n - (1+x)^{2n+1}}{1-x} \\
&= \frac{2^{n+1}(1+x)^n}{1-x} - \frac{(1+x)^{2n+1}}{1-x} \\
&= 2^{n+1}(1+x)^n \sum_{s=0}^{\infty} x^s - (1+x)^{2n+1} \sum_{s=0}^{\infty} x^s \\
&= 2^{n+1} \sum_{r=0}^{n} \binom{n}{r} x^r \sum_{s=0}^{\infty} x^s - \sum_{r=0}^{2n+1} \binom{2n+1}{r} x^r \sum_{s=0}^{\infty} x^s
\end{aligned}$$

The coefficient of x^n on the RHS is given by

$$2^{n+1} \sum_{r=0}^{n} \binom{n}{r} - \sum_{r=0}^{2n+1} \binom{2n+1}{r} = 2^{n+1} \cdot 2^n - 2^{2n} = 2^{2n}$$

Thus $\sum_{r=0}^{n} 2^r \binom{2n-r}{n} = 2^{2n}$, as desired.

As a byproduct, it follows from equation (7.4) that the weighted sum $\sum_{r=0}^{n} 2^r \binom{2n-r}{n}$ is a square.

Banach's Matchbox Problem

Formula (7.4) is closely related to the well-known **matchbox problem**, developed by the eminent Polish mathematician Stefan Banach (1892–1945):

Banach used to carry two identical matchboxes in his pocket. Each contained n matches originally. Every time he needed to light a match, he would select a box at random. After taking out a match, he would put the box back in his pocket even if it was empty. When the box chosen was empty, he would discard it, and reach for the other box. Find the probability that exactly r matches were left in the second box at that time, where $0 \le r \le n$.

Let A and B denote the two boxes. Initially, $|A| = n = |B|$. When box A is empty, there are r matches left in box B; so $n - r$ matches must have been taken from B; in other words, when the chosen box is empty, $n + (n - r) = (2n - r)$ matches must have been taken by then. Since the outcomes are equally likely, there are 2^{2n-r} different ways of choosing $2n - r$ matches. But the n matches from A and $n - r$ matches from B can be selected in $\binom{n+(n-r)}{n} = \binom{2n-r}{n}$ different ways.

Thus the probability that n matches are selected from A and $n - r$ from B is given by $\frac{1}{2^{2n-r}}\binom{2n-r}{n}$; that is, the probability that exactly r matches are left in box B when the empty box A is selected is $\frac{1}{2^{2n-r}}\binom{2n-r}{n}$, where $0 \le r \le n$.

Since the sum of the probabilities is one, it follows that $\sum_{r=0}^{n} \frac{1}{2^{2n-r}}\binom{2n-r}{n} = 1$. That is, $\sum_{r=0}^{n} 2^r \binom{2n-r}{n} = 2^{2n}$, as desired.

Additional Properties

The combinatorial formula

$$\binom{n}{2} + \binom{n+1}{2} = n^2 \tag{7.5}$$

coupled with the fact that

$$\sum_{k=r}^{m} \binom{k}{r} = \binom{m+1}{r+1} \tag{7.6}$$

can be used to develop a formula for $\sum_{i=1}^{n} i^2$. Since $i^2 = \binom{i}{2} + \binom{i+1}{2}$, we have:

$$\begin{aligned}\sum_{i=1}^{n} i^2 &= \sum_{i=1}^{n} \binom{i}{2} + \sum_{i=1}^{n} \binom{i+1}{2} \\ &= \binom{n+1}{3} + \binom{n+2}{3}\end{aligned}$$

This implies that $\sum_{i=1}^{n} i^2$ can be computed as the sum of two consecutive elements on the ascending diagonal 3; see the loops in Figure 7.2.

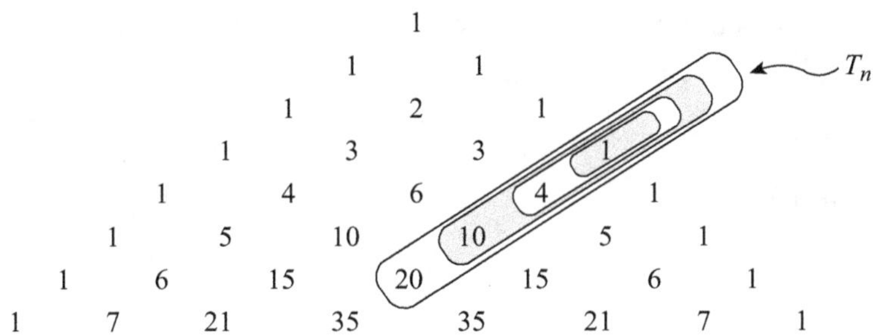

Figure 7.2

Clearly, $\sum_{i=1}^{n} i^2 = \binom{n+1}{3} + \binom{n+2}{3} = \frac{n(n+1)(2n+1)}{6}$, as expected.

Figurate Numbers Revisited

- Consider any three consecutive tetrahedral numbers $T_n = \binom{n+2}{3}$ on ascending diagonal 3; see Figure 7.2. The sum of the two ends and four times the middle number is a cube. For example, $20+4\cdot 10+4 = 64 = 4^3$.

  More generally,

$$\begin{aligned}T_n+4T_{n-1}+T_{n-2} &= \binom{n+2}{3}+4\binom{n+1}{3}+\binom{n}{3} \\ &= \frac{(n+2)(n+1)n}{6}+4\frac{(n+1)n(n-1)}{6}+\frac{n(n-1)(n-2)}{6} \\ &= \frac{n}{6}(6n^2)=n^3\end{aligned}$$

  Thus

$$\binom{n}{3}+4\binom{n+1}{3}+\binom{n+2}{3}=n^3 \tag{7.7}$$

Using the same technique as before, combinatorial formulas (7.6) and (7.7) can be used to develop a formula for $\sum_{i=1}^{n} i^3$:

$$i^3 = \binom{i}{3} + 4\binom{i+1}{3} + \binom{i+2}{3}$$

$$\sum_{i=1}^{n} i^3 = \sum_{i=1}^{n}\binom{i}{3} + 4\sum_{i=1}^{n}\binom{i+1}{3} + \sum_{i=1}^{n}\binom{i+2}{3}$$

$$= \binom{n+1}{4} + 4\binom{n+2}{4} + \binom{n+3}{4} \tag{7.8}$$

Thus $\sum_{i=1}^{n} i^3$ can be read using three consecutive elements on the descending diagonal 4 with proper weights.

For example, $\sum_{i=1}^{4} i^3 = \binom{5}{4} + 4\binom{6}{4} + \binom{7}{4} = 5 + 4 \cdot 15 + 35 = 100$; see Figure 7.3.

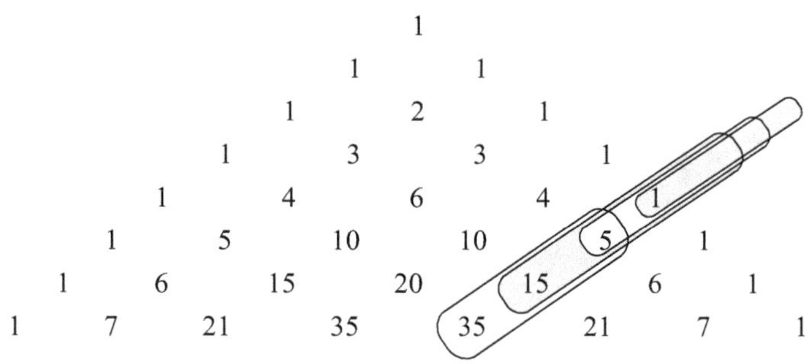

Figure 7.3

- Every tetrahedral number T_n is the sum of the triangular numbers on ascending diagonal 2. For example, $10 = 6 + 3 + 1$ and $20 = 10 + 6 + 3 + 1$; see Figure 7.3. This is true in general since

$$\sum_{i=1}^{n} t_k = \sum_{i=1}^{n}\binom{k+1}{2} = \binom{n+2}{3} = T_n$$

Returning to formulas (7.5) and (7.7), we have:

$$n = \binom{n}{1}$$

$$n^2 = \binom{n}{2} + \binom{n+1}{2}$$

$$n^3 = \binom{n}{3} + 4\binom{n+1}{3} + \binom{n+2}{3}$$

Following this pattern, we may conjecture that

$$n^4 = \binom{n}{4} + a\binom{n+1}{4} + b\binom{n+2}{4} + \binom{n+3}{4}$$

for some integers a and b. Using symmetry, we may even gamble that $a = b$. So

$$n^4 = \binom{n}{4} + a\binom{n+1}{4} + a\binom{n+2}{4} + \binom{n+3}{4}$$

Suppose we let $n = 2$. Then:

$$\begin{aligned} 16 &= \binom{2}{4} + a\binom{3}{4} + a\binom{4}{4} + \binom{5}{4} \\ &= 0 + 0 + a + 5 \\ a &= 11 \end{aligned}$$

Thus we conjecture that

$$\binom{n}{4} + 11\binom{n+1}{4} + 11\binom{n+2}{4} + \binom{n+3}{4} = n^4 \qquad (7.9)$$

Fortunately, this can be confirmed algebraically:

$$\begin{aligned} \text{LHS} &= \binom{n}{4} + 11\binom{n+1}{4} + 11\binom{n+2}{4} + \binom{n+3}{4} \\ &= \frac{(n+3)(n+2)(n+1)n}{24} + 11\frac{(n+2)(n+1)n(n-1)}{24} \\ &\quad + 11\frac{(n+1)n(n-1)(n-2)}{24} + \frac{n(n-1)(n-2)(n-3)}{24} \\ &= \frac{n}{24}[(n+2)(n+1)(12n-8) + (n-1)(n-2)(12n+8)] \\ &= \frac{n}{6}\left[(3n-2)(n^2+3n+2) + (3n+2)(n^2-3n+2)\right] \\ &= \frac{n}{6} \cdot 6n^3 = n^4 \end{aligned}$$

as desired. It can be confirmed using PMI also.

For example, $\binom{5}{4} + 11\binom{6}{4} + 11\binom{7}{4} + \binom{8}{4} = 625 = 5^4$; see Figure 7.4.

As before, it follows from formula (7.9) that

$$\sum_{i=1}^{n} i^4 = \binom{n+1}{5} + 11\binom{n+2}{5} + 11\binom{n+3}{5} + \binom{n+4}{5} \qquad (7.10)$$

For example, $\sum_{i=1}^{5} i^4 = \binom{6}{5} + 11\binom{7}{5} + 11\binom{8}{5} + \binom{9}{5} = 979$.

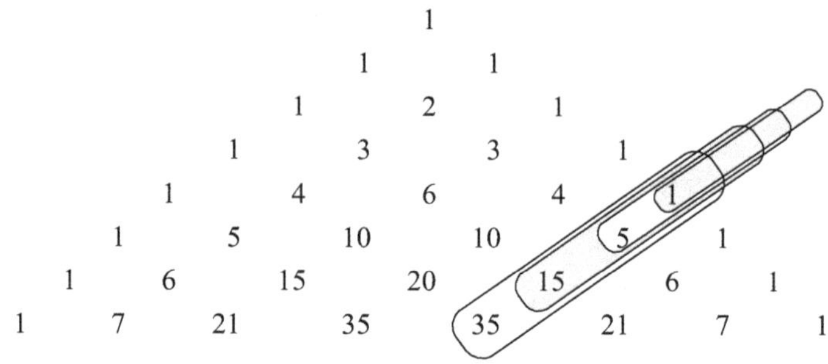

Figure 7.4

- Using the same technique we employed to develop formula (7.9), we can develop a similar one for n^5:

$$\binom{n}{5} + 26\binom{n+1}{5} + 66\binom{n+2}{5} + 26\binom{n+3}{5} + \binom{n+4}{5} = n^5 \tag{7.11}$$

In other words, rising diagonal 5 can be used to compute n^5.

For example, $\binom{6}{5} + 26\binom{7}{5} + 66\binom{8}{5} + 26\binom{9}{5} + \binom{10}{5} = 7{,}776 = 6^5$; see Figure 7.5.

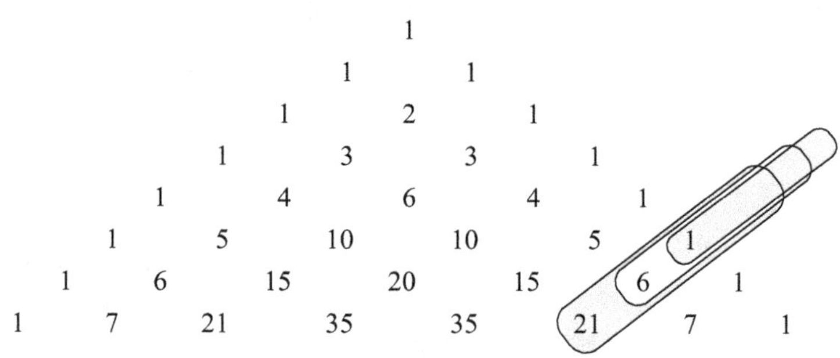

Figure 7.5

A Surprising Dividend

The left-hand sides of formulas (7.5), (7.7), (7.9), and (7.11) for the first four powers of n are summarized in Table 7.1. The various binomial coefficients $\binom{m}{r}$ in this array can be used to form the triangular array in Figure 7.6. It contains a hidden treasure: The row sums are factorials. For example, $1 + 11 + 11 + 1 = 24 = 4!$.

That each row sum is $n!$ was proposed as a problem in 1993 by D. Callan of the University of Wisconsin at Whitewater, by adding pre- and post-subscripts to

Table 7.1

$$\binom{n}{1}$$
$$\binom{n}{2} + \binom{n+1}{2}$$
$$\binom{n}{3} + 4\binom{n+1}{3} + \binom{n+2}{3}$$
$$\binom{n}{4} + 11\binom{n+1}{4} + 11\binom{n+2}{4} + \binom{n+3}{4}$$
$$\binom{n}{5} + 26\binom{n+1}{5} + 66\binom{n+2}{5} + 26\binom{n+3}{5} + \binom{n+4}{5}$$

n \ j	1	2	3	4	5	Row sums
1	1					1
2	1	1				2
3	1	4	1			6
4	1	11	11	1		24
5	1	26	66	26	1	120

↑ $n!$ (Row sums)

Figure 7.6

each entry in Figure 7.6; see Figure 7.7. Each element is a linear combination of its northern neighbor and its western neighbor with the subscripts as coefficients. For example, $26 = 4 \cdot 1 + 2 \cdot 11$.

$${}_1 1_1$$
$${}_1 1_2 \quad {}_2 1_1$$
$${}_1 1_3 \quad {}_2 4_2 \quad {}_3 1_1$$
$${}_1 1_4 \quad {}_2 11_3 \quad {}_3 11_2 \quad {}_4 11_1$$
$${}_1 1_5 \quad {}_2 26_4 \quad {}_3 66_3 \quad {}_4 26_2 \quad {}_4 1_1$$

Figure 7.7

Notice that the sum of the subscripts at each entry in row n is $n + 1$, where $n \geq 1$. Consequently, when the elements in row $n + 1$ are added, each entry in row n is counted $n + 1$ times. So, the sum of the elements in row $n + 1$ is $n + 1$ times that of the elements in row n. But the sum of the elements in row 1 is $1 = 1!$ and that in row 2 is $2 = 2!$. Thus, by PMI, the row sum in row n is $n!$, as M. P. Eisner of Mount Vernon College, Washington, D.C. established in 1994. The problem of finding the row sum in the array is the standard problem in combinatorial enumeration of counting permutations by the number of ascents in the permutation;[†] see Chapter 22. (The permutation $p_1 p_2 \cdots p_n$ of $\{1, 2, \cdots, n\}$ has j ascents if $p_k < p_{k+1}$ exactly j times.)

Eulerian Polynomials

The entries in row n in Figure 7.7 (without the subscripts) are the coefficients in *Eulerian polynomials* E_n, published by Euler in 1755 in his book *Intitutiones Calculi Differentialis*. The entries, known as *Eulerian coefficients*, are denoted by $\left\langle {n \atop j} \right\rangle$. The first six Eulerian polynomials are:

$$E_1(x) = x \qquad E_4(x) = x^4 + 11x^3 + 11x^2 + x$$

$$E_2(x) = x^2 + x \qquad E_5(x) = x^5 + 26x^4 + 66x^3 + 26x^2 + x$$

$$E_3(x) = x^3 + 4x^2 + x \qquad E_6(x) = x^6 + 57x^5 + 302x^4 + 302x^3 + 57x^2 + x$$

The *Eulerian numbers* $\left\langle {n \atop j} \right\rangle$ are often defined recursively as follows:

$$\left\langle {n \atop 1} \right\rangle = 1 = \left\langle {n \atop n} \right\rangle$$

$$\left\langle {n \atop j} \right\rangle = (n - j + 1)\left\langle {n-1 \atop j-1} \right\rangle + j\left\langle {n-1 \atop j} \right\rangle \tag{7.12}$$

where $\left\langle {n \atop j} \right\rangle = 0$ if $n < j$ and $j > 1$.

For example,

$$\left\langle {3 \atop 2} \right\rangle = 2 \cdot \left\langle {2 \atop 1} \right\rangle + 2 \cdot \left\langle {2 \atop 2} \right\rangle = 2 \cdot 1 + 2 \cdot 1 = 4,$$

and

$$\left\langle {5 \atop 3} \right\rangle = 3 \cdot \left\langle {4 \atop 2} \right\rangle + 3 \cdot \left\langle {4 \atop 3} \right\rangle = 3 \cdot 11 + 3 \cdot 11 = 66.$$

Next, we will show that

$$\left\langle {n \atop j} \right\rangle = \left\langle {n \atop n-j+1} \right\rangle \tag{7.13}$$

That is, each row is symmetric. For example, $\left\langle {5 \atop 2} \right\rangle = 26 = \left\langle {5 \atop 4} \right\rangle = \left\langle {5 \atop 5-2+1} \right\rangle$.

[†] See R. L. Graham et al., *Concrete Mathematics*, Addison-Wesley, Waltham, MA, 1990.

We will establish property (7.13) using PMI. First, since $\left\langle {n \atop 1} \right\rangle = \left\langle {n \atop n} \right\rangle$, it is true when $j = 1$.

Now, assume that it is true for an arbitrary integer $j < n$. Then $\left\langle {n-1 \atop n-j} \right\rangle = \left\langle {n-1 \atop j} \right\rangle$ and $\left\langle {n-1 \atop n-j+1} \right\rangle = \left\langle {n-1 \atop j-1} \right\rangle$. So, by the recurrence relation (7.12),

$$\begin{aligned}\left\langle {n \atop n-j+1} \right\rangle &= j\left\langle {n-1 \atop j-1} \right\rangle + (n-j+1)\left\langle {n-1 \atop n-j+1} \right\rangle \\ &= j\left\langle {n-1 \atop j} \right\rangle + (n-j+1)\left\langle {n-1 \atop j-1} \right\rangle \\ &= \left\langle {n \atop j} \right\rangle\end{aligned}$$

Thus, by PMI, property (7.13) holds for every j, where $1 \leq j \leq n$.

Points of Intersection of the Diagonals of a Convex *n*-gon

Next, we would like to find the total number of points $f(n)$ at which the diagonals of a convex n-gon intersect in its interior, where no three diagonals intersect inside the n-gon and $n \geq 3$. Clearly,

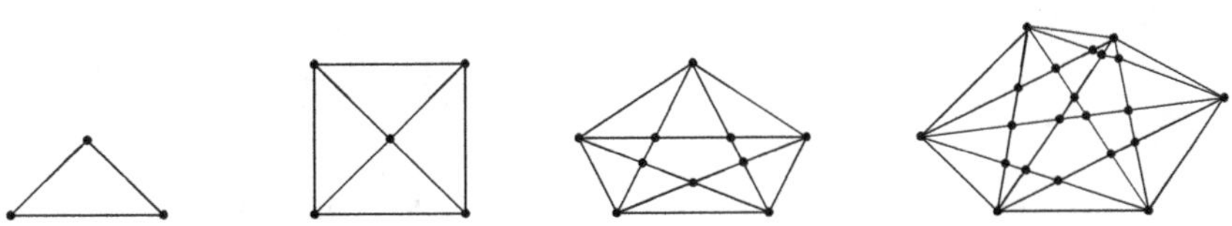

Figure 7.8

$f(3) = 0, f(4) = 1, f(5) = 5$, and $f(6) = 15$; see Figure 7.8. More generally, $f(n) = \binom{n}{4}$, where $n \geq 3$. Consequently, $f(n)$ is given by the entries on the rising diagonal 4. See Figure 7.9.

Number of Regions Formed Inside a Convex *n*-gon by Its Diagonals

Next, we turn to a closely related problem: Find the number $g(n)$ of non-overlapping regions formed inside a convex n-gon by its diagonals such that no three diagonals meet in the interior, where $n \geq 3$. See Figure 7.8. Table 7.2 summarizes the first few values of $g(n)$.

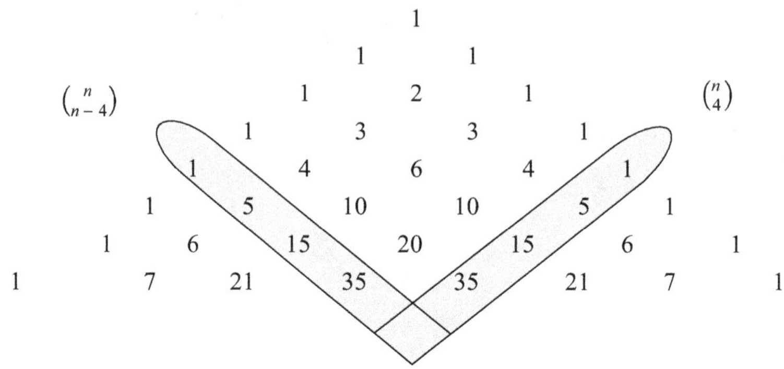

Figure 7.9

Table 7.2

n	3	4	5	6	7	$\cdots$
$g(n)$	1	4	11	25	50	$\cdots$

Notice an intriguing pattern:

$$g(3) = 1 = 1 + 0 + 0 = \binom{2}{2} + \binom{2}{3} + \binom{2}{4}$$

$$g(4) = 4 = 3 + 1 + 0 = \binom{3}{2} + \binom{3}{3} + \binom{3}{4}$$

$$g(5) = 11 = 6 + 4 + 1 = \binom{4}{2} + \binom{4}{3} + \binom{4}{4}$$

More generally, we predict that $g(n) = \binom{n-1}{2} + \binom{n-1}{3} + \binom{n-1}{4}$, where $n \geq 3$.

This formula can be established using the well-known *Euler's formula*[†] for a planar graph:[‡] Let v denote the number of vertices, e the number of edges, and r the number of regions (including the outside region) formed on the plane by the graph; then $v - e + r = 2$. Consequently, if r^* denotes the number of interior regions, then $r^* = e - v + 1$.

In the case of the convex n-gon, there are $\binom{n}{4}$ interior points and n vertices; so $v = \binom{n}{4} + n$. Since there are exactly four edges at each interior point and $n - 1$ edges at each vertex of the n-gon, there is a total of $4\binom{n}{4} + n(n-1)$ edges; *but* each edge is counted twice in this total; so $e = 2\binom{n}{4} + \binom{n}{2}$.

[†] See D. I. A. Cohen, *Basic Techniques of Combinatorial Theory*, Wiley, New York, 1978.

[‡] See Chapter 9.

Thus, the total number of distinct regions inside the n-gon is given by

$$\begin{aligned} g(n) &= \left[4\binom{n}{4} + \binom{n}{2}\right] - \left[\binom{n}{4} + n\right] + 1 \\ &= \binom{n}{4} + \binom{n}{2} - n + 1 \\ &= \binom{n-1}{4} + \binom{n-1}{3} + \binom{n-1}{2} \end{aligned}$$

Consequently, the value of $g(n)$ can be computed by adding the three binomial coefficients in row $n-1$ beginning at ascending diagonal 2, as shown in Figure 7.10.

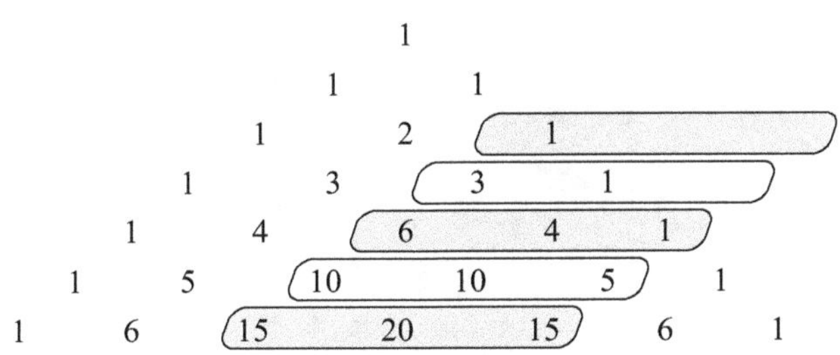

Figure 7.10

For example, $g(8) = \binom{7}{2} + \binom{7}{3} + \binom{7}{4} = 21 + 35 + 35 = 91$.

For the curious-minded, we add that $g(n)$ can be defined recursively: $g(n) = g(n-1) + \binom{n-1}{3} + (n-2)$, where $n \geq 3$ and $g(3) = 1$.

- What can we say about the sum of the squares of consecutive odd integers on the rising diagonal 1? For example, we have:

$$1^2 = 1 = T_1 \qquad 1^2 + 3^2 = 10 = T_3$$

$$1^2 + 3^2 + 5^2 = 35 = T_5 \qquad 1^2 + 3^2 + 5^2 + 7^2 = 84 = T_7$$

where $T_n = \binom{n+2}{3}$ denotes the nth tetrahedral number; see Figure 7.11.

More generally, we conjecture that $\sum_{i=1}^{n}(2i-1)^2 = \binom{2n+1}{3} = T_{2n-1}$; this can be confirmed by expanding the sum on the LHS. Likewise, $\sum_{i=1}^{n}(2i)^2 = \binom{2n+2}{3} = T_{2n}$; see Figure 7.12.

Since tetrahedral numbers are cumulative sums of triangular numbers, it follows that they are indeed sums of squares of odd positive integers or

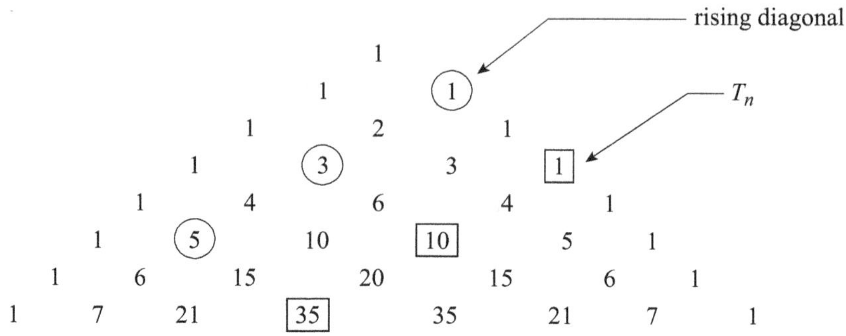

Figure 7.11

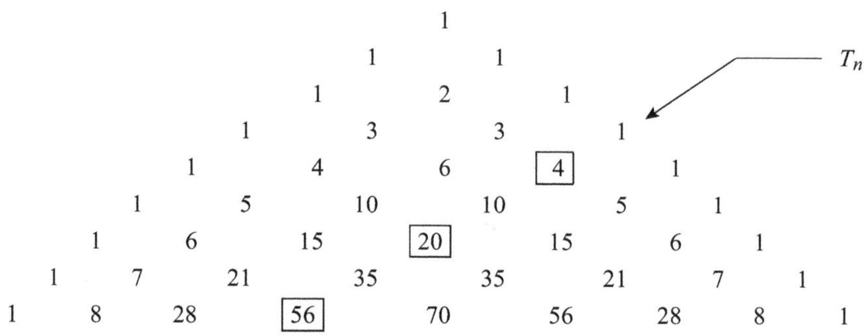

Figure 7.12

even positive integers:

$$\begin{aligned}1+3+6+10+15+21+\cdots &= (1+3)+(6+10)+(15+21)+\cdots\\ &= 2^2+4^2+6^2+8^2+\cdots\end{aligned}$$

and

$$\begin{aligned}1+3+6+10+15+21+\cdots &= 1+(3+6)+(10+15)+(21+28)+\cdots\\ &= 1^2+3^2+5^2+7^2+\cdots\end{aligned}$$

Lagrange's Identity Revisited

- Lagrange's identity can be interpreted slightly differently. Suppose we build a triangular array in which each element equals the square of the corresponding element in Pascal's triangle. The nth row sum of the resulting array is $\binom{2n}{n} = (n+1)C_n$. See Figure 7.13.

	Row sums
1	$1 = \binom{0}{0}$
1 1	$2 = \binom{2}{1}$
1 4 1	$6 = \binom{4}{2}$
1 9 9 1	$20 = \binom{6}{3}$
1 16 36 16 1	$70 = \binom{8}{4}$
1 25 100 100 25 1	$252 = \binom{10}{5}$

Figure 7.13

A Hockey Stick Property

In 1982, M. Boscarol of Free University, Taranto, Italy, developed the following property of binomial coefficients, which we call *Boscarol's identity*:

$$\sum_{i=0}^{m} \frac{\binom{n+i}{i}}{2^{n+i}} + \sum_{j=0}^{n} \frac{\binom{n+m-j}{m}}{2^{n+m-j}} = 2 \tag{7.14}$$

where $m, n \geq 0$. Its meaning and relevance become clear if we relate the identity to Pascal's triangle.

An Interpretation of Boscarol's Identity

Consider the path $\binom{n}{0} + \binom{n+1}{1} + \binom{n+2}{2} + \cdots$ along the southeast diagonal originating at the binomial coefficient $\binom{n}{0}$. Stop at any point $\binom{n+m}{m}$. Then continue in the northeast direction until the end $\binom{m}{m}$ of the diagonal. See Figure 7.14. Counting the turning point $\binom{n+m}{m}$ twice, divide each binomial coefficient $\binom{r}{s}$ along the path by 2^r. The sum of the resulting ratios is 2.

$$\begin{array}{ccccccccc}
\binom{n}{0} & & & & & & & & \binom{m}{m} \\
 & \binom{n+1}{1} & & & & & & \binom{n+m-2}{m} & \\
 & & \binom{n+2}{2} & & & & \binom{n+m-1}{m} & & \\
 & & & \ddots & & \iddots & & & \\
 & & & & \binom{n+m}{m} & & & &
\end{array}$$

Figure 7.14

For example, consider the path in Figure 7.15. Then

$$\frac{1}{16} + \frac{5}{32} + \frac{15}{64} + \frac{35}{128} + \frac{35}{128} + \frac{20}{64} + \frac{10}{32} + \frac{4}{16} + \frac{1}{8} = 2$$

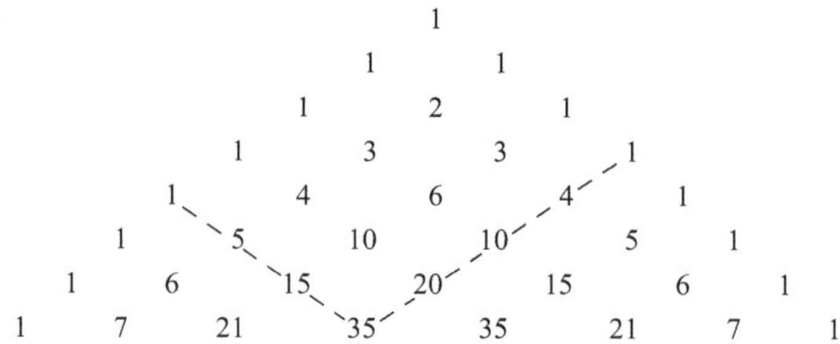

Figure 7.15

Similarly,

$$\frac{\binom{3}{0}}{2^3}+\frac{\binom{4}{1}}{2^4}+\frac{\binom{5}{2}}{2^5}+\frac{\binom{5}{2}}{2^5}+\frac{\binom{4}{2}}{2^4}+\frac{\binom{3}{2}}{2^3}+\frac{\binom{2}{2}}{2^2}=2$$

Since the path formed by the binomial coefficients in the above algorithm takes the shape of a hockey stick, we take the liberty of naming it a *hockey stick property* of binomial coefficients.

Next, we present a proof of Boscarol's identity.

A Proof of Boscarol's Identity

In order to establish Boscarol's identity (7.14), we need the following result, which follows by a repeated application of Pascal's identity.

Let a, b, and c be any integers such that $a < b$ and $c \geq 0$. Then

$$2\sum_{r=a}^{b}\binom{c}{r} = \sum_{r=a+1}^{b}\binom{c+1}{r} + \binom{c}{a} + \binom{c}{b} \tag{7.15}$$

We are now ready to prove Boscarol's identity using PMI. It is true when $m = n = 0$. So assume that $n > 0$. Let $k = n - j$. Then identity (7.14) becomes

$$\sum_{i=0}^{m}\frac{\binom{n+i}{i}}{2^{n+i}}+\sum_{k=0}^{n}\frac{\binom{m+k}{m}}{2^{m+k}}=2$$

That is,

$$\sum_{i=0}^{m}\frac{\binom{n+i}{i}}{2^{n+i}}+\sum_{j=0}^{n}\frac{\binom{m+j}{m}}{2^{m+j}}=2 \tag{7.16}$$

Since the sum on the LHS is symmetric in m and n, we can assume without any loss of generality, that $m \leq n$.

By Corollary 3.1,

$$2 = \frac{\sum_{k=0}^{m} \binom{m}{k}}{2^{m-1}} = \frac{2\sum_{k=0}^{m} \binom{m}{k}}{2^m}$$

Since $\binom{m}{k} = 0$ for $k < 0$, the sum may start at $k = m - n \leq 0$:

$$2 = \frac{2 \sum_{k=m-n}^{m} \binom{m}{k}}{2^m}$$

Since $m > m - n$, using identity (7.15), the RHS can be written as follows:

$$\text{RHS} = \frac{\sum_{k=m-n+1}^{m} \binom{m+1}{k} + \binom{m}{m-n} + \binom{m}{m}}{2^m}$$

$$= \frac{2 \sum_{k=m-n+1}^{m} \binom{m+1}{k}}{2^{m+1}} + \frac{\binom{m}{m-n} + \binom{m}{m}}{2^m}$$

Again, $m > m - n + 1$; so we can apply identity (7.15) again. Applying it r times, we get:

$$\text{RHS} = \frac{2 \sum_{k=m-n+r}^{m} \binom{m+r}{k}}{2^{m+r}} + \sum_{j=0}^{n-1} \frac{\binom{m+j}{m-n+j} + \binom{m+j}{m}}{2^{m+j}}$$

In fact, we can continue applying identity (7.15) until $m - n + r < m$; that is, until $r < n$. Then:

$$\text{RHS} = \frac{2\binom{m+n}{n}}{2^{m+n}} + \sum_{j=0}^{n-1} \frac{\binom{m+j}{m-n+j} + \binom{m+j}{m}}{2^{m+j}}$$

$$= \sum_{j=0}^{n} \frac{\binom{m+j}{m-n+j} + \binom{m+j}{m}}{2^{m+j}} = \sum_{s=0}^{n} \frac{\binom{m+s}{m-n+s}}{2^{m+s}} + \sum_{j=0}^{n} \frac{\binom{m+j}{m}}{2^{m+j}}$$

Let $i = m - n + s$. Since $m - n < 0$, we can restrict the first sum to nonnegative values of i. Thus

$$\text{RHS} = \sum_{i=0}^{m} \frac{\binom{n+i}{i}}{2^{n+i}} + \sum_{j=0}^{n} \frac{\binom{m+j}{m}}{2^{m+j}}$$

That is,

$$\sum_{i=0}^{m} \frac{\binom{n+i}{i}}{2^{n+i}} + \sum_{j=0}^{n} \frac{\binom{m+j}{m}}{2^{m+j}} = 2$$

as desired. ■

A Moscow Mathematical Olympiad Problem

The next example presents an elegant application of Pascal's triangle. It is based on the following problem† from the eighth Moscow Mathematical Olympiad of 1945 for ninth and tenth graders:

There are 2^{1000} travelers at the apex of the network of roads in Figure 7.16. At each intersection, exactly half go in the southwest direction and the other half in the southeast direction. Find the number of people reaching each intersection in row 1000.

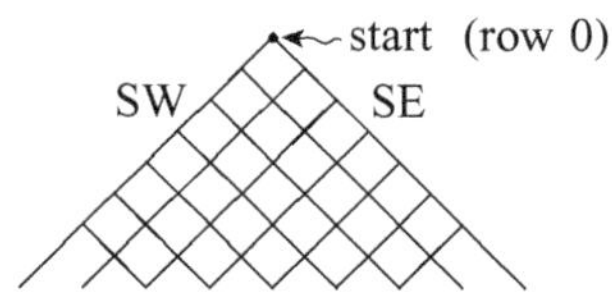

Figure 7.16

Let $f(n, r)$ denote the number of travelers reaching intersection r in row n, where $0 \leq r \leq n \leq 1000$. Since there are 2^{1000} people at the apex, $f(0, 0) = 2^{1000}$. Clearly, $f(1, 0) = 2^{999} = f(1, 1)$.

More generally,

$$f(n, 0) = \frac{f(n-1, 0)}{2} \tag{7.17}$$

and

$$f(n, n) = \frac{f(n-1, n-1)}{2} \tag{7.18}$$

See Figure 7.17. Since $f(0, 0) = 2^{1000}$, it follows that both $f(n, 0)$ and $f(n, n)$ are integers. Since one-half of the people $f(n-1, r-1)$ at intersection $r-1$ and that of the people $f(n-1, r)$ at intersection r, both in row $n-1$, go to intersection r in row n, it follows that

$$f(n, r) = \frac{f(n-1, r-1) + f(n-1, r)}{2} \tag{7.19}$$

where $1 \leq r \leq n-1$. See Figure 7.17.

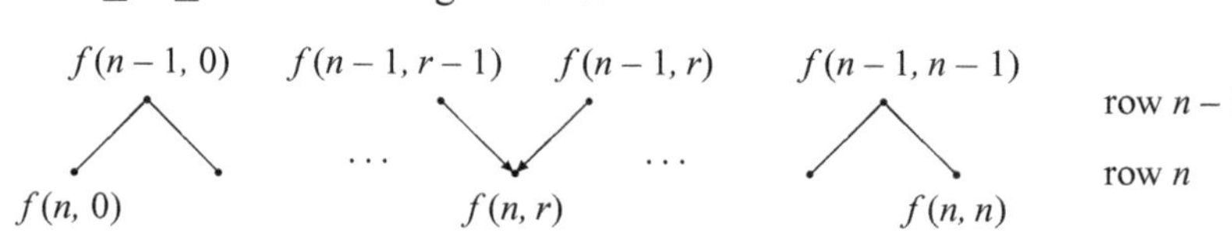

Figure 7.17

†V. A. Uspenskii, *Pascal's Triangle*, University of Chicago Press, Chicago, 1974.

For example,

$$f(2,0) = \frac{f(1,0)}{2} = 2^{998}$$

$$f(2,1) = \frac{f(1,0)+f(1,1)}{2} = \frac{2^{999}+2^{999}}{2} = 2^{999}$$

Similarly,

$$f(2,2)=2^{998} \quad f(3,0)=2^{997} \quad f(3,1)=3\cdot 2^{997} \quad f(3,2)=3\cdot 2^{997} \quad f(3,3)=2^{997}$$

Notice that

$$\begin{aligned} f(3,0)+f(3,1)+f(3,2)+f(3,3) &= 2^{997}+3\cdot 2^{997}+3\cdot 2^{997}+2^{997} = 2^{1000} \\ &= f(2,0)+f(2,1)+f(2,2) \\ &= f(1,0)+f(1,1) = f(0,0) \end{aligned}$$

as expected.

If the number of people at each intersection in row $n-1$ is even, then it follows by properties (7.17), (7.18), and (7.19) that the number of travelers at each intersection in row n is also even. Since those numbers in rows 0 and 1 are even, it follows by PMI that the number of travelers at every intersection in each row is even.

The values $f(0,0), f(1,r), f(2,r)$, and $f(3,r)$ manifest a fascinating pattern:

$$f(0,0) = \binom{0}{0}2^{1000-0} \qquad f(1,0) = \binom{1}{0}2^{1000-1} \qquad f(1,1) = \binom{1}{1}2^{1000-1}$$

$$f(2,0) = \binom{2}{0}2^{1000-2} \qquad f(2,1) = \binom{2}{1}2^{1000-2} \qquad f(2,2) = \binom{2}{2}2^{1000-2}$$

$$f(3,0) = \binom{3}{0}2^{1000-3} \qquad f(3,1) = \binom{3}{1}2^{1000-3} \qquad f(3,2) = \binom{3}{2}2^{1000-3}$$

More generally, we conjecture that

$$f(n,r) = \binom{n}{r}2^{1000-n} \tag{7.20}$$

where $0 \le r \le n \le 1000$. Clearly, this formula works when $n = 0$. Now, assume that it works for an arbitrary integer n, where $0 \le n < 1000$. Then, by the recurrence relation (7.19),

$$\begin{aligned} f(n+1,r) &= \frac{f(n,r-1)+f(n,r)}{2} = \frac{\binom{n}{r-1}2^{1000-n}+\binom{n}{r}2^{1000-n}}{2} \\ &= \left[\binom{n}{r-1}+\binom{n}{r}\right]2^{1000-(n+1)} = \binom{n+1}{r}2^{1000-(n+1)} \end{aligned}$$

Thus, by PMI, formula (7.20) works for every nonnegative integer $n \le 1000$.

In particular,

$$f(1000, r) = \binom{1000}{r} 2^{1000-1000} = \binom{1000}{r}$$

In other words, the number of travelers reaching intersection r in row 1000 is $\binom{1000}{r}$, where $0 \le r \le 1000$.

The total number of travelers in row n is given by

$$\sum_{r=0}^{n} f(1000, r) = \sum_{r=0}^{n} \binom{1000}{r} = 2^{1000}$$

as expected.

More generally, if there are 2^m people at the apex, then $\binom{m}{r}$ of them will reach intersection r in row m, where $0 \le r \le m$.

Extracting Square Roots

In 1989, I. M. Richards of Penwith Sixth Form College, Penzance, Cornwall, England, discovered an intriguing way of evaluating the principal square roots of positive integers using the CBCs.

To see how to achieve this, recall from Chapter 2 that

$$\frac{1}{\sqrt{1-4x}} = \sum_{n=0}^{\infty} \binom{2n}{n} x^n \tag{7.21}$$

This power series converges when $|4x| < 1$; that is, when $|x| < \frac{1}{4}$.

Let $x = \frac{y-1}{4y}$, where $y > 0$. Then:

$$\frac{1}{\sqrt{1-4x}} = \frac{1}{\sqrt{1 - \frac{4(y-1)}{4y}}} = \sqrt{y}$$

Thus

$$\sqrt{y} = \sum_{n=0}^{\infty} \binom{2n}{n} x^n \tag{7.22}$$

where $x = \frac{y-1}{4y}$. This expansion can be employed to compute $\sqrt{y}$ to any desired accuracy, provided $\frac{y-1}{4y} < \frac{1}{4}$.

For example, let $y = 2$. Then $x = \frac{1}{8}$ and

$$\begin{aligned}\sqrt{2} &= \sum_{n=0}^{\infty} \binom{2n}{n} \left(\frac{1}{8}\right)^n \\ &= 1 + 2\left(\frac{1}{8}\right) + 6\left(\frac{1}{8}\right)^2 + 20\left(\frac{1}{8}\right)^3 + \cdots\end{aligned}$$

Golden Ratio and the CBCs

Suppose we let $y = 5$ in equation (7.22). Then $x = \frac{1}{5}$ and

$$\begin{aligned}\sqrt{5} &= \sum_{n=0}^{\infty} \binom{2n}{n} \left(\frac{1}{5}\right)^n \qquad (7.23) \\ &= 1 + \sum_{n=1}^{\infty} \binom{2n}{n} \left(\frac{1}{5}\right)^n \\ \sqrt{5} - 1 &= \sum_{n=1}^{\infty} \binom{2n}{n} \left(\frac{1}{5}\right)^n \\ \frac{\sqrt{5} - 1}{2} &= \frac{1}{2} \sum_{n=1}^{\infty} \binom{2n}{n} \left(\frac{1}{5}\right)^n\end{aligned}$$

But $\frac{1}{2}\binom{2n}{n} = \binom{2n-1}{n-1}$. Therefore,

$$\frac{\sqrt{5} - 1}{2} = \sum_{n=1}^{\infty} \binom{2n-1}{n-1} \left(\frac{1}{5}\right)^n \qquad (7.24)$$

That is,

$$\frac{1}{\alpha} = 1 + 3\left(\frac{1}{5}\right) + 10\left(\frac{1}{5}\right)^2 + 35\left(\frac{1}{5}\right)^3 + \cdots$$

where $\alpha = \frac{1+\sqrt{5}}{2}$ denotes the golden ratio.[†]

Notice that the coefficients $\binom{2n-1}{n-1}$ in the expansion (7.24) form the vertical column to the left of the central column; see Figure 7.18.

[†]For an extensive discussion of the golden ratio, see the author's *Fibonacci and Lucas Numbers with Applications.*

$$
\begin{array}{ccccccccccccc}
 & & & & & & 1 & & & & & & \\
 & & & & & \boxed{1} & & 1 & & & & & \\
 & & & & 1 & & 2 & & 1 & & & & \\
 & & & 1 & & \boxed{3} & & 3 & & 1 & & & \\
 & & 1 & & 4 & & 6 & & 4 & & 1 & & \\
 & 1 & & 5 & & \boxed{10} & & 10 & & 5 & & 1 & \\
1 & & 6 & & 15 & & 20 & & 15 & & 6 & & 1 \\
 & & & & & \uparrow & & & & & & & \\
 & & & & & \binom{2n-1}{n-1} & & & & & & &
\end{array}
$$

Figure 7.18

Clever substitutions for y could lead to series that converge faster. For example, let $y = \frac{5}{4}$. Then $x = \frac{1}{20}$ and

$$
\begin{aligned}
\sqrt{5} &= 2\sum_{n=0}^{\infty} \binom{2n}{n} \left(\frac{1}{20}\right)^n \\
&= 2\left[1 + 2\left(\frac{1}{20}\right) + 6\left(\frac{1}{20}\right)^2 + 20\left(\frac{1}{20}\right)^3 + 70\left(\frac{1}{20}\right)^4 + \cdots\right]
\end{aligned}
$$

The first five terms in this expansion yields $\frac{1}{\alpha} \approx 0.6179375$, which is not that far from the correct value $\frac{1}{\alpha} = 0.6180339887\ldots$.

Pascal's Triangle and π

A year after Richards published his work on extracting square roots, A. Fearnehough of Portsmouth Sixth Form College showed how the CBCs can be used to develop an interesting formula for π.

To this end, consider the infinite sum $\sum_{n=0}^{\infty} \frac{1}{(2n+1)\binom{2n}{n}}$, which can be rewritten as $\sum_{n=0}^{\infty} \frac{(n!)^2}{(2n+1)!}$. A repeated application of integration by parts yields

$$
\int_0^1 x^n(1-x)^n \, dx = \frac{(n!)^2}{(2n+1)!}
$$

Therefore,

$$\sum_{n=0}^{\infty} \frac{1}{(2n+1)\binom{2n}{n}} = \sum_{n=0}^{\infty} \int_0^1 x^n(1-x)^n \, dx$$

$$= \int_0^1 \sum_{n=0}^{\infty} x^n(1-x)^n \, dx = \int_0^1 \frac{1}{1-x(1-x)} \, dx$$

$$= \int_0^1 \frac{dx}{x^2-x+1} = \int_0^1 \frac{dx}{\left(x-\frac{1}{2}\right)^2 + \left(\frac{\sqrt{3}}{2}\right)^2}$$

$$= \frac{2}{\sqrt{3}} \int_{-\pi/6}^{\pi/6} d\theta = \frac{2\pi}{3\sqrt{3}}$$

where we have used the substitution $x - \frac{1}{2} = \frac{\sqrt{3}}{2} \tan\theta$.

Thus

$$\sum_{n=0}^{\infty} \frac{1}{(2n+1)\binom{2n}{n}} = \frac{2\pi}{3\sqrt{3}}$$

Soon after Fearnehough discovered this formula, Richards developed a similar one using expansion (7.21), which converges when $|x| < 1/4$:

$$\frac{1}{\sqrt{1-4x}} = \sum_{n=0}^{\infty} \binom{2n}{n} x^n$$

Changing x to x^2, this yields

$$\frac{1}{\sqrt{1-4x^2}} = \sum_{n=0}^{\infty} \binom{2n}{n} x^{2n}$$

where $|x| < 1/2$. So

$$\int_0^{1/4} \frac{dx}{\sqrt{1-4x^2}} = \sum_{n=0}^{\infty} \binom{2n}{n} \int_0^{1/4} x^{2n} \, dx$$

$$\frac{1}{2} \arcsin 2x \Big]_{x=0}^{x=1/2} = \sum_{n=0}^{\infty} \frac{\binom{2n}{n}}{2n+1} x^{2n+1} \Big]_{x=0}^{x=1/4}$$

This yields

$$\frac{1}{2} \cdot \frac{\pi}{3} = \sum_{n=0}^{\infty} \frac{\binom{2n}{n}}{2n+1} \left(\frac{1}{4}\right)^{2n+1}$$

That is,

$$\sum_{n=0}^{\infty} \frac{\binom{2n}{n}}{(2n+1)16^n} = \frac{2\pi}{3}$$

Suppose we would like to evaluate the sum $\sum\limits_{n=0}^{\infty} \frac{1}{\binom{2n}{n}}$. Since $\frac{1}{\binom{2n}{n}} = \frac{(n!)^2}{(2n)!}$, it follows from the above discussion that

$$\frac{1}{\binom{2n}{n}} = (2n+1) \int_0^1 x^n (1-x)^n \, \mathrm{d}x$$

Therefore,

$$\begin{aligned}
\sum_{n=0}^{\infty} \frac{1}{\binom{2n}{n}} &= \sum_{n=0}^{\infty} (2n+1) \int_0^1 x^n (1-x)^n \, \mathrm{d}x \\
&= 2 \int_0^1 \sum_{n=0}^{\infty} n x^n (1-x)^n \, \mathrm{d}x + \int_0^1 \sum_{n=0}^{\infty} x^n (1-x)^n \, \mathrm{d}x \\
&= 2 \int_0^1 \sum_{n=0}^{\infty} n x^n (1-x)^n \, \mathrm{d}x + \frac{2\pi}{3\sqrt{3}}
\end{aligned}$$

Now, consider

$$\begin{aligned}
\int_0^1 \sum_{n=0}^{\infty} n x^n (1-x)^n \, \mathrm{d}x &= \int_0^1 x(1-x) \left[\sum_{n=0}^{\infty} n (x - x^2)^{n-1} \right] \mathrm{d}x \\
&= \int_0^1 (x - x^2) \frac{\mathrm{d}x}{(1 - x + x^2)^2} \\
&= \int_0^1 \frac{x \mathrm{d}x}{(x^2 - x + 1)^2} - \int_0^1 \frac{x^2 \, \mathrm{d}x}{(x^2 - x + 1)^2}
\end{aligned}$$

Using the substitution $x - \frac{1}{2} = \frac{\sqrt{3}}{2}\tan\theta$ in both integrals on the RHS, we have:

$$I_1 = \int_0^1 \frac{x\mathrm{d}x}{(x^2 - x + 1)^2}$$

$$= \int_{-\pi/6}^{\pi/6} \frac{(\frac{1}{2} + \frac{\sqrt{3}}{2}\tan\theta)\frac{\sqrt{3}}{2}\sec^2\theta\,\mathrm{d}\theta}{\frac{9}{16}\sec^4\theta}$$

$$= \frac{\sqrt{3}}{4} \cdot \frac{16}{9} \int_{-\pi/6}^{\pi/6} (1 + \sqrt{3}\tan\theta)\cos^2\theta\,\mathrm{d}\theta$$

$$= \frac{4\sqrt{3}}{9} \int_{-\pi/6}^{\pi/6} \left[\frac{1 + \cos 2\theta}{2} + \sqrt{3}\frac{\sin 2\theta}{2}\right]\mathrm{d}\theta$$

$$= \frac{2\sqrt{3}}{9}\left[\theta + \frac{1}{2}\sin 2\theta - \frac{\sqrt{3}}{2}\cos 2\theta\right]_{-\pi/6}^{\pi/6}$$

$$= \frac{2\sqrt{3}}{9}\left[\left(\frac{\pi}{6} + \frac{1}{2}\cdot\frac{\sqrt{3}}{2} - \frac{\sqrt{3}}{2}\cdot\frac{\sqrt{3}}{2}\right) - \left(-\frac{\pi}{6} - \frac{1}{2}\cdot\frac{\sqrt{3}}{2} - \frac{\sqrt{3}}{2}\cdot\frac{\sqrt{3}}{2}\right)\right]$$

$$= \frac{2\sqrt{3}}{9}\left(\frac{\pi}{3} + \frac{\sqrt{3}}{2}\right) = \frac{2\pi\sqrt{3}}{27} + \frac{1}{3}$$

and

$$I_2 = \int_0^1 \frac{x^2\mathrm{d}x}{(x^2 - x + 1)^2}$$

$$= \frac{\sqrt{3}}{2} \cdot \frac{16}{9} \cdot \frac{1}{4} \int_{-\pi/6}^{\pi/6} (1 + 2\sqrt{3}\tan\theta + 3\tan^2\theta)\cos^2\theta\,\mathrm{d}\theta$$

$$= \frac{2\sqrt{3}}{9} \int_{-\pi/6}^{\pi/6} \left[\frac{1 + \cos 2\theta}{2} + \sqrt{3}\sin 2\theta + 3\left(\frac{1 - \cos 2\theta}{2}\right)\right]\mathrm{d}\theta$$

$$= \frac{2\sqrt{3}}{18}\left[\theta + \frac{\sin 2\theta}{2} - \sqrt{3}\cos 2\theta + 3\theta - \frac{3}{2}\sin 2\theta\right]_{-\pi/6}^{\pi/6}$$

$$= \frac{\sqrt{3}}{9}\left[4\theta - \sqrt{3}\cos 2\theta - \sin 2\theta\right]_{-\pi/6}^{\pi/6}$$

$$= \frac{\sqrt{3}}{9}\left[(4\theta - \sqrt{3}\cos 2\theta - \sin 2\theta) - (-4\theta - \sqrt{3}\cos 2\theta + \sin 2\theta)\right]_{\theta=\pi/6}$$

$$= \frac{\sqrt{3}}{9}\left[8\theta - 2\sin 2\theta\right]_{\theta=\pi/6}$$

$$= \frac{\sqrt{3}}{9}\left(8\cdot\frac{\pi}{6} - 2\cdot\frac{\sqrt{3}}{2}\right) = \frac{\sqrt{3}}{9}\left(4\cdot\frac{\pi}{3} - \sqrt{3}\right)$$

$$= \frac{4\pi\sqrt{3}}{27} - \frac{1}{3}$$

Therefore,

$$\sum_{n=0}^{\infty}\frac{1}{\binom{2n}{n}} = 2I_1 - 2I_2 + \frac{2\pi}{3\sqrt{3}}$$

$$= \frac{4\pi\sqrt{3}}{27} + \frac{2}{3} - \frac{8\pi\sqrt{3}}{27} + \frac{2}{3} + \frac{2\pi}{3\sqrt{3}}$$

$$= \frac{4}{3} - \frac{4\pi\sqrt{3}}{27} + \frac{2\pi}{3\sqrt{3}}\cdot\frac{3\sqrt{3}}{3\sqrt{3}}$$

$$= \frac{4}{3} + \frac{2\pi\sqrt{3}}{27}$$

8

Pascal's *m*-ary Triangles

In the two preceding chapters, we discussed a number of properties of the amazing Pascal's triangle. This chapter continues the investigation of additional properties of the array, through a geometric lens using the congruence relation as a powerful tool.

Fractal-like Patterns

Suppose we replace each even number in Pascal's triangle by a black dot and each odd number by an open circle. Figure 8.1 shows the resulting binary pattern, *Pascal's binary triangle* (PBT).

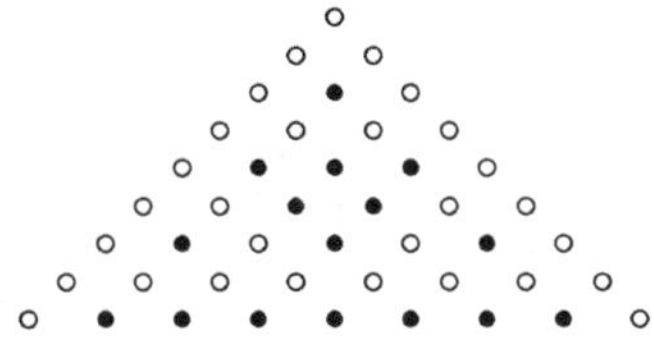

Figure 8.1 Pascal's Binary Triangle

Recall from Theorem 4.7 that every entry in row n is odd if and only if n is a Mersenne number, where $n > 1$. Consequently, if n is a Mersenne number, then row n consists of open circles only. There are exactly six such rows n, where $0 \le n \le 100$; namely, rows 0, 1, 3, 7, 31, and 63.

PBT and Even Perfect Numbers

In addition to its aesthetic beauty, the binary triangle contains a hidden treasure. To see it, consider the centrally located triangles ∇_n pointing downward and having their bases in row 2^n, where $n \geq 1$. Since the base of ∇_n contains $2^n - 1$ zeros, the number of zeros inside ∇_n equals

$$N = \frac{(2^n - 1)(2^n - 1 + 1)}{2} = 2^{n-1}(2^n - 1)$$

which is a perfect number if n and $2^n - 1$ are primes.

For example, the base of ∇_4 lies in row 32 and it contains $2^5 - 1 = 31$ zeros. So the number of zeros inside ∇_4 equals

$$31 + 30 + \cdots + 2 + 1 = \frac{31 \cdot 32}{2} = 496$$

which is the third perfect number.

More generally, since every number $N = 2^{n-1}(2^n - 1)$ is represented by ∇_n, every even perfect number $N = 2^{p-1}(2^p - 1)$ is represented by ∇_p, where p and $2^p - 1$ are primes; in other words, even perfect numbers are represented geometrically by a subsequence of the sequence $\{\nabla_n\}$, as observed in 1956 by Alan L. Brown of South Orange, New Jersey.

In 1979, Ko-Wei Lih of Academica Sinica, Tapei, Taiwan, asked if the binary triangle contains the binary word 1101 or 1011 in any row. Suppose, for instance, it contains the word 1011. Then it must also contain the array in Figure 8.2. This contradicts the Hoggatt–Hansell hexagon property in Chapter 6, since $0 \cdot 0 \cdot 0 \neq 1 \cdot 1 \cdot 1$, as shown in 1980 by M. Mays of West Virginia University. Likewise, the binary triangle cannot contain the pattern 1101 also.

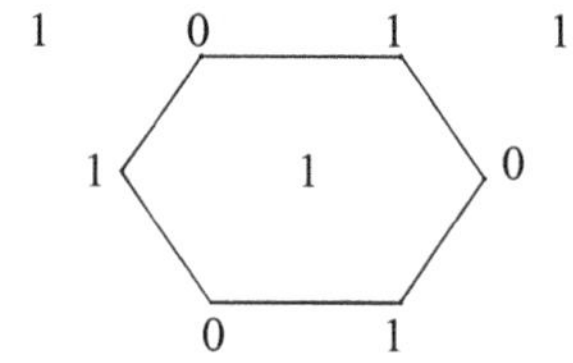

Figure 8.2

In fact, using Lucas' theorem,† J. G. Mauldon of Amherst College, Amherst, Massachusetts, proved a more general result pertaining to Pascal's p-ary triangle: *The string* 10ab *or its reverse occurs in Pascal's p-ary triangle if and only if* $a(2a + b) \equiv 0 \pmod{p}$. With $a = 1 = b$, $a(2a + b) \not\equiv 0 \pmod{p}$, so 10$ab$ or ba01 will *not* occur in Pascal's p-ary triangle.

†See Theorem 4.10.

PBT can be used to extract several properties of Pascal's triangle, as P. Häggmark of Sweden did in 1977. First, notice that row 2^k has the form $1\underbrace{00\cdots00}_{M_k \text{ zeros}}1$. So rows 2^k through $2^{k+1}-1 = M_{k+1}$ consist of two identical copies of rows 0 through M_k, with an inverted triangle of 0s in between them. Row 2^{k+1} has the form $1\underbrace{000\cdots000}_{M_{k+1} \text{ zeros}}1$. The pattern now repeats itself.

Let $0 \le j < 2^k$. Then row $2^k + j$ consists of two identical copies of row j, separated by 2^k zeros between them. Consequently, if N_r denotes the number of 1s in row r, then

$$N_{2^k+j} = 2N_j \tag{8.1}$$

This implies that the number of odd numbers in row n of Pascal's triangle is 2^m, where m is the number of 1s in n_{two}.

For example, let $n = 11 = 2^3 + 2 + 1 = 1011_{\text{two}}$. So $N_{11} = N_{2^3+2+1} = 2N_{2+1} = 2 \cdot 2N_1 = 2 \cdot 2 \cdot 2 = 2^3$.

Next, we compute the relative frequency of 1s in rows 0 through M_k.

Relative Frequency of 1s in Rows 0 through M_k

Let $T_{0,n}$ denote the number of 1s in rows 0 through n. Then $T_{0,1} = 3$; $T_{0,3} = 3T_{0,1} = 3^2$; and $T_{0,7} = 3T_{0,3} = 3^3$. More generally, $T_{0,M_k} = 3^k$; this can be confirmed using PMI.

The total number of bits in rows 0 through M_k is given by

$$\sum_{i=0}^{M_k}(i+1) = \sum_{j=1}^{M_k+1} j = \frac{(M_k+1)(M_k+2)}{2} = \frac{2^k(2^k+1)}{2}$$

So the relative frequency of 1s in rows 0 through M_k is given by

$$\frac{2 \cdot 3^k}{2^k(2^k+1)} = \frac{2 \cdot 3^k}{2^k(4^k+2^k)} = \frac{2}{(4/3)^k + (2/3)^k)}$$

This approaches 0 as $k \to \infty$, so nearly all binomial coefficients are even.

Next we present a close relationship between PBT and *Fermat numbers* f_n; they are defined by $f_n = 2^{2^n} + 1$, where $n \ge 0$. The Fermat numbers $f_0 = 3$, $f_1 = 5, f_2 = 17, f_3 = 257$, and $f_4 = 65{,}537$ are the only known Fermat primes. Since $641|f_5$, it follows that f_5 is composite.†

† See the author's *Elementary Number Theory with Applications*.

PBT and Fermat Numbers

To see the relationship, we consider each row of PBT as a binary number. Let B_k denote the decimal value of row k. For example, $B_5 = 110011_{\text{two}} = 51$ and $B_{13} = \underbrace{110011}\,00\,\underbrace{110011}_{\text{two}} = 13{,}107$.

More generally, it follows from equation (8.1) that $B_{2^k+j} = 2^{2^k}B_j + B_j = \left(2^{2^k}+1\right)B_j$. For example,

$$\begin{aligned} B_{13} &= B_{2^3+2^2+2^0} \\ &= \left(2^{2^3}+1\right)B_{2^2+2^0} = \left(2^{2^3}+1\right)\left(2^{2^2}+1\right)B_0 \\ &= \left(2^{2^3}+1\right)\left(2^{2^2}+1\right)\left(2^{2^0}+1\right) = f_3f_2f_0 \end{aligned}$$

In general, every number B_k is the product of a nonnegative number of distinct Fermat numbers. Conversely, the binary version of every Fermat number and every product of distinct Fermat numbers will appear as a row in PBT.

For example, $f_2f_1 = \left(2^{2^2}+1\right)\left(2^{2^1}+1\right) = 85 = 1010101_{\text{two}}$. Notice that 1010101 is row $2^2 + 2^1 =$ in PBT.

Table 8.1 shows rows 1 through 10 in PBT, their decimal values B_k, their factorizations, and factorizations in terms of Fermat numbers.

Table 8.1

Row k	(Row k)$_{\text{two}}$	B_k	B_k in factored form	B_k in terms of Fermat numbers
1	11	3	3	f_0
2	101	5	5	f_1
3	1111	15	$5 \cdot 3$	f_1f_0
4	10001	17	17	f_2
5	110011	51	$17 \cdot 3$	f_2f_0
6	1010101	85	$17 \cdot 5$	f_2f_1
7	11111111	255	$17 \cdot 5 \cdot 3$	$f_2f_1f_0$
8	100000001	257	257	f_3
9	1100000011	771	$257 \cdot 3$	f_3f_0
10	10100000101	1285	$257 \cdot 5$	f_3f_1

The above discussion provides a simple algorithm to identify all odd (and hence even) binomial coefficients in row n of Pascal's triangle, as the following example illustrates.

Example 8.1 Find the integers r for which $\binom{42}{r}$ is odd.

Solution We have $42 = 32 + 8 + 2$. Correspondingly,

$$\begin{aligned} B_{42} &= \left(2^{32}+1\right)\left(2^{8}+1\right)\left(2^{2}+1\right) \\ &= 2^{42}+2^{40}+2^{34}+2^{32}+2^{10}+2^{8}+2^{2}+2^{0} \end{aligned}$$

Consequently, $\binom{42}{r}$ is odd when $r = 0, 2, 8, 10, 32, 34, 40$, and 42. (You may confirm this using Lucas' theorem in Chapter 4.) ■

The close link between the number of odd binomial coefficients in row n and Fermat numbers has a delightful application to abstract algebra and geometry.

Ruler and Compass Constructibility

In 1796, Gauss established that a regular n-gon can be constructed with a ruler and compass† if and only if n is the product of 2^s and a nonnegative number of distinct Fermat primes. Suppose n is odd. Since there are only five known Fermat primes, there are only $\sum_{r=1}^{5} \binom{5}{r} = 31$ known possible choices for n. Consequently, the only known odd integers n for which a regular n-gon is constructible with a ruler and compass are the ones with $n = B_k$, where $1 \le k \le 31$.

For example, let $k = 10 = 8 + 2$. Then $n = (2^8 + 1)(2^2 + 1) = 257 \cdot 5 =1{,}285$ and a regular 1285-gon is constructible.

Returning to Pascal's binary pattern, obviously it can be extended to higher modulii m. For example, Figure 8.3 shows the spectacular patterns corresponding to $m = 3, 4, 5, 6, 7, 8, 9$, and 10.

Locating Multiples of Primes in Pascal's Triangle

Recall from Corollary 2.2 (or Theorem 4.1) that p is a prime if and only if $p \mid \binom{p}{r}$, where $0 < r < p$; that is, if and only if p divides every interior element in row p of Pascal's triangle. Also $p \nmid \binom{2p}{p}$, but $p \mid \binom{2p}{r}$ and hence $p \mid \binom{2p}{p+r}$, where $0 < r < p$. Thus, row $2p$ contains $(p-1)+(p-1) = 2p-2$ elements divisible by p. Since

$$\begin{aligned} \binom{p+k}{r} &= \frac{(p+k)!}{r!(p+k-r)!} \\ &= \frac{(p+k)(p+k-1)\cdots p \cdots [p+k-(r-1)]}{r!} \end{aligned}$$

† See I. N. Herstein, *Topics in Algebra*, 2nd edition, Xerox, Lexington, MA, 1975.

it follows that $p \mid \binom{p+k}{r}$ if and only if $r < p$ and $p > p+k-r$; that is, if and only if $k < r < p$.

For example, let $p = 23$. then $23 \mid \binom{23+k}{r}$ if and only if $k < r < 23$. In particular, let $k = 4$; then $23 \mid \binom{27}{r}$ for $4 < r < 23$; that is, $\binom{27}{5}, \binom{27}{6}, \cdots, \binom{27}{22}$ are divisible by 23.

In 1989, L. Cannon of Utah State University, Logan, Utah, investigated the binomial coefficients that are multiples of a given number. To this end, let R_n denote row n in Pascal's triangle.

In particular, suppose we would like to find all multiples of 1001. Notice that $1001 = 7 \cdot 11 \cdot 13$. Every interior element in R_{11} is divisible by 11 and every interior element in R_{13} is divisible by 13; see the triangular areas in Figure 8.4. From the above discussion, R_4 contains 12 elements divisible by $p = 7$. But $7 \nmid \binom{14}{7}$; that is, 7 does not divide the CBC $\binom{14}{7} = 3{,}432$. So the twelve elements are divided

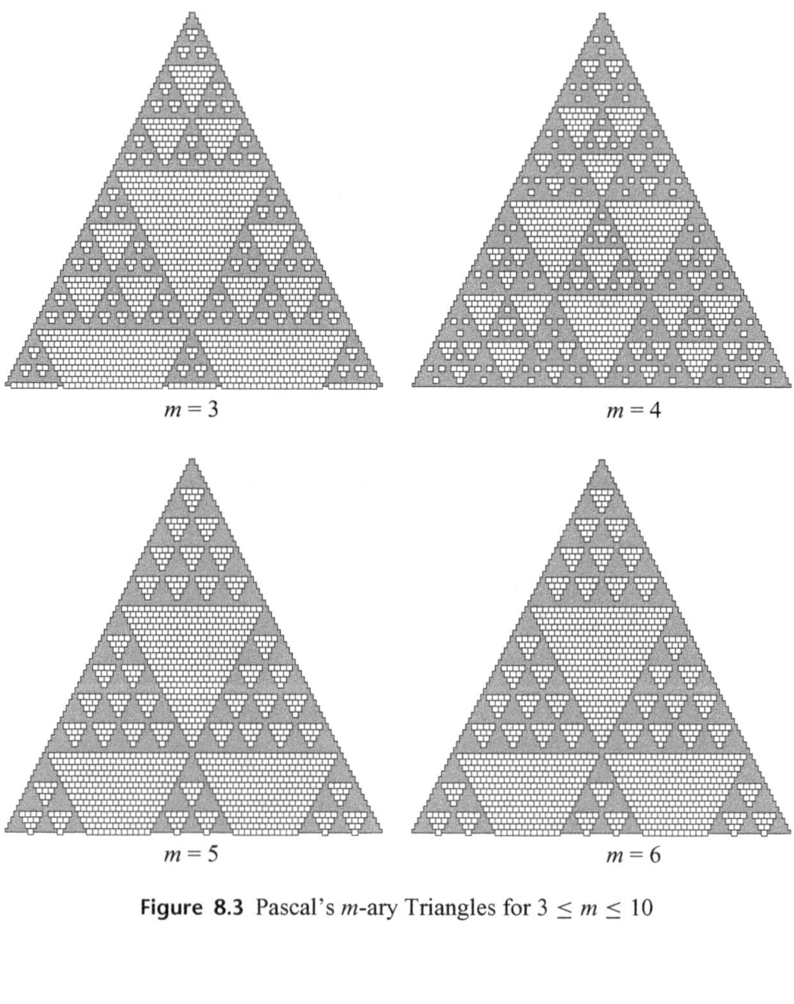

Figure 8.3 Pascal's m-ary Triangles for $3 \le m \le 10$

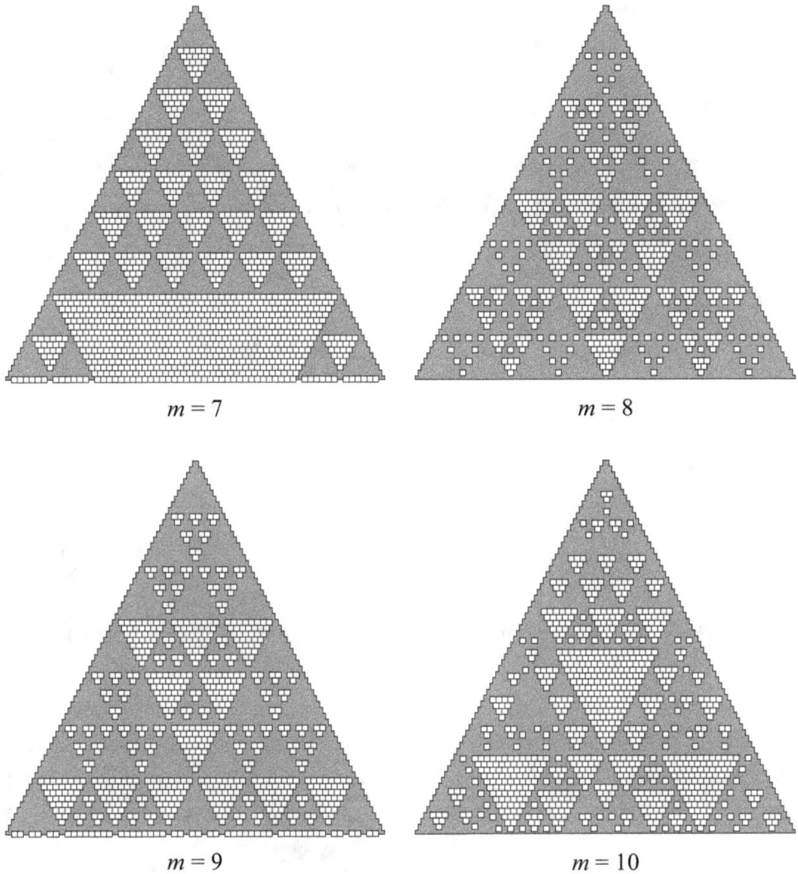

Figure 8.3 Continued

into two blocks of six each, separated by 3,432; see the two triangular arrays in Figure 8.4.

Pascal's *p*-ary Triangle

The binomial coefficients that are multiples of a prime p are precisely those that are congruent to 0 modulo p. Consequently, reducing R_p modulo p yields $p - 1$ consecutive zeros with a 1 at each end:

$$1 \quad 0 \quad 0 \quad \cdots \quad 0 \quad 0 \quad 1 \quad \leftarrow \quad R_p \quad \text{modulo} \quad p$$

Since no binomial coefficient above R_p has p as a factor, $\binom{n}{r} \not\equiv 0 \pmod{p}$ for $0 \le n < p$; in other words, no binomial coefficient when reduced modulo p yields

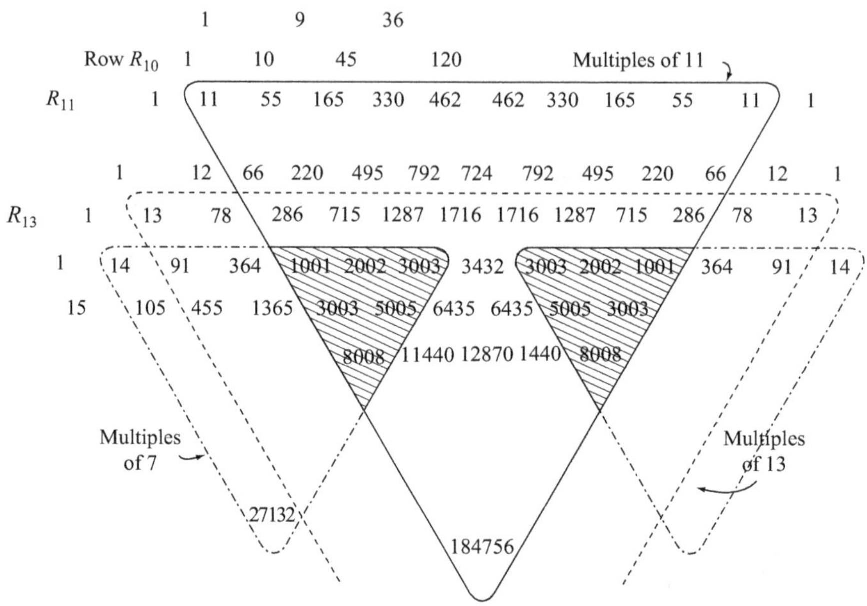

Figure 8.4 The Two Shaded Triangular Areas Consist of Multiples of $7 \cdot 11 \cdot 13 = 1001$

a zero; that is, no row above R_p contains a zero. Since R_p consists of two exterior 1s and a string of $p-1$ zeros, it follows by Pascal's identity that R_{p-1} when reduced modulo p consists of two distinct elements, 1 and $p-1$, and they alternate. Figure 8.5 shows array T_p when rows R_0 through R_{p-1} are reduced modulo p.

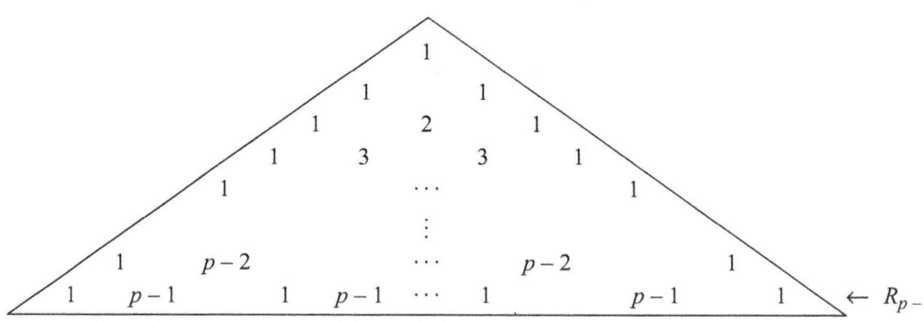

Figure 8.5 Triangular Array T_p of Residues Modulo p

Using Pascal's identity, the string of $p-1$ zeros in R_p yields a triangular array O of zeros; see the dotted triangle in Figure 8.6. The trapezoid with bases R_p and R_{2p-1} contains two identical copies of T_p. R_{2p} contains two strings of $p-1$ zeros, separated by a 2.

Continuing this intriguing pattern downward, R_{2p} gives rise to three non-zero triangular blocks: two duplicates of T_p, and one block of $2T_p$ whose entries are

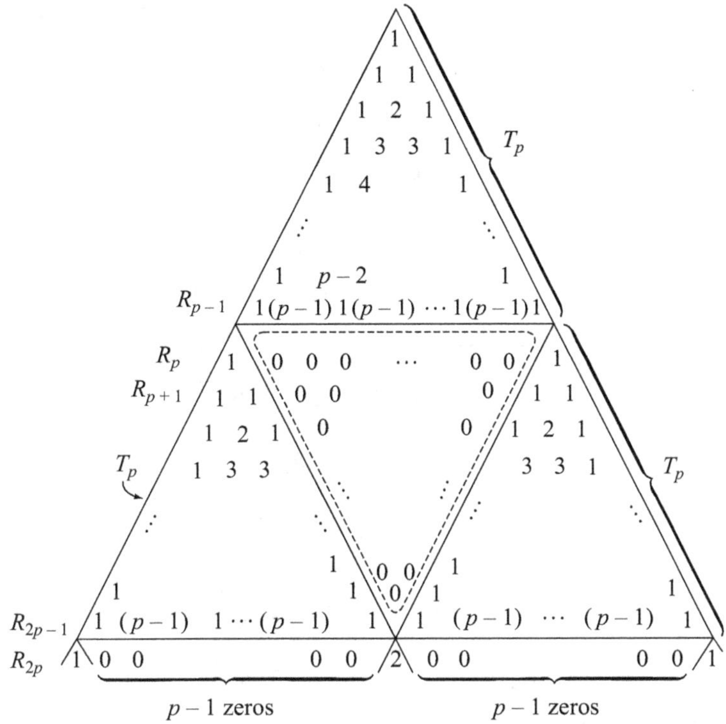

Figure 8.6 Pascal's p-ary Triangle

twice modulo p of the corresponding entries in T_p; they are separated by two copies of T_0. Clearly, this pattern continues. Figure 8.7 shows the resulting array for $p = 5$.

Figure 8.8 shows the structure of Pascal's 5-ary triangle. (The same pattern occurs for any prime p.) Notice that the multiples of 5 occur in triangular blocks T_0 down toward R_{25}, which contains a string of $5 \cdot 4 + 4 \cdot 1 = 24 = 5^2 - 1$ zeros.

Interestingly, the triangular array T_{25} displays the same structure as T_5 under the correspondence $1 \quad \leftrightarrow \quad T_5$; see Figure 8.9. In other words, the structure of T_5 is magnified by a factor of 5 in T_{25}.

Note that the distribution of arrays O in Figure 8.8 also shows an intriguing pattern (see Figure 8.10), where each T_0 represents a triangular block of multiples of 5. Rows R_0 through R_{24} contribute $1 + 2 + 3 + 4 = 10$ triangular blocks of multiples of 5, where each block represents $4 + 3 + 2 + 1 = 10$ binomial coefficients, for a total of 100 multiples of 5.

More generally, rows R_0 through R_{p^2-1} give rise to $1+2+3+\cdots+(p-1) = \frac{p(p-1)}{2}$ triangular blocks of multiples of p, each block containing $(p - 1)+ (p - 2) + \cdots + 2 + 1 = \frac{p(p-1)}{2}$ multiples of p.

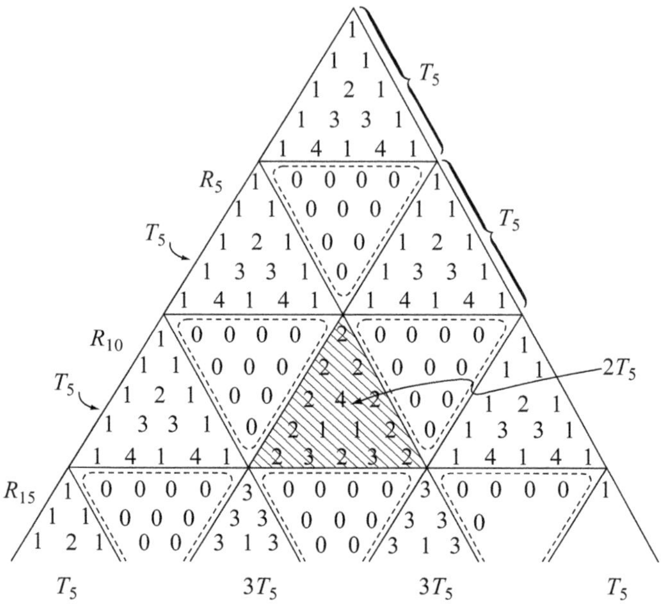

Figure 8.7 Pascal's 5-ary Triangle

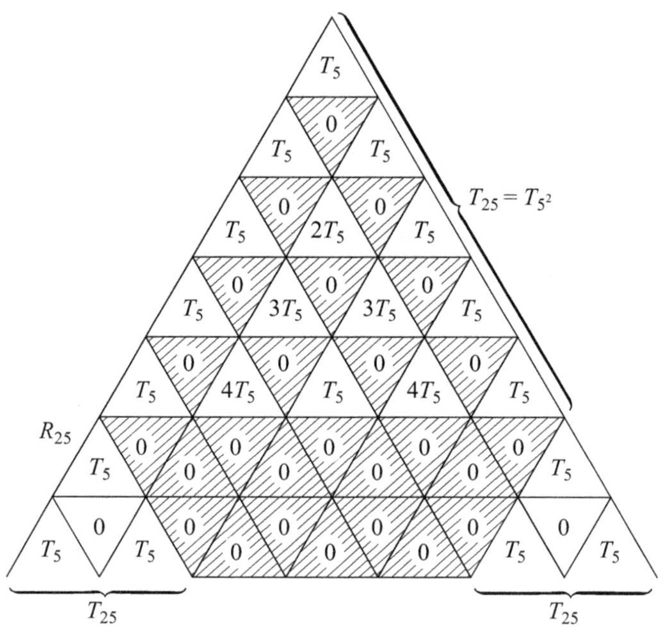

Figure 8.8 Pascal's 5-ary Triangle

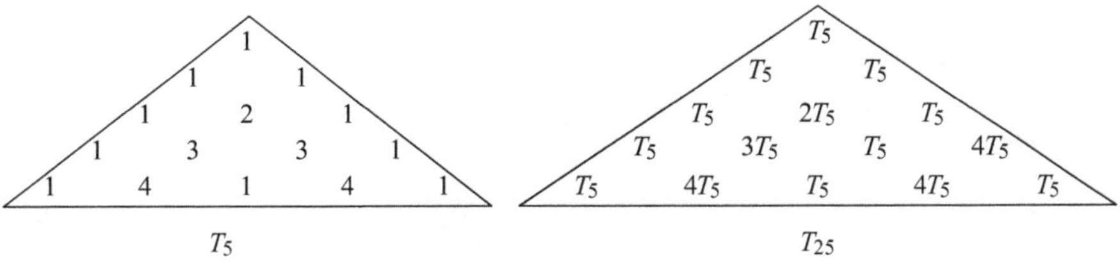

Figure 8.9

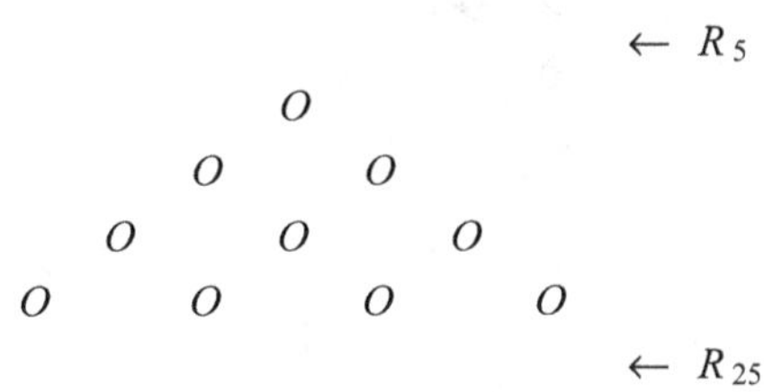

Figure 8.10

Since R_{25} contains a string of $5^2 - 1$ zeros, a larger triangular block of multiples of 5 can be seen beginning at R_{25} and continuing downward. It accounts for

$$(5^2 - 1) + (5^2 - 2) + \cdots + 2 + 1 = \frac{(5^2 - 1)5^2}{2} = 300$$

triangular arrays O, for a total of 3,000 multiples of 5.

In general, beginning with R_{p^2}, Pascal's p-ary triangle produces a large triangular block containing

$$(p^2 - 1) + (p^2 - 2) + \cdots + 2 + 1 = \frac{(p^2 - 1)p^2}{2}$$

arrays of $(p - 1) + (p - 2) + \cdots + 2 + 1 = \frac{p(p-1)}{2}$ multiples of p each, a total of $\frac{(p+1)(p-1)^2p^3}{4}$ multiples of p. Obviously, this interesting pattern continues, creating larger and larger triangular blocks, as Figure 8.11 shows.

An Application

Finally, suppose we would like to locate in Pascal's triangle the first occurrence of a positive multiple of a product of distinct primes. For instance, suppose we would like to find the smallest positive multiple of $2001 = 3 \cdot 23 \cdot 29$. Clearly,

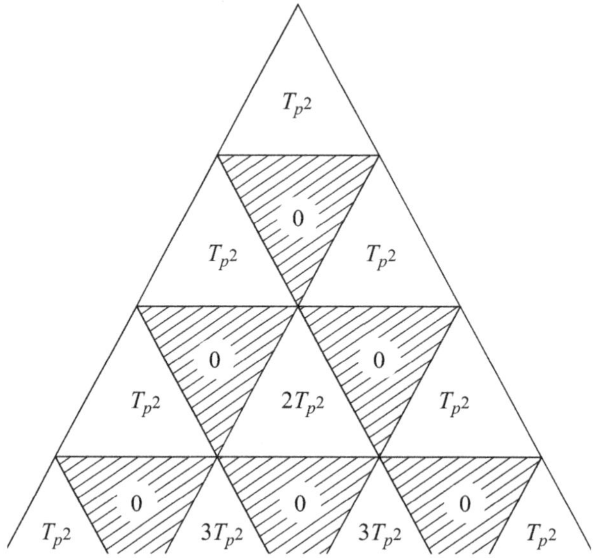

Figure 8.11 Pascal's p-ary Triangle

2001 occurs in R_{2001} and is the smallest positive multiple of 2001. So, what is the smallest positive multiple of 2001 without going to R_{2001}?

Since $29 \mid 2001$, all 28 interior elements in R_{29} are positive multiple of 29. Likewise, each interior element in R_{23} is a multiple of 23. Since $27 = 3^3$, each interior entry in R_{27} is a multiple of 3. Now, locate the triangular arrays of positive multiples of 3, 23, and 29. It follows from the above discussion that the smallest block of multiples of $3 \cdot 23 \cdot 29 = 2001$ is given by the area in Pascal's triangle where the three triangular blocks overlap: namely, the shaded area T in Figure 8.12.

There are $16 + 15 + \cdots + 2 + 1 = 136$ such multiples in T. The smallest of them is $\binom{29}{7} = 1{,}560{,}780 = 780 \cdot 2001$ and the largest is $\binom{44}{22} = 210{,}409{,}892{,}370 = 105152370 \cdot 2001$.

C. T. Long of Washington State University, Pullman, Washington, also investigated Pascal's p-ary triangle and arrived at some interesting results. Let k, n, and m be nonnegative integers, where $n \geq k \geq 0$ and $m \geq 1$. Let $\Delta_{n,k}$ denote the triangular array, where the apex consists of the binomial coefficient $\binom{np^m}{kp^m}$, and the last row begins with $\binom{np^m+p^m-1}{kp^m}$ and ends in $\binom{np^m+p^m-1}{kp^m+p^m-1}$:

$$\Delta_{n,k} = \begin{array}{ccccc} & & \binom{np^m}{kp^m} & & \\ & \cdot^{\cdot^{\cdot}} & & \ddots & \\ \binom{np^m+p^m-1}{kp^m} & & \cdots & & \binom{np^m+p^m-1}{kp^m+p^m-1} \end{array}$$

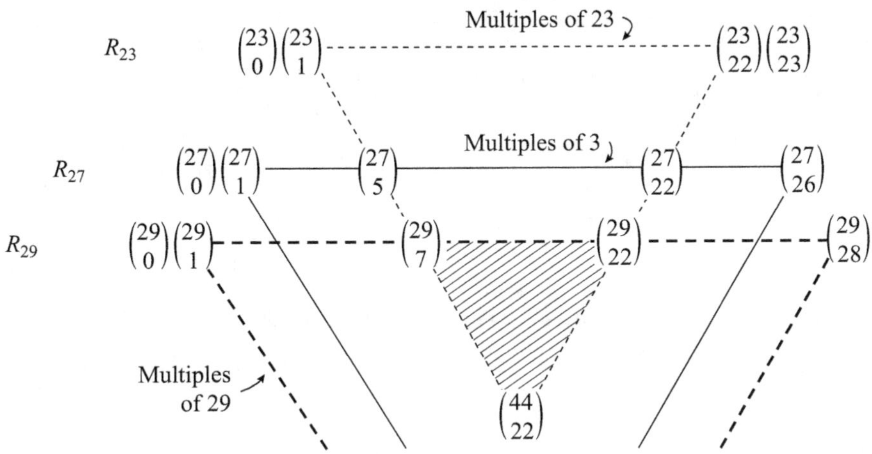

Shaded triangle contains 136 mutiples of $3 \cdot 23 \cdot 29 = 2001$.

Figure 8.12 The Smallest Block of Positive Multiples of $3 \cdot 23 \cdot 29 = 2001$

For example, when $p = 2$ and $m = 1$:

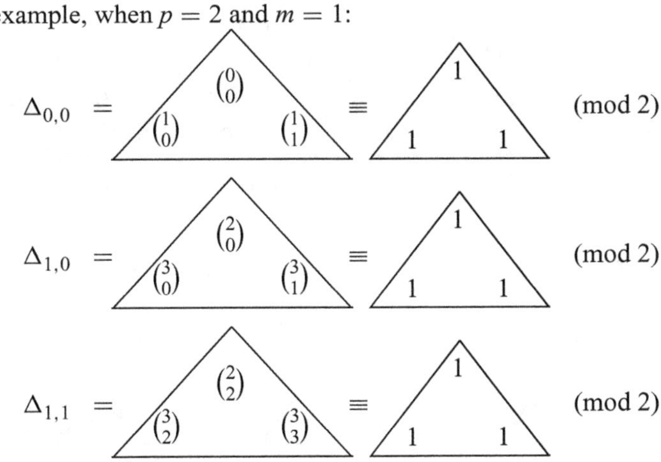

Consequently, Pascal's binary triangle (see Figure 8.13) can be partitioned in terms of $\Delta_{n,k}$, as in Figure 8.14, which exhibits an interesting pattern:

$$\Delta_{n,k} + \Delta_{n,k+1} = \Delta_{n+1,k+1}$$

where addition denotes element-wise addition modulo 2.

For example,

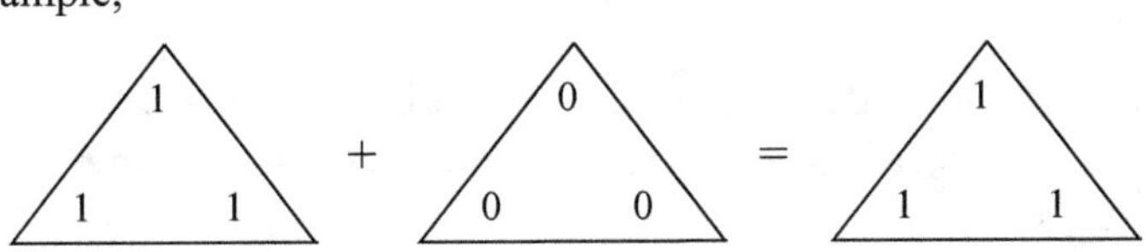

1
1 1
1 0 1
1 1 1 1
1 0 0 0 1
1 1 0 0 1 1
1 0 1 0 1 0 1
1 1 1 1 1 1 1 1

Figure 8.13 Pascal's Binary Triangle

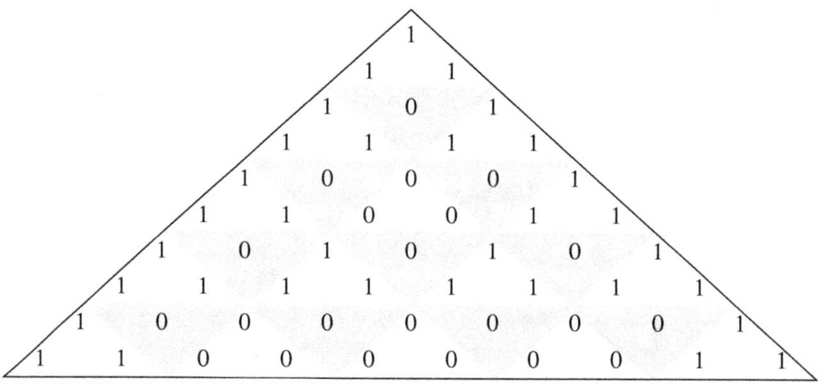

Figure 8.14

and

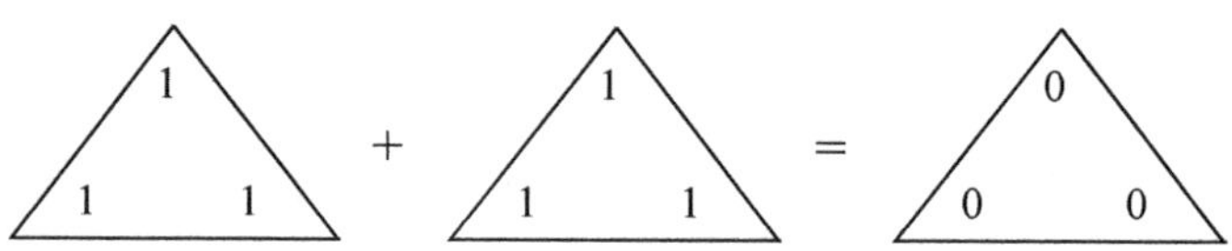

In Figure 8.14, suppose we suppress those triangles (pointing downward) with a single zero inside and make the substitution

$$\begin{matrix} & 0 & \\ 0 & & 0 \end{matrix} \quad \longleftrightarrow \quad 0 \quad \text{and} \quad \begin{matrix} & 1 & \\ 1 & & 1 \end{matrix} \quad \longleftrightarrow \quad 1$$

This transformation yields Pascal's binary triangle in Figure 8.13.

Suppose, we choose $m = 2$. Then Pascal's binary triangle can be partitioned, as in Figure 8.15.

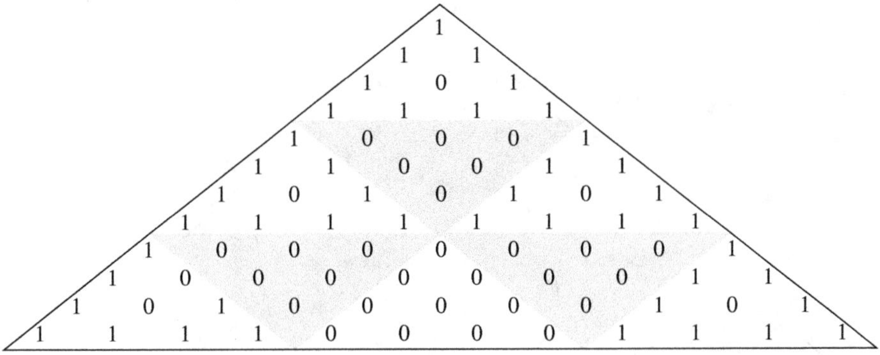

Figure 8.15

Once again, if we suppress those triangles pointing downward, then the correspondences

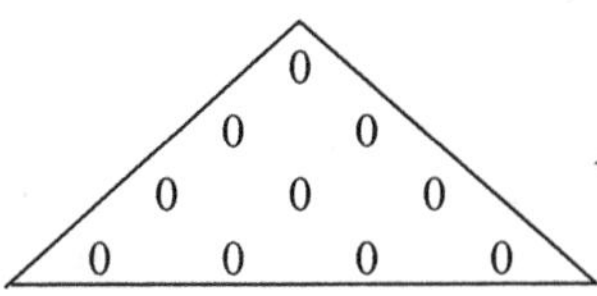

$\longleftrightarrow$ 0 and
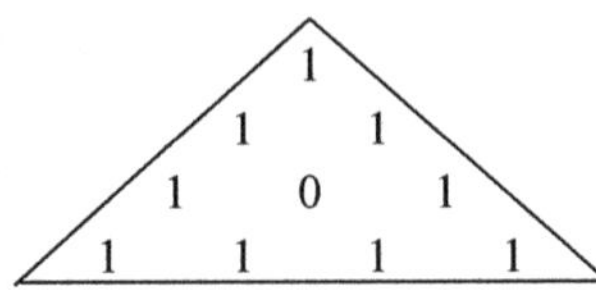

$\longleftrightarrow$ 1

yield Pascal's binary triangle as before – see Figure 8.16.

$$
\begin{array}{ccccccccccc}
 & & & & & 1\cdot\Delta & & & & & \\
 & & & & 1\cdot\Delta & & 1\cdot\Delta & & & & \\
 & & & 1\cdot\Delta & & 2\cdot\Delta & & 1\cdot\Delta & & & \\
 & & 1\cdot\Delta & & 3\cdot\Delta & & 3\cdot\Delta & & 1\cdot\Delta & & \\
 & 1\cdot\Delta & & 4\cdot\Delta & & 6\cdot\Delta & & 4\cdot\Delta & & 1\cdot\Delta & \\
1\cdot\Delta & & 5\cdot\Delta & & 10\cdot\Delta & & 10\cdot\Delta & & 5\cdot\Delta & & 1\cdot\Delta
\end{array}
$$

Figure 8.16

where $\Delta = \Delta_{0,0}$ and multiplication denotes multiplication modulo 2.

On the other hand, suppose we let $p = 3$. Then *Pascal's ternary triangle* looks as in Figure 8.17. As expected, that, too, exhibits a fractal behavior.

Notice that the top nine rows of the ternary triangle are given by

$\Delta_2 =$
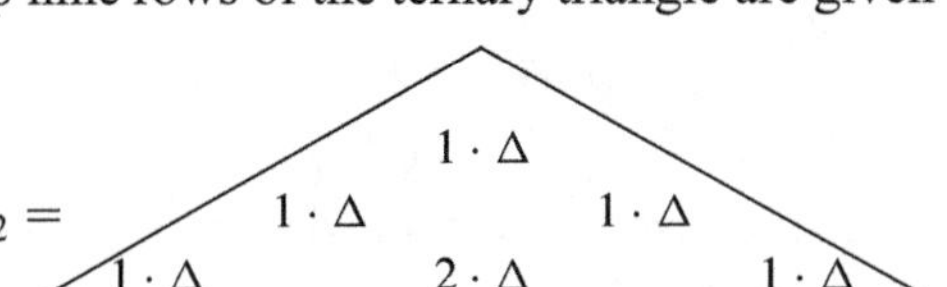

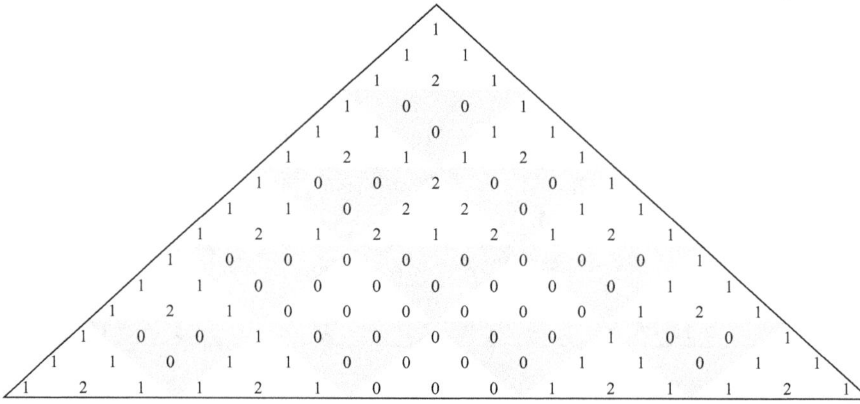

Figure 8.17 Pascal's Ternary Triangle

and the top 27 rows are given by

$$\Delta_3 = \begin{array}{ccccc} & & 1\cdot\Delta_2 & & \\ & 1\cdot\Delta_2 & & 1\cdot\Delta_2 & \\ 1\cdot\Delta_2 & & 2\cdot\Delta_2 & & 1\cdot\Delta_2 \end{array}$$

By letting $m = 1$, as above, the top three rows of the ternary triangle are given by

$$\Delta = \Delta_{0,0} = \begin{array}{ccccc} & & 1 & & \\ & 1 & & 1 & \\ 1 & & 2 & & 1 \end{array}$$

If we delete the shaded triangles (pointing downward) consisting of zeros, the ternary triangle displays the interesting Pascal-like pattern in Figure 8.18, where multiplication denotes multiplication modulo 3.

$$\begin{array}{ccccccccccc} & & & & & 1\cdot\Delta & & & & & \\ & & & & 1\cdot\Delta & & 1\cdot\Delta & & & & \\ & & & 1\cdot\Delta & & 2\cdot\Delta & & 1\cdot\Delta & & & \\ & & 1\cdot\Delta & & 3\cdot\Delta & & 3\cdot\Delta & & 1\cdot\Delta & & \\ & 1\cdot\Delta & & 4\cdot\Delta & & 6\cdot\Delta & & 4\cdot\Delta & & 1\cdot\Delta & \\ 1\cdot\Delta & & 5\cdot\Delta & & 10\cdot\Delta & & 10\cdot\Delta & & 5\cdot\Delta & & 1\cdot\Delta \end{array}$$

Figure 8.18

These results are summarized in the following theorem, which is an application of the Lucas' theorem in Theorem 4.10.

Theorem 8.1 (Long, 1981) Let p be a prime, $0 \leq k \leq n$, and $m \geq 1$. Then:

$$\Delta_{n,k} = \begin{array}{ccccc} & & \binom{n}{k}\binom{0}{0} & & \\ & \binom{n}{k}\binom{1}{0} & & \binom{n}{k}\binom{1}{1} & \\ & & \vdots & & \\ \binom{n}{k}\binom{p^m-1}{0} & & \cdots & & \binom{n}{k}\binom{p^m-1}{p^m-1} \end{array}$$

where multiplication is multiplication modulo p and $\Delta_{n,k} + \Delta_{n,k+1} = \Delta_{n+1,k+1}$, where addition denotes element-wise addition modulo p. Finally, every element in Pascal's triangle that is not in any $\Delta_{n,k}$ is congruent to zero modulo p.

Proof:

1) The elements of $\Delta_{n,k}$ are the binomial coefficients $\binom{np^m+r}{kp^m+s}$, where $0 \leq s \leq r < p^m$. By Lucas' theorem, $\binom{np^m+r}{kp^m+s} \equiv \binom{n}{k}\binom{r}{s} \pmod p$, as desired.
2) Again, by Lucas' theorem,

$$\begin{aligned}\binom{np^m+r}{kp^m+s} + \binom{np^m+r}{(k+1)p^m+s} &\equiv \binom{n}{k}\binom{r}{s} + \binom{n}{k+1}\binom{r}{s} \pmod p \\ &\equiv \left[\binom{n}{k} + \binom{n}{k+1}\right]\binom{r}{s} \pmod p \\ &\equiv \binom{n+1}{k+1}\binom{r}{s} \equiv \binom{(n+1)p^m+r}{(k+1)p^m+s} \pmod p\end{aligned}$$

Consequently, $\Delta_{n,k} + \Delta_{n,k+1} = \Delta_{n+1,k+1}$.
3) Finally, the entries of Pascal's triangle not included in any $\Delta_{n,k}$ form triangles $\nabla_{n,k}$, where

$$\nabla_{n,k} = \begin{array}{ccc} \binom{np^m}{kp^m+1} & \cdots & \binom{np^m}{kp^m+p^m-1} \\ & \vdots & \\ & \binom{np^m+p^m-2}{kp^m+p^m-1} & \end{array}$$

where the elements are reduced modulo p. Thus, every element in $\nabla_{n,k}$ is of the form $\binom{np^m+r}{kp^m+s}$, where $0 \leq r < s \leq p^m - 1$. By Lucas' theorem, since $r < s$,

$$\begin{aligned}\binom{np^m+r}{kp^m+s} &\equiv \binom{n}{k}\binom{r}{s} \pmod{\text{p}} \\ &\equiv \binom{n}{k} \cdot 0 \equiv 0 \pmod 2\end{aligned}$$

This completes the proof. ∎

A Greatest Common Divisor Property

The next theorem gives the greatest common divisor (gcd) of the binomial coefficients in the triangular array

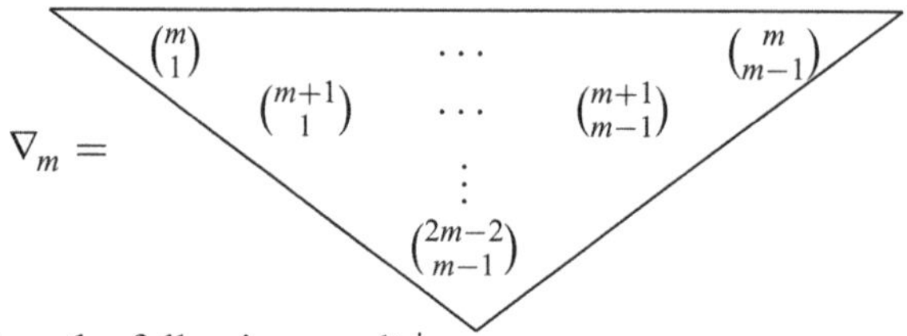

The proof requires the following result.[†]

Lemma 8.1 Let p be a prime and $0 \le k \le n$. If $p^f \parallel \binom{n}{k}$, then f is the number of carries needed when k is added to $n-k$ in base p, where $p^f \parallel b$ means $p^f \mid b$, but $p^{f+1} \nmid b$. ■

We are now ready to present the theorem.

Theorem 8.2 (Long, 1981) Let p be a prime, d the gcd of all elements in ∇_m, and D the gcd of the three corner elements $\binom{m}{1}$, $\binom{m}{m-1}$, and $\binom{2m-2}{m-1}$. Then:

1) If $m = p$, then $d = D = p$;
2) If $m = p^\alpha$, then $d = p$ and $D = p^\alpha$, where $\alpha > 1$ is an integer; and
3) $d = 1$ and $D = m$ for all other integers $m \ge 2$.

Proof:

1) Since $\binom{m}{1} = \binom{p}{1} = p$, $d \mid D$, and $D \mid \binom{m}{1}$, it suffices to show that $p \mid d$. Consider an arbitrary element $\binom{p+k}{h}$ of ∇_p, where $0 \le k \le p-2$ and $k+1 \le h \le p-1$. By Lucas' theorem, since $k < h$, $\binom{p+k}{h} \equiv \binom{k}{h} \equiv 0 \pmod{p}$. Thus p divides every element of ∇_p, so $p \mid d$, as desired.
2) The elements of ∇_{p^α} are of the form $\binom{p^\alpha+k}{h}$, where $0 \le k \le p-2$ and $k+1 \le h \le p-1$. Again, by Lucas' theorem, since $k < h$, $\binom{p^\alpha+k}{h} \equiv \binom{k}{h} \equiv 0 \pmod{p}$. Thus, $p \mid d$ and $d \mid D$.

   On the other hand, $p \parallel \binom{p^\alpha}{p^{\alpha-1}}$, since the number of carries we make when $p^{\alpha-1}$ is added to $p^\alpha - p^{\alpha-1} = p^{\alpha-1}(p-1)$ in base p is one. This implies that $d \mid p$. Thus, $p \mid d$ and $d \mid p$, so $d = p$.

   Besides, $\binom{p^\alpha}{1} = \binom{p^\alpha}{p^\alpha-1} = p^\alpha$. Also $p^\alpha \parallel \binom{2p^\alpha-2}{p^\alpha-1}$, since $p^\alpha - 1 = \sum_{i=0}^{\alpha-1}(p-1)p^i$. So, there are α carries when $p^\alpha - 1$ is added to $p^\alpha - 1$ in base p. Therefore, $D = p^\alpha$, as desired.

[†]See C. T. Long, *Elementary Introduction to the Theory of Numbers*, 2nd edition, Heath, Lexington, MA, 1972.

3) In this case, m is *not* a prime power. Since $\binom{m}{1} = \binom{m}{m-1} = m$, it follows that $D \mid m$. Thus, it remains to show that $m \mid D$. To this end, let $m = \prod_{i=1}^{r} p_i^{\alpha_i}$ be the canonical decomposition of m. Let $m = kp$, where k is an integer and $p \nmid k$. Since

$$kp^{\alpha} - 1 = (k-1)p^{\alpha} + p^{\alpha} - 1 = (k-1)p^{\alpha} + \sum_{i=0}^{\alpha-1}(p-1)p$$

it follows that the number of carries made in adding $kp^{\alpha} - 1$ to $kp^{\alpha} - 1$ in base p is at least α. Therefore, $p^{\alpha} \mid \binom{2kp^{\alpha}-2}{kp^{\alpha}-1}$. Thus $p_i^{\alpha_i} \mid \binom{2m-2}{m-1}$, where $1 \le i \le r$. So $m \mid \binom{2m-2}{m-1}$.

Thus, $m \mid \binom{m}{1}, m \mid \binom{m}{m-1}$, and $m \mid \binom{2m-2}{m-1}$. Consequently, $m \mid D$; so $D = m$.

The proof that $d = 1$ is a bit complicated, so we skip it. ■

We now turn to a non-numeric triangular array that reminds us of Pascal's fractal-like patterns.

Sierpiński Triangle

The *Sierpiński triangle* is a fascinating fractal, discovered in 1915 by the outstanding Polish mathematician Waclaw Sierpiński (1882–1969). Although non-numeric, the array has a striking resemblance to Pascal's ternary triangle in Figure 8.3.

To construct the Sierpiński triangle, we begin with an equilateral triangle; see Figure 8.19a. Join the midpoints of the three sides. This yields four congruent equilateral triangles. Now, color the central triangle pointing downward, leaving the three triangles at the vertices; see Figure 8.19b. Repeat this procedure with each of the three triangles. This yields nine smaller and congruent equilateral triangles; see Figure 8.19c. Continuing this algorithm with each of them yields 27 even smaller and congruent equilateral triangles.

Obviously, this iterative algorithm can be continued indefinitely. Figure 8.19d shows the Sierpiński triangle after four iterations.

(a) (b) (c) (d)

Figure 8.19 Sierpiński Triangles

Let s_n denote the number of triangles after n iterations. Then $s_n = 3s_{n-1}$, where $s_0 = 1$ and $n \geq 1$.

Next, we compute the area of the Sierpiński triangle.

Area of the Sierpiński triangle

Let A_n denote the area of the Sierpiński triangle after n iterations. Then $A_n = \frac{3}{4}A_{n-1}$, where $A_0 = 1$ and $n \geq 1$.

It follows by iteration that $A_n = \left(\frac{3}{4}\right)^n$. Consequently, the area A after an infinite number of iterations is given by

$$A = \lim_{n\to\infty} A_n = \lim_{n\to\infty} \left(\frac{3}{4}\right)^n = 0$$

That is, after infinitely many iterations, the triangle consists of an infinite number of points.

Perimeter of the Sierpiński triangle

Let p denote the perimeter of the original triangle and p_n the perimeter of the triangle after n iterations, where $p = p_0$. Then:

$$p_0 = 3^1 \left(\frac{1}{2}\right)^0 p \qquad p_2 = 3^3 \left(\frac{1}{2}\right)^2 p$$

$$p_1 = 3^2 \left(\frac{1}{2}\right)^1 p \qquad p_3 = 3^4 \left(\frac{1}{2}\right)^2 p$$

More generally, $p_n = 3^{n+1}\left(\frac{1}{2}\right)^n p$, where $n \geq 0$. As n gets larger, so does the perimeter. As the iterative process is continued indefinitely, the perimeter becomes infinite.

9

Pascal Graphs

In 1983, N. Deo and M. J. Quinn of Washington State University, Pullman, Washington, introduced the *Pascal matrix*, a triangular array closely related to PBT, investigated in the preceding chapter. They developed the Pascal matrix in their pursuit of graphs with the following properties for computer networks:

- The design must be simple and recursive, so when a new vertex is added the entire network does not have to be reconfigured.
- The graph must contain a central vertex adjacent to all other vertices.
- For the sake of reliability, there must exist several paths between every pair of vertices; some of these paths must be of short lengths to reduce communication delays.
- The graphs should have good cohesion and connectivity.

Complete graphs K_n do satisfy these properties, but they are ruled out because of the expense involved in their implementation. With this mind, Deo and Quinn introduced a family of graphs, called *Pascal graphs*, using Pascal matrices.

We begin our discussion with a definition of Pascal matrices.

Pascal Matrices

A symmetric binary matrix $M_n = (m_{ij})_{n \times n}$ is a *Pascal matrix* of *order n* if:

- Its main diagonal consists of zeros; that is, $m_{ii} = 0$ for every n; and
- Its lower triangle (and hence upper triangle also) consists of the top $n - 1$ rows of PBT.

For example,

$$M_2 = \begin{bmatrix} 0 & 1 \\ 1 & 0 \end{bmatrix} \text{ and } M_3 = \begin{bmatrix} 0 & 1 & 1 \\ 1 & 0 & 1 \\ 1 & 1 & 0 \end{bmatrix}$$

We are now ready to present Pascal graphs.

Pascal Graphs

A graph with adjacency matrix M_n is a *Pascal graph* G_n.

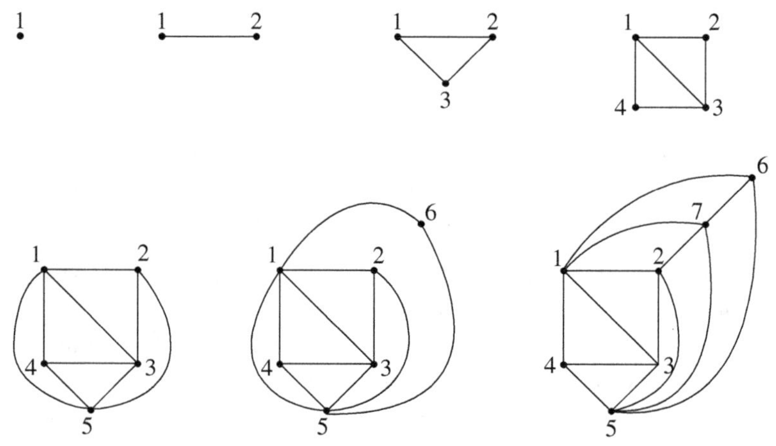

Figure 9.1 Pascal Graphs G_1 through G_7

Figure 9.1 shows the first seven Pascal graphs. Their corresponding adjacency matrices are:

$$M_1 = [0] \quad M_2 = \begin{bmatrix} 0 & 1 \\ 1 & 0 \end{bmatrix} \quad M_3 = \begin{bmatrix} 0 & 1 & 1 \\ 1 & 0 & 1 \\ 1 & 1 & 0 \end{bmatrix} \quad M_4 = \begin{bmatrix} 0 & 1 & 1 & 1 \\ 1 & 0 & 1 & 0 \\ 1 & 1 & 0 & 1 \\ 1 & 0 & 1 & 0 \end{bmatrix}$$

$$M_5 = \begin{bmatrix} 0 & 1 & 1 & 1 & 1 \\ 1 & 0 & 1 & 0 & 1 \\ 1 & 1 & 0 & 1 & 1 \\ 1 & 0 & 1 & 0 & 1 \\ 1 & 1 & 1 & 1 & 0 \end{bmatrix} \quad M_6 = \begin{bmatrix} 0 & 1 & 1 & 1 & 1 & 1 \\ 1 & 0 & 1 & 0 & 1 & 0 \\ 1 & 1 & 0 & 1 & 1 & 0 \\ 1 & 0 & 1 & 0 & 1 & 0 \\ 1 & 1 & 1 & 1 & 0 & 1 \\ 1 & 0 & 0 & 0 & 1 & 0 \end{bmatrix}$$

$$M_7 = \begin{bmatrix} 0 & 1 & 1 & 1 & 1 & 1 & 1 \\ 1 & 0 & 1 & 0 & 1 & 0 & 1 \\ 1 & 1 & 0 & 1 & 1 & 0 & 0 \\ 1 & 0 & 1 & 0 & 1 & 0 & 0 \\ 1 & 1 & 1 & 1 & 0 & 1 & 1 \\ 1 & 0 & 0 & 0 & 1 & 0 & 1 \\ 1 & 1 & 0 & 0 & 1 & 1 & 0 \end{bmatrix}$$

We now pursue some properties of Pascal graphs.

Properties of Pascal Graphs

First, since $m_{ii} = 0$ for every i, G_n is loop-free. M_n is a submatrix of M_{n+1}. Consequently, G_n is a subgraph of G_{n+1} for every $n \geq 1$.

Lemma 9.1 G_n is a subgraph of G_{n+1} for every $n \geq 1$. ■

Before we investigate G_n further, we would like to introduce a family of graphs, called *planar graphs*.

Planar Graphs

A graph is *planar* if it can be drawn in the plane in such a way that no two edges meet at a non-vertex. Such a drawing is a *planar representation* of the graph.

For example, the graph in Figure 9.2 is planar; Figure 9.3 shows its planar representation. The complete graph K_5 and the *complete bipartite graph* $K_{3,3}$ in Figure 9.4 are *non-planar*.

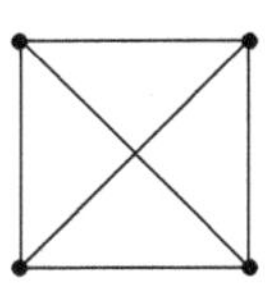

Figure 9.2

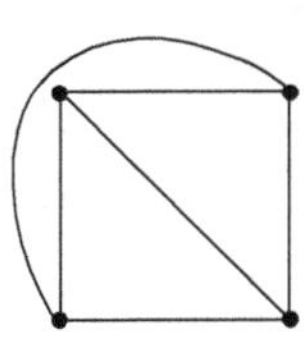

Figure 9.3

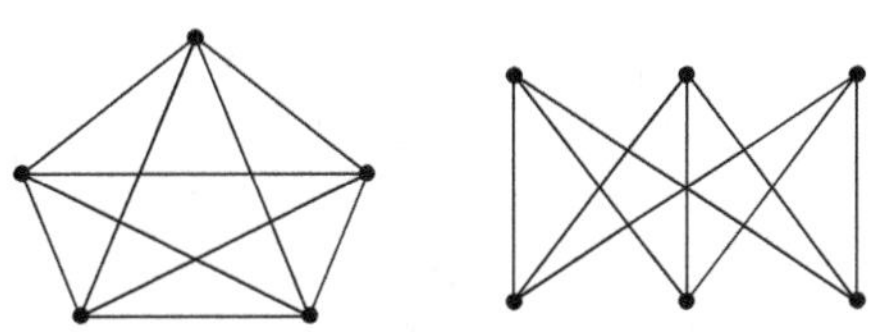

K_5 $K_{3,3}$

Figure 9.4

In 1930, the Polish mathematician Kazimierz Kuratowski (1896–1980) established a test for determining the planarity of a graph: A graph is planar if and only if it does not contain a subgraph that is 'homeomorphic' to K_5 or $K_{3,3}$.[†]

[†]Deleting an edge $x - y$ from a graph and adding a vertex z together with the edges $x - z$ and $z - y$ is called an *elementary subdivision*. Two graphs are *homeomorphic* if one can be obtained from the other by a sequence of elementary subdivisions.

Planarity is an extremely useful concept in the real world. For example, it plays a pivotal role in the design of circuit boards.

The next result identifies Pascal graphs that are planar.

Theorem 9.1 G_n is planar if and only if $1 \leq n \leq 7$.

Proof: Suppose $n \leq 7$. Then it follows from Figure 9.1 that G_n is planar for every $n \leq 7$.

On the other hand, suppose $n \geq 8$. For instance, let $n = 8$. Figure 9.5 shows the corresponding Pascal graph G_8. It contains $K_{3,3}$ as a subgraph; see Figure 9.6. Consequently, G_8 is non-planar. So, by Lemma 9.1, G_n is non-planar for $n \geq 8$.

$$M_8 = \begin{bmatrix} 0 & 1 & 1 & 1 & 1 & 1 & 1 & 1 \\ 1 & 0 & 1 & 0 & 1 & 0 & 1 & 0 \\ 1 & 1 & 0 & 1 & 1 & 0 & 0 & 1 \\ 1 & 0 & 1 & 0 & 1 & 0 & 0 & 0 \\ 1 & 1 & 1 & 1 & 0 & 1 & 1 & 1 \\ 1 & 0 & 0 & 0 & 1 & 0 & 1 & 0 \\ 1 & 1 & 0 & 0 & 1 & 1 & 0 & 1 \\ 1 & 0 & 1 & 0 & 1 & 0 & 1 & 0 \end{bmatrix}$$

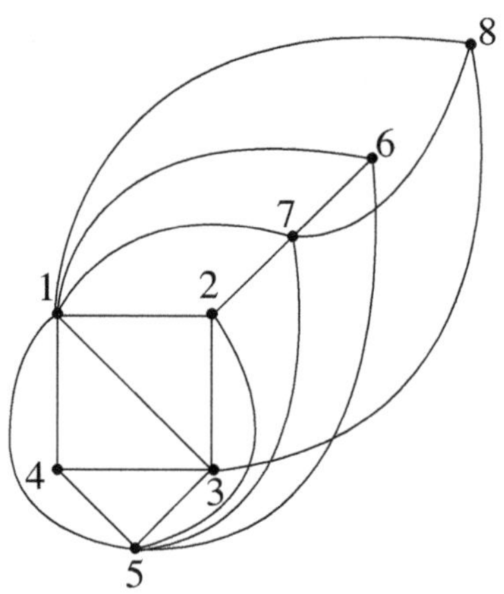

Figure 9.5 Pascal Graph G_8

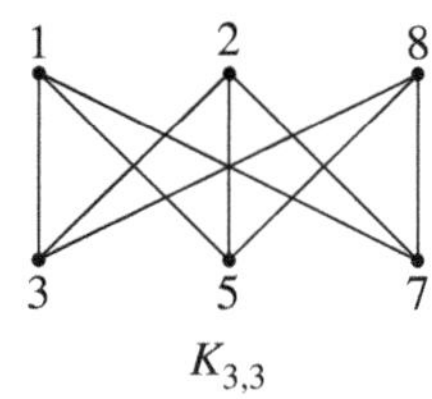

Figure 9.6 $K_{3,3}$

Thus, G_n is planar if and only if $1 \leq n \leq 7$. ■

The next property of G_n is related to Hamiltonian graphs.

Theorem 9.2 Vertex v_1 is adjacent to all other vertices in G_n and vertex v_i is adjacent to v_{i+1}, where $i \geq 1$.

Proof: By the definition of the Pascal matrix M_n, $m_{ij} \equiv \binom{i-2}{j-1} \pmod 2$, where $i > j \geq 1$. When $i \geq 2$, $m_{i1} \equiv \binom{i-2}{0} = 1 \pmod 2$, so v_1 is adjacent to every vertex v_i, where $i \geq 2$.

Let $i \geq 1$. Then $m_{i+1,i} \equiv \binom{i-1}{i-1} = 1 \pmod 2$, so v_i is adjacent to v_{i+1} for every $i \geq 1$. ■

Since v_i is adjacent to v_{i+1} for every $i \geq 1$ and v_n is adjacent to v_1 in G_n, it follows that G_n contains a Hamiltonian cycle; namely, $v_1 - v_2 - v_3 - \cdots - v_n - v_1$. Accordingly, we have the following result.

Corollary 9.1 G_n is Hamiltonian. ■

The next result employs the Lucas' theorem in Theorem 4.10.

Lemma 9.2 Let n be a positive integer and $k = 2^n + 1$. Then v_k is adjacent to all v_i, where $1 \leq i < 2k$ and $i \neq k$.

Proof:

Case 1 Let $1 \leq i < k$. Then $m_{ki} \equiv \binom{k-2}{i-1} \equiv \binom{2^n-1}{i-1} \pmod 2$. Since $i < k$, $i - 1 < 2^n$. Let

$$i - 1 = c_0 + c_1 \cdot 2 + c_2 \cdot 2^2 + \cdots + c_{n-1} \cdot 2^{n-1}$$

where each $c_i = 0$ or 1. We also have $2^n - 1 = 1 + 1 \cdot 2 + 1 \cdot 2^2 + \cdots + 1 \cdot 2^{n-1}$. So, by Lucas' theorem,

$$\binom{2^n - 1}{i - 1} \equiv \binom{1}{c_0}\binom{1}{c_1}\binom{1}{c_2} \cdots \binom{1}{c_{n-1}} \pmod 2$$

But $\binom{1}{c_i} = 1$ for every i. Consequently, $\binom{2^n-1}{i-1} \equiv 1 \pmod 2$.

Case 2 Let $k < i < 2k$. Then:

$$\begin{aligned} m_{ki} &= m_{ik} \\ &\equiv \binom{i-2}{k-1} \equiv \binom{i-2}{2^n} \pmod 2 \end{aligned}$$

Let

$$i - 2 = d_0 + d_1 \cdot 2 + d_2 \cdot 2^2 + \cdots + d_{n-1} \cdot 2^{n-1} + 1 \cdot 2^n$$

Then, by Lucas' theorem,

$$\binom{i-2}{2^n} \equiv \binom{d_0}{0}\binom{d_1}{0}\binom{d_2}{0} \cdots \binom{d_{n-1}}{0}\binom{1}{1} \equiv 1 \pmod 2$$

Thus, $m_{ki} = 1$ for every i, where $1 \leq i < 2k$ and $i \neq k$. In other words, v_k is adjacent to all such vertices v_i. ■

For example, let $n = 1$ and $k = 3$. Notice that in Pascal graph G_6, v_3 is adjacent to v_1, v_2, v_4, and v_5. In G_7 also, v_3 is adjacent to v_1, v_2, v_4, and v_5.

We will use this lemma in Theorem 9.6.

The next lemma gives some additional information about adjacency relationships in a Pascal graph.

Lemma 9.3 No two even-numbered vertices in G_n are adjacent.

Proof: Let v_i and v_j be any two distinct vertices in G_n, where i and j are even integers, and $i > j$. (Recall that G_n is loop-free.) $m_{ij} \equiv \binom{i-2}{j-1} \pmod{2}$. Since $i - 2$ is even and $j - 1$ is odd, it follows that $\binom{i-2}{j-1} \equiv 0 \pmod{2}$, so $m_{ij} = 0$; that is, v_i and v_j are not adjacent vertices. ■

For example, consider the graph G_7 in Figure 9.1. Notice that vertices v_2 and v_4, v_4 and v_6, and v_6 and v_2 are *not* adjacent.

Theorem 9.3 Let v_i be adjacent to v_j, where j is even and $|i - j| > 1$. Then i is odd and v_i is adjacent to v_{j-1}.

Proof: Since j is even, it follows by Lemma 9.3 that i is odd. It remains to show that v_i is adjacent to v_{j-1}.

Let $i > j$. Since $i - 1$ is even, by Lemma 9.3, $m_{i-1,j} = 0$. Since v_i is adjacent to v_j, $m_{ij} = 1$. But

$$m_{ij} \equiv m_{i-1,j-1} + m_{i-1,j} \pmod{2}$$

$$1 \equiv m_{i-1,j-1} + 0 \pmod{2}$$

$$1 \equiv m_{i-1,j-1} \pmod{2}$$

Thus, by Lemma 9.3, $m_{i,j-1} \equiv m_{i-1,j-2} + m_{i-1,j-1} \equiv 0 + 1 \equiv 1 \pmod{2}$.

Similarly, it can be shown that $m_{i,j-1} = 1$ when $i < j$ also. Thus, in both cases, $m_{i,j-1} = 1$; so v_i is adjacent to v_{j-1}. ■

For example, consider the graph G_6 in Figure 9.1. Let $j = 4$ and $|i - j| > 1$. Then $i = 1$. Notice that v_1 is adjacent to v_4 and v_3. On the other hand, let $j = 2$, so $i = 5$. Vertex v_5 is adjacent to both v_2 and v_1.

Notice that the complete graph K_n has $\binom{n}{2} = \frac{n(n-1)}{2} = O(n^2)$ edges.* The number of edges in K_n increases at a rate of n^2. But the following theorem shows that the number of edges in G_n increases at a much lower rate.

Theorem 9.4 Let e_n denote the number of edges in G_n. Then $e_n \leq \lfloor (n-1)^{\lg 3} \rfloor$.†

*A function f is of *order at most* n^2 if there is a positive constant C and a positive integer n_0 such that $|f(n) \leq Cn^2$ for every $n \geq n_0$. We then write $f(n) = O(n^2)$; read this as $f(n)$ is *big-oh* of n^2.

†$\lg x = \log_2 x$.

Proof (by PMI): It follows from Figure 9.1 that:

$$e_1 = 0 \le 0 = \lfloor 0^{\lg 3} \rfloor$$

$$e_2 = 1 \le 1 = \lfloor 1^{\lg 3} \rfloor$$

$$e_3 = 3 \le 3 = \lfloor 2^{\lg 3} \rfloor$$

Now, assume that the result is true for every G_n, where $1 \le n \le 2^k + 1$ and $k \ge 1$. We will show that it is also true for G_n, where $2^k + 2 \le n \le 2^{k+1} + 1$.

Let $n - 1 = 2^k + r$, where $1 \le r \le 2^k$. Then:

$$\begin{aligned} e_n &= e_{2^k+1} + e_{r+1} \\ &\le \lfloor (2^k)^{\lg 3} \rfloor + \lfloor r^{\lg 3} \rfloor \\ &\le \lfloor (2^k)^{\lg 3} + r^{\lg 3} \rfloor \\ &\le \lfloor (2^k + r)^{\lg 3} \rfloor \\ &= \lfloor (n-1))^{\lg 3} \rfloor \end{aligned}$$

where we have used the facts that $\lfloor x + y \rfloor \ge \lfloor x \rfloor + \lfloor y \rfloor$ and $(x+y)^m \ge x^m + y^m$.

Thus, by induction, the result is true for every Pascal graph G_n. ■

For example, $e_6 = 11 \le \lfloor 5^{\lg 3} \rfloor$ and $e_7 = 13 \le \lfloor 6^{\lg 3} \rfloor$.

We now note also that Pascal graphs do not enjoy the fewest possible number of edges satisfying the desired structural properties, but they are useful in designing practical networks. For example, the edge $v_2 - v_7$ in G_7 is redundant.

Next, we turn to determinants $| M_n |$ of Pascal matrices M_n.

Determinants of Pascal Matrices

First, notice that:

$$|M_4| = \begin{vmatrix} 0 & 1 & 1 & 1 \\ 1 & 0 & 1 & 0 \\ 1 & 1 & 0 & 1 \\ 1 & 0 & 1 & 0 \end{vmatrix} = 0$$

$$|M_6| = \begin{vmatrix} 0 & 1 & 1 & 1 & 1 & 1 \\ 1 & 0 & 1 & 0 & 1 & 0 \\ 1 & 1 & 0 & 1 & 1 & 0 \\ 1 & 0 & 1 & 0 & 1 & 0 \\ 1 & 1 & 1 & 1 & 0 & 1 \\ 1 & 0 & 0 & 0 & 1 & 0 \end{vmatrix} = 0 \quad |M_8| = \begin{vmatrix} 0 & 1 & 1 & 1 & 1 & 1 & 1 & 1 \\ 1 & 0 & 1 & 0 & 1 & 0 & 1 & 0 \\ 1 & 1 & 0 & 1 & 1 & 0 & 0 & 1 \\ 1 & 0 & 1 & 0 & 1 & 0 & 0 & 0 \\ 1 & 1 & 1 & 1 & 0 & 1 & 1 & 1 \\ 1 & 0 & 0 & 0 & 1 & 0 & 1 & 0 \\ 1 & 1 & 0 & 0 & 1 & 1 & 0 & 1 \\ 1 & 0 & 1 & 0 & 1 & 0 & 1 & 0 \end{vmatrix} = 0$$

More generally, we have the following result.

Theorem 9.5 Let n be an even integer ≥ 4. Then $|M_n| = 0$.

Proof: From the theory of determinants, it suffices to show that the even-numbered rows of M_n are linearly dependent.

By Lemma 9.3, no two even-numbered vertices in G_n are adjacent. So such vertices are adjacent to odd-numbered vertices only. To show that the even-numbered rows in M_n are linearly dependent, we shall first create a reduced Pascal matrix M' by deleting the odd-numbered rows and the even-numbered columns of M_n (see Figure 9.7) and then show that the determinant of the reduced Pascal matrix M' is zero.

$$M_8 = \begin{bmatrix} 0 & 1 & 1 & 1 & 1 & 1 & 1 & 1 \\ 1 & 0 & 1 & 0 & 1 & 0 & 1 & 0 \\ 1 & 1 & 0 & 1 & 1 & 0 & 0 & 1 \\ 1 & 0 & 1 & 0 & 1 & 0 & 0 & 0 \\ 1 & 1 & 1 & 1 & 0 & 1 & 1 & 1 \\ 1 & 0 & 0 & 0 & 1 & 0 & 1 & 0 \\ 1 & 1 & 0 & 0 & 1 & 1 & 0 & 1 \\ 1 & 0 & 1 & 0 & 1 & 0 & 1 & 0 \end{bmatrix} \qquad M' = \begin{bmatrix} 1 & 1 & 1 & 1 \\ 1 & 1 & 1 & 0 \\ 1 & 0 & 1 & 1 \\ 1 & 1 & 1 & 1 \end{bmatrix}$$

Figure 9.7

By Theorem 9.2, v_1 is adjacent to all other vertices. Let $k = 2^{\lfloor \lg(n-1) \rfloor} + 1$. By Lemma 9.2, vertex v_k is adjacent to all other v_i, where $1 \leq i < k$ or $k+1 < i < 2k$. Thus, columns 1 and $\lfloor k/2 \rfloor + 1$ of M' consist of only 1s.

Since M' contains two identical columns, $|M'| = 0$. Consequently, $|M_n| = 0$, as desired. ■

Before we state the next result, notice that:

$$|M_1| = |0| = 0 \qquad |M_2| = \begin{vmatrix} 0 & 1 \\ 1 & 0 \end{vmatrix} = 2$$

$$|M_5| = \begin{vmatrix} 0 & 1 & 1 & 1 & 1 \\ 1 & 0 & 1 & 0 & 1 \\ 1 & 1 & 0 & 1 & 1 \\ 1 & 0 & 1 & 0 & 1 \\ 1 & 1 & 1 & 1 & 0 \end{vmatrix} = 0 \qquad |M_7| = \begin{vmatrix} 0 & 1 & 1 & 1 & 1 & 1 & 1 \\ 1 & 0 & 1 & 0 & 1 & 0 & 1 \\ 1 & 1 & 0 & 1 & 1 & 0 & 0 \\ 1 & 0 & 1 & 0 & 1 & 0 & 0 \\ 1 & 1 & 1 & 1 & 0 & 1 & 1 \\ 1 & 0 & 0 & 0 & 1 & 0 & 1 \\ 1 & 1 & 0 & 0 & 1 & 1 & 0 \end{vmatrix} = 2$$

More generally, we have the following result. We omit its proof for brevity.

Theorem 9.6 Let n be an odd integer. Then $|M_n|$ is even. ■

Finally, we add an interesting observation made by Deo and Quinn:

$$|M_n| = \begin{cases} 2 & \text{for } n = 3, 7, 11, 23, 43, 87 \\ 0 & \text{for all other } n, \text{ where } 4 \leq n \leq 86 \end{cases}$$

The integers n, where $|M_n| = 2$, seem to follow an interesting pattern:

$$t_0 = 3 = t_1$$

$$t_n = 2^n + t_{n-2}, \qquad n \geq 2$$

These observations await confirmations or counterexamples.

10

Maclaurin's Series

This chapter extends the definition of $\binom{n}{r}$ to negative integers and fractions n. In the process, we will encounter the binomial coefficient $\binom{n+r-1}{r}$.

A Definition of $\binom{-n}{r}$

The quantity $\binom{-n}{r}$ has no known combinatorial interpretation. Nonetheless, it can still be assigned a meaningful value. To this end, we change n to $-n$ in the binomial coefficient $\binom{n}{r} = \frac{n(n-1)\cdots(n-r+1)}{r!}$:

$$\binom{-n}{r} = \frac{-n(-n-1)\cdots(-n-r+1)}{r!} = (-1)^r \frac{n(n+1)\cdots(n+r-1)}{r!}$$
$$= (-1)^r \binom{n+r-1}{r} \tag{10.1}$$

For example, $\binom{-7}{3} = (-1)^3\binom{9}{3} = -84$, $\binom{-3}{7} = (-1)^7\binom{9}{7} = -36$, and $\binom{-1}{2} = (-1)^2\binom{2}{2} = 1$.

Using definition (10.1), we can extend Pascal's triangle in the opposite direction, as in Figure 10.1.[†] Consequently, we can employ row $-n$ to expand $(1+x)^{-n}$.

For example, using row -4, $(1+x)^{-4} = 1 - 4x + 10x^2 - 20x^3 + 35x^4 - \cdots$.

Using definition (10.1), Newton extended the binomial theorem to negative integral exponents in 1664 or 1665.

[†] An extended Pascal's triangle can be found in Lucas' *Théorie des Numbres*.

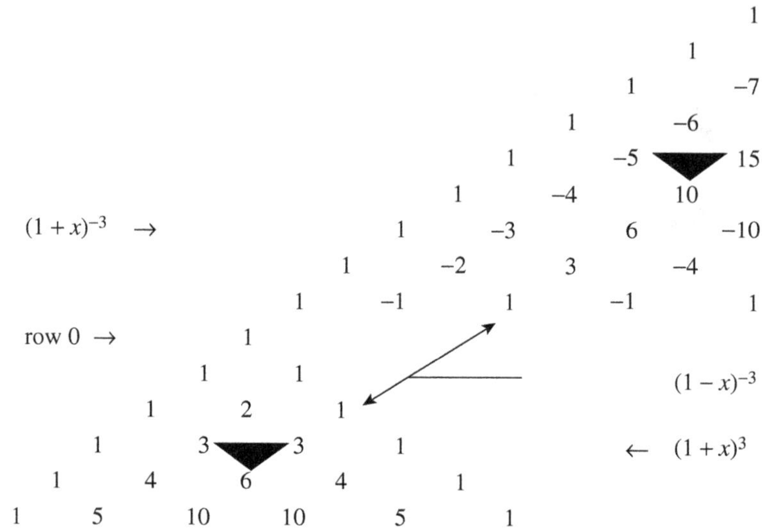

Figure 10.1 Extended Pascal's Triangle

Maclaurin's Series

The binomial coefficient $\binom{n+r-1}{r}$ is the coefficient of x^n in the **Maclaurin's expansion** for $(1-x)^{-n}$. This series is named after the Scottish mathematician Colin Maclaurin (1698–1746):

$$(1-x)^{-n} = \sum_{r=0}^{\infty} \binom{-n}{r}(-x)^r = \sum_{r=0}^{\infty} \binom{n+r-1}{r} x^r$$

$$= 1 + \binom{n}{1}x + \binom{n+1}{2}x^2 + \binom{n+2}{3}x^3 + \cdots$$

So, row $-n$ in the extended array can be used to expand $(1-x)^{-n}$. For example:

$$(1-x)^{-3} = 1 + 3x + 6x^2 + 10x^3 + 15x^4 + \cdots = \sum_{n=0}^{\infty} t_{n+1}x^n$$

where t_k denotes the kth triangular number; see the northeast diagonal 2 in the upper half and southwest diagonal 2 in the lower half in Figure 10.1. In fact, the southwest diagonal n gives the skeleton for the expansion of $(1-x)^{-n}$.

The following example, an application of Maclaurin's series, shows that tetrahedral numbers $T_n = \binom{n+2}{3}$ occur in Maclaurin's expansions also.

Example 10.1 Find the coefficient of x^n in $\left[1+2x+3x^2+\cdots(n+1)x^n+\cdots\right]^2$.

Solution We have

$$(1-x)^{-2} = 1 + 2x + 3x^2 + \cdots$$

Therefore,

$$(1-x)^{-4} = (1 + 2x + 3x^2 + \cdots)^2$$

So, the desired coefficient is the coefficient of x^n in the expansion of $(1-x)^{-4}$; namely,

$$\binom{4+n-1}{n} = \binom{n+3}{n} = \binom{n+3}{3} = T_{n+1}$$ ■

Since

$$(1 + x + x^2 + \cdots)^4 = \left[(1-x)^{-1}\right]^4 = (1-x)^{-4}$$

it follows from Example 10.1 that the coefficient of x^n in the expansion of $(1 + x + x^2 + \cdots)^4$ is also T_{n+1}.

A Prime Characterization

In 1956, L. E. Clarke of the University College of the Gold Coast, Australia, generalized Piza's theorem (Theorem 4.9), providing a necessary and sufficient condition for a positive integer $n \geq 2$ to be a prime. The proof of the generalization, given by E. M. Wright of King's College, Aberdeen, Scotland, employs Maclaurin's series.

Theorem 10.1 (Clarke, 1956) An integer $n \geq 2$ is a prime if and only if $\binom{m}{n} \equiv \lfloor m/n \rfloor \pmod{n}$ for every integer m.

Proof: We have:

$$(1-x)^{-n-1} = \sum_{r=0}^{\infty} \binom{n+r}{r} x^r = \sum_{m=n}^{\infty} \binom{m}{n} x^{m-n}$$

We also have

$$\sum_{m=n}^{\infty} \lfloor m/n \rfloor x^{m-n} = (1 + x + x^2 + \cdots + x^{n-1})(1 + 2x^n + 3x^n + \cdots)$$

$$= \frac{1-x^n}{1-x} \cdot \frac{1}{(1-x^n)^2} = (1-x)^{-1}(1-x^n)^{-1}$$

Thus $\binom{m}{n} \equiv \lfloor m/n \rfloor \pmod{n}$ if and only if $(1-x)^{-n-1} \equiv (1-x)^{-1}(1-x^n)^{-1} \pmod{n}$; that is, if and only if

$$(1-x)^n \equiv 1 - x^n \pmod{n} \tag{10.2}$$

This congruence is true by Theorem 4.2, when n is a prime.

Conversely, suppose n is composite. Let p be a prime factor of n. Then congruence (10.2) implies that $n|\binom{n}{p}$, so $p|(n-1)(n-2)\cdots(n-p+1)$. Then $(-1)(-2)\cdots[-(p-1)] \equiv 0 \pmod{p}$; that is, $(p-1)!(-1)^{p-1} \equiv 0 \pmod{p}$. By Wilson's theorem, this implies $(-1)^p \equiv 0 \pmod{p}$, which is a contradiction. Thus, n must be a prime. ■

For example, $\binom{12}{7} = 792 \equiv 1 = \pmod{7}$, whereas $\binom{13}{6} = 1{,}716 \equiv 0 \;\neq 2 \equiv \lfloor 13/6 \rfloor \pmod{6}$.

A Definition of $\binom{p/q}{r}$

We now extend the definition of the binomial coefficient to fractional numbers $n = \frac{p}{q}$:

$$\begin{aligned}\binom{p/q}{r} &= \frac{p}{q}\left(\frac{p}{q}-1\right)\cdots\left(\frac{p}{q}-r+1\right)\\ &= \frac{p(p-q)(p-2q)\cdots[p-(r-1)q]}{q^r\, r!}\end{aligned} \tag{10.3}$$

For example,

$$\binom{3/4}{5} = \frac{3(3-1\cdot 4)(3-2\cdot 4)(3-3\cdot 4)}{4^5\, 5!} = -\frac{117}{8192}$$

and

$$\begin{aligned}\binom{1/2}{n} &= \frac{\frac{1}{2}\left(\frac{1}{2}-1\right)\left(\frac{1}{2}-2\right)\cdots\left(\frac{1}{2}-n+1\right)}{n!}\\ &= \frac{(-1)^{n-1}\,1(2\cdot 1-1)(2\cdot 2-1)\cdot[2(n-1)-1]}{2^n\, n!}\\ &= (-1)^{n-1}\frac{1\cdot 3\cdot 5\cdots(2n-3)}{2^n\, n!} = (-1)^{n-1}\frac{(2n-2)!}{2^n\, n![2\cdot 4\cdot 6\cdots(2n-2)]}\\ &= (-1)^{n-1}\frac{(2n-2)!}{2^n\, n!\, 2^{n-1}\,(n-1)!} = (-1)^{n-1}\frac{(2n-2)!}{2^{2n-1}\, n!\,(n-1)!}\\ &= \frac{(-1)^{n-1}}{2^{2n-1}\, n}\binom{2n-2}{n-1} = \frac{(-1)^{n-1}}{2^{2n-1}}C_{n-1}\end{aligned}$$

where C_k denotes the kth Catalan number. In particular, $\binom{1/2}{2} = -\frac{1}{8}$ and $\binom{1/2}{3} = -\frac{1}{16}$.

Thus:

$$
\begin{aligned}
(1+x)^{1/2} &= \sum_{n=0}^{\infty}\binom{1/2}{n}x^n = 1+\sum_{n=1}^{\infty}\binom{1/2}{n}x^n \\
&= 1+\sum_{n=1}^{\infty}\frac{(-1)^{n-1}}{n\cdot 2^{2n-1}}\binom{2n-2}{n-1}x^n = 1+\sum_{n=1}^{\infty}\frac{(-1)^{n-1}C_{n-1}}{2^{2n-1}}x^n
\end{aligned}
$$

So

$$(1+4x)^{1/2} = 1-2x\sum_{n=0}^{\infty}(-1)^n C_n x^n$$

It now follows that:

$$(1-x)^{1/2} = 1-\sum_{n=1}^{\infty}\frac{C_{n-1}}{2^{2n-1}}x^n$$

$$(1-4x)^{1/2} = 1-2x\sum_{n=0}^{\infty}C_n x^n$$

We will use the latter power series expansion in Chapter 17.

Using the expansions of $\sqrt{1+x}$ and $\sqrt{1-x}$, it can be shown that

$$\frac{\sqrt{1+x}-\sqrt{1-x}}{\sqrt{1+x}+\sqrt{1-x}} = \frac{x}{2}+\sum_{n=1}^{\infty}\frac{1\cdot 3\cdot 5\cdots(2n-1)}{2\cdot 4\cdot 6\cdots(2n)}\cdot\frac{x^{2n+1}}{2n+2}$$

Formula (10.3) enables us to extend the binomial theorem to fractional exponents, as Newton did in 1664, or shortly thereafter, for his study of calculus:

$$
\begin{aligned}
(x+y)^{m/n} &= x^{m/n}(1+y/x)^{m/n} \\
&= x^{m/n}\sum_{r=0}^{\infty}\binom{m/n}{r}(y/x)^r \qquad (10.4)
\end{aligned}
$$

In particular, $(1+x)^{m/n} = \sum\limits_{r=0}^{\infty}\binom{m/n}{r}x^r$.

This expansion can be employed to approximate $\sqrt[n]{1+x}$, where $|x|<1$. For example,

$$\sqrt[5]{244} = 3\sqrt[5]{\left(1+\frac{1}{243}\right)} = 3\sum_{r=0}^{\infty}\binom{1/5}{r}\left(\frac{1}{243}\right)^r \approx 2.9975349$$

By changing p to $-p$ in formula (10.3), $\binom{-p/q}{r}$ also can be defined:

$$\binom{-p/q}{r} = \frac{-p(-p-q)(-p-2q)\cdots[-p-(r-1)q]}{q^r\, r!}$$
$$= (-1)^r \frac{p(p+q)(p+2q)\cdots[p+(r-1)q]}{q^r\, r!} \tag{10.5}$$

For example,

$$\binom{-2/3}{5} = (-1)^5 \frac{2(2+1\cdot 3)(2+2\cdot 3)(2+3\cdot 3)(2+4\cdot 3)}{3^5\, 5!} = -\frac{308}{729}$$

and

$$\binom{-1/2}{n} = (-1)^n \frac{1\cdot(1+2\cdot 1)(1+2\cdot 2)(1+2\cdot 3)\cdots[1+2(n-1)]}{2^n\, n!}$$
$$= (-1)^n \frac{1\cdot 3\cdot 5\cdot 7\cdots(2n-1)}{2^n\, n!} = (-1)^n \frac{(2n-1)!}{2^n\, n!\,[2\cdot 4\cdot 6\cdots(2n-2)]}$$
$$= (-1)^n \frac{(2n-1)!}{2^{2n-1}\, n!(n-1)!} = \frac{(-1)^n}{2^{2n}}\binom{2n}{n} = \left(-\frac{1}{4}\right)^n \binom{2n}{n}$$

As can be expected, we can use formula (10.5) to extend Maclaurin's series to negative fractional exponents:

$$(1-x)^{-p/q} = \sum_{r=0}^{\infty} (-1)^r \binom{-p/q}{r} x^r$$
$$= \sum_{r=0}^{\infty} \frac{p(p+q)(p+2q)\cdots[p+(r-1)q]}{r!} \left(\frac{x}{q}\right)^r \tag{10.6}$$

For example,

$$(1-x)^{-1/3} = \sum_{r=0}^{\infty} \frac{1(1+1\cdot 3)(1+2\cdot 3)\cdots[1+3(n-1)]}{n!} \left(\frac{x}{3}\right)^n$$
$$= 1 + \frac{1}{1!}\left(\frac{x}{3}\right) + \frac{1\cdot 4}{2!}\left(\frac{x}{3}\right)^2 + \frac{1\cdot 4\cdot 7}{3!}\left(\frac{x}{3}\right)^3 + \cdots$$

A CBC Identity

We now invoke Vandermonde's identity

$$\sum_{r=0}^{n} \binom{r}{k}\binom{s}{n-k} = \binom{r+s}{n}$$

to develop the remarkable property about the CBCs we encountered in Example 5.3. Letting $r = -1/2 = s$, this identity yields

$$\sum_{k=0}^{n}\binom{-1/2}{k}\binom{-1/2}{n-k} = \binom{-1}{n} = (-1)^n$$

That is:

$$\sum_{k=0}^{n}\left(-\frac{1}{4}\right)^k\binom{2k}{k}\cdot\left(-\frac{1}{4}\right)^{n-k}\binom{2n-2k}{n-k} = (-1)^n$$

$$\left(-\frac{1}{4}\right)^n\sum_{k=0}^{n}\binom{2k}{k}\binom{2n-2k}{n-k} = (-1)^n$$

$$\sum_{k=0}^{n}\binom{2k}{k}\binom{2n-2k}{n-k} = 4^n \qquad (10.7)$$

Notice that identity (10.7) can be rewritten as $\sum_{k=0}^{n}(k+1)(n-k+1)C_k C_{n-k} = 4^n$.

Using Maclaurin series for $(1-x)^{-1/2}$, the next example identifies the sequence $\{a_n\}$ that has a unique property.†

Example 10.2 Find the sequence $\{a_n\}$ of real numbers such that $a_0 = 1$ and $\sum_{k=0}^{n} a_k a_{n-k} = 1$.

Solution Let $a(x) = \sum_{k=0}^{\infty} a_n x^n$ be the generating function of the sequence $\{a_n\}$. Then $\sum_{k=0}^{n} a_k a_{n-k}$ is the coefficient of x^n in $a^2(x) = \left(\sum_{n=0}^{\infty} a_n x^n\right)^2$. By virtue of the recurrence relation satisfied by a_n, this yields:

$$a^2(x) = \sum_{n=0}^{\infty} x^n = \frac{1}{1-x}$$

$$a(x) = (1-x)^{-1/2} = \sum_{n=0}^{\infty}(-1)^n\binom{-1/2}{n}x^n$$

$$a_n = (-1)^n\binom{-1/2}{n}$$

$$= (-1)^n\frac{\left(-\frac{1}{2}\right)\left(-\frac{3}{2}\right)\left(-\frac{5}{2}\right)\cdots\left(-\frac{2n-1}{2}\right)}{n!}$$

†R. P. Stanley, *Enumerative Combinatorics*, Vol. 1, Wadsworth and Brooks/Cole, Monterey, California.

$$= \frac{1 \cdot 3 \cdot 5 \cdots (2n-1)}{2^n\, n!} = \frac{(2n)!}{2^n\, n![2 \cdot 4 \cdot 6 \cdots (2n)]}$$

$$= \frac{(2n)!}{2^n\, n!\, 2^n\, n!} = \frac{(2n)!}{4^n\, n!\, n!} = \frac{n+1}{4^n} C_n \qquad \blacksquare$$

Reversing the order, we would like to define the sequence $\{a_n\}$ recursively, where $a_n = \frac{1}{4^n}\binom{2n}{n}$. First, notice that $a_0 = 1$.

Next, using identity (10.7),

$$a_k a_{n-k} = \frac{1}{4^k}\binom{2k}{k} \cdot \frac{1}{4^{n-k}}\binom{2n-2k}{n-k} = \frac{1}{4^n}\binom{2k}{k}\binom{2n-2k}{n-k}$$

$$\sum_{k=0}^{n} a_k a_{n-k} = \frac{1}{4^n}\sum_{k=0}^{n}\binom{2k}{k}\binom{2n-2k}{n-k} = \frac{1}{4^n} \cdot 4^n = 1$$

Thus, the sequence $\{a_n\}$ can be defined recursively as follows, where $n \geq 1$:

$$a_0 = 1$$

$$a_n = \frac{1}{2} - \frac{1}{2}\sum_{k=0}^{n-1} a_k a_{n-k}$$

For example,

$$a_1 = \frac{1}{2} - \frac{1}{2}\sum_{k=1}^{0} a_k a_{1-k} = \frac{1}{2} - \frac{1}{2} \cdot 0 = \frac{1}{2}$$

and

$$a_2 = \frac{1}{2} - \frac{1}{2}\sum_{k=1}^{1} a_k a_{2-k} = \frac{1}{2} - \frac{1}{2} \cdot a_1 \cdot a_1 = \frac{1}{2} - \frac{1}{2} \cdot \frac{1}{2} \cdot \frac{1}{2} = \frac{3}{8}$$

Notice that $a_0 a_2 + a_1 a_1 + a_2 a_0 = 1 \cdot \frac{3}{8} + \frac{1}{2} \cdot \frac{1}{2} + \frac{3}{8} \cdot 1 = 1$, as required in Example 10.2.

Let $a_n = \frac{1 \cdot 3 \cdot 5 \cdots (2n-1)}{2 \cdot 4 \cdot 6 \cdots (2n)}$. Then using the fact that $(1-x)^{-1} = (1-x)^{-1/2}(1-x)^{-1/2}$, and equating the coefficients of x^{2n+1}, it can be shown that $a_{2n+1} + a_1 a_{2n} + a_2 a_{2n-1} + \cdots + a_n a_{n+1} = \frac{1}{2}$. Likewise, $a_n + a_{n-1} b_1 + a_{n-2} b_2 + \cdots + b_n = \frac{(n+1)(n+2)}{2}$, where $b_n = \frac{5 \cdot 7 \cdot 9 \cdots (2n+3)}{2 \cdot 4 \cdot 6 \cdots (2n)}$.

The next example, proposed in 1989 by D. Doster of Choate Rosemary Hall of Wallingford, Connecticut, is an application of the binomial expansion for $(1+x)^{-1/2}$. The solution is based on the one given by C. Hill, when he was a student at Colorado State University, Fort Collins, Colorado.

Example 10.3 Consider the sequence $\{p_n\}$, defined recursively by $p_n = \frac{6n-1}{4n} p_{n-1} - \frac{2n-1}{4n} p_{n-2}$, where $p_0 = 1$ and $p_1 = 5/4$. Evaluate $\lim_{n\to\infty} p_n$.

Solution We have:

$$\begin{aligned} p_n - p_{n-1} &= \frac{2n-1}{4n}(p_{n-1} - p_{n-2}) \\ &= \frac{2n-1}{4n} \cdot \frac{2n-3}{4(n-1)} \cdot \frac{2n-5}{4(n-2)} \cdots \frac{3}{4 \cdot 2} \cdot \frac{1}{4} \\ &= \frac{\left(-\frac{1}{2}\right)\left(-\frac{3}{2}\right)\left(-\frac{5}{2}\right)\cdots\left(-\frac{2n-1}{2}\right)}{n!}\left(-\frac{1}{2}\right)^n \\ &= \binom{-1/2}{n}\left(-\frac{1}{2}\right)^n, \quad n \geq 2 \end{aligned}$$

Notice that $\binom{-1/2}{1}\left(-\frac{1}{2}\right)^1 = \frac{1}{4} = p_1 - p_0$ and $\binom{-1/2}{0}\left(-\frac{1}{2}\right)^0 = 1 = p_0$. Therefore,

$$p_n = \sum_{i=1}^{n}(p_i - p_{i-1}) + p_0 = \sum_{r=0}^{n}\binom{-1/2}{r}\left(-\frac{1}{2}\right)^r, \quad n \geq 0.$$

By the Maclaurin's expansion, $(1+x)^{-1/2} = \sum\limits_{r=0}^{\infty}\binom{-1/2}{r}x^r$, where $|x| < 1$. Consequently,

$$\lim_{n\to\infty} p_n = \sum_{r=0}^{\infty}\binom{-1/2}{r}\left(-\frac{1}{2}\right)^r = \left(1 - \frac{1}{2}\right)^{-1/2} = \sqrt{2} \qquad \blacksquare$$

The next example, proposed in 1963 by J. S. Frame of Michigan State University, is to evaluate the sum of an infinite series involving the ubiquitous CBC. The solution, given in the following year by A. Weinmann of the University of Leicester, England, involves $\binom{-p/q}{r}$, integral calculus, the golden ratio $\alpha = \frac{1+\sqrt{5}}{2}$, and the infinite sum $\frac{1}{1^2} - \frac{1}{2^2} + \frac{1}{3^2} - \cdots = \frac{\pi^2}{12}$.

Example 10.4 Evaluate the sum $\sum\limits_{r=0}^{\infty}\binom{2k}{k}(-16)^{-k}(2k+1)^{-2}$.

Solution Let $S = \sum\limits_{r=0}^{\infty}\binom{2k}{k}(-16)^{-k}(2k+1)^{-2}$. Then:

$$\begin{aligned} S &= \sum_{r=0}^{\infty}\binom{-1/2}{k}\left(\frac{1}{4}\right)^k \int_0^{\infty} e^{-(2k+1)t}t\,\mathrm{d}t \\ &= \int_0^{\infty} \frac{te^{-t}}{\left(1 + \frac{1}{4}e^{-2t}\right)^{1/2}}\,\mathrm{d}t \end{aligned}$$

Using the substitution $e^{-t} = \frac{z^2-1}{z}$ and integration by parts, this yields $S = 2(A + B - C)$, where

$$A = \int_1^{\alpha}\frac{\log z}{z+1}\mathrm{d}z, \quad B = \int_1^{\alpha}\frac{\log z}{z-1}\mathrm{d}z, \quad \text{and} \quad C = \int_1^{\alpha}\frac{\log z}{z}\mathrm{d}z = \frac{1}{2}(\log\alpha)^2$$

Let

$$D = \int_1^{\alpha} \frac{\log z}{z+1} dz$$

$$= -\frac{1}{1^2} + \frac{1}{2^2} - \frac{1}{3^2} - \cdots = -\frac{\pi^2}{12}$$

Since $\alpha^2 = \alpha + 1$, using integration by parts, we have:

$$B = \log \alpha \cdot \log(\alpha - 1) - \int_1^{\alpha} \frac{\log(z-1)}{z} dz$$

$$= -(\log \alpha)^2 - D - \int_1^{\alpha - 1} \frac{\log t}{t+1} dt$$

$$= -2C - D - \int_1^{\alpha} \frac{\log v}{v(v+1)} dv, \text{ where } v = \frac{1}{t}.$$

$$= -2C - D - C + A = -D - 3C + A$$

But

$$A + B = \int_1^{\alpha} \frac{\log z}{z+1} dz + \int_1^{\alpha} \frac{\log z}{z-1} dz$$

$$= \int_1^{\alpha} \frac{2z \log z}{z^2 - 1} dz$$

$$= \frac{1}{2} \int_1^{\alpha} \frac{\log(t+1)}{t} dt, \text{ where } z^2 = t + 1.$$

$$= \frac{1}{2} \log(\alpha + 1) \cdot \log \alpha - \frac{1}{2} D - \frac{1}{2} A, \text{ by parts}$$

$$= 2C - \frac{1}{2} D - \frac{1}{2} A$$

Thus $A + B = 2C - \frac{1}{2}D - \frac{1}{2}A$ and $A - B = 3C + D$. Adding, $A = 2C + \frac{1}{5}D$, so $D = 2(A + B) - 2C = -\frac{6}{5}D = \frac{\pi^2}{10}$. ■

11

Fibonacci and Lucas Numbers

This chapter presents two of the most intriguing classes of positive integers, *Fibonacci numbers* and *Lucas numbers*. They

are the two shining stars in the vast array of integer sequences. They have fascinated both amateurs and professional mathematicians for centuries, and they continue to charm us with their beauty, their abundant applications, and their ubiquitous habit of occurring in totally surprising and unrelated places. They continue to be a fertile ground for creative amateurs and mathematicians alike.[†]

Fibonacci numbers are named after Leonardo Fibonacci (*c*.1170–1250), the most outstanding Italian mathematician of the Middle Ages. However, according to P. Singh of Raj Narain College, Bihar, India, Fibonacci numbers and their recursive definition were known in India several centuries before Fibonacci; they were given by Virahanka between A.D. 600 and 800; by Gopala prior to A.D. 1135, and by Acharya Hemachandra about A.D. 1150, about 50 years before Fibonacci. Hemachandra discovered them while counting the number of rhythmic patterns of length n made up of single-beat and double-beat notes. Fibonacci numbers also occur as a special case of a formula established by Narayana Pandit (A.D. 1156).

There is a dazzling puzzle behind these numbers. It appears as a problem in Chapter 12 in Fibonacci's famous work, *Liber Abaci*, published in 1202. This introduced the Fibonacci sequence to Europe. So, we begin with the puzzle.

[†]See the author's *Fibonacci and Lucas Numbers with Applications* for an extensive discussion of Fibonacci and Lucas numbers, and their myriad applications.

Fibonacci's Rabbits

Suppose we are given two newborn rabbits, one male and the other female. Find the number of pairs of rabbits at the end of the year if:

- Each baby pair takes a month to mature;
- Each adult pair produces a mixed (one male and the other female) pair every month; and
- All rabbits are immortal.

For convenience, assume that the original pair of rabbits was born on January 1. It takes a month to become mature; so there is one adult pair in February, but no baby pairs. On March 1, this adult pair is two months old, so it produces a mixed baby pair, a total of 2 pairs. Continue like this; there will be a total of 3 pairs on April 1, 5 pairs on May 1, 8 pairs on June 1, and so on.

Fibonacci Numbers

The numbers 1, 1, 2, 3, 5, 8, $\cdots$ are the *Fibonacci numbers* F_n, a name given to them by Lucas. They display a fascinating property: $F_n = F_{n-1} + F_{n-2}$, where $n \geq 3$. (Thus, there will be 144 pairs of rabbits at the end of the year.) Thus Fibonacci numbers can be defined recursively: $F_n = F_{n-1} + F_{n-2}$, where $F_1 = 1 = F_2$ and $n \geq 3$.

Fibonacci numbers occur quite frequently in numerous places, including nature. For example, the most common number of petals on flowers is the Fibonacci number 5. The spiral arrangements of seeds on a sunflower provide a striking occurrence of Fibonacci numbers; the numbers of clockwise spirals and counterclockwise spirals are usually adjacent Fibonacci numbers (see Figures 11.1 and 11.2). The scale patterns on pine cones and pineapples, and the arrangement of leaves on some trees also follow the Fibonacci pattern.

Lucas Numbers

Closely related to Fibonacci numbers are the well-known *Lucas numbers*, named after Lucas. They follow the same Fibonacci pattern, except that the two initial seeds are 1 and 3. Accordingly, the various Lucas numbers L_n are $1, 3, 4, 7, 11, \cdots$. They too are often defined recursively: $L_n = L_{n-1} + L_{n-2}$, where $L_1 = 1$, $L_2 = 3$, and $n \geq 3$.

Using the recursive definitions, both Fibonacci and Lucas numbers can be extended to zero and negative subscripts: $F_0 = 0$, $F_{-n} = (-1)^{n-1}F_n$; and $L_0 = 2$, $L_{-n} = (-1)^n L_n$.

Figure 11.1 Sunflower: A Sunflower Head. (*Source*: Runk/Schoenberger/Grant Heilman Photography, Inc.)

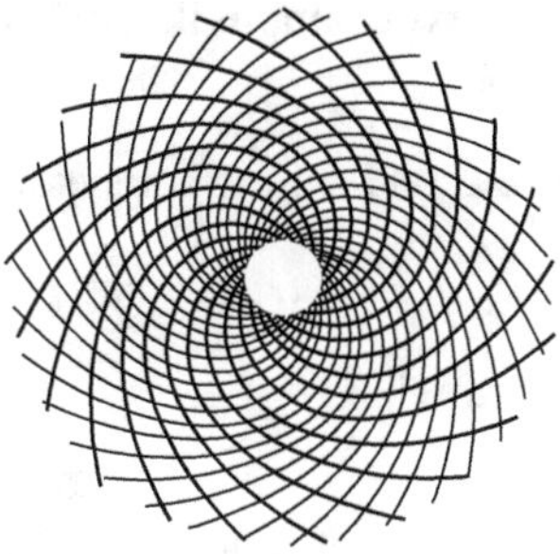

Figure 11.2 The Spiral Pattern in a Sunflower. (*Source*: H. E. Huntley, *The Divine Proportion*, Mincola, NY: Dover, 1970. Reproduced with permission of Dover Publications.)

Binet's Formulas

Both Fibonacci numbers and Lucas numbers can be defined explicitly, using *Binet's formulas*:

$$F_n = \frac{\alpha^n - \beta^n}{\alpha - \beta} \quad \text{and} \quad L_n = \alpha^n + \beta^n$$

where $\alpha = \frac{1+\sqrt{5}}{2}$ and $\beta = \frac{1-\sqrt{5}}{2}$ are the solutions† of the quadratic equation $x^2 = x + 1$. These two formulas are named after the French mathematician and astronomer Jacques Philippe Marie Binet (1786–1865), who found the formula for F_n in 1843. It was, however, discovered in 1718 by De Moivre, using generating functions.

Fibonacci and Lucas Identities

Fibonacci and Lucas numbers satisfy a plethora of interesting identities. A few of them are:

$F_{2n} = F_n L_n$

$F_{n+1} + F_{n-1} = L_n$

$F_{n+2} - F_{n-2} = 5F_n$

$L_{n+1}L_{n-1} - L_n^2 = 5(-1)^{n-1}$

$F_{n+1}F_{n-1} - F_n^2 = (-1)^n$ (Cassini's formula)

$F_{n+1}^2 + F_n^2 = F_{2n+1}$ (Lucas, 1876)

$F_{n+1}^2 - F_n^2 = F_{2n}$ (Lucas, 1876)

They can be established using Binet's formulas or PMI.

†The number $\alpha = \frac{1+\sqrt{5}}{2}$ is the well-known *golden ratio* and $\beta = -1/\alpha$.

The next example illustrates how to find a generating function for Fibonacci numbers.

Example 11.1 Find a generating function for Fibonacci numbers.

Solution Let $g(x) = \sum_{n=0}^{\infty} F_n x^n$ be the generating function of the Fibonacci sequence. Then:

$$xg(x) = \sum_{n=2}^{\infty} F_{n-1}x^n$$

$$x^2g(x) = \sum_{n=3}^{\infty} F_{n-2}x^n$$

$$g(x) - xg(x) - x^2g(x) = F_1x + (F_2 - F_1)x^2 + \sum_{n=3}^{\infty}(F_n - F_{n-1} - F_{n-2})x^n = x$$

$$(1 - x - x^2)g(x) = x$$

$$g(x) = \frac{x}{1 - x - x^2}$$

Thus

$$\frac{x}{1 - x - x^2} = \sum_{n=0}^{\infty} F_n x^n$$ ■

Similarly, it can be shown that

$$\frac{2 - x}{1 - x - x^2} = \sum_{n=0}^{\infty} L_n x^n$$

Binomial Coefficients and Fibonacci Numbers

Curiously enough, binomial coefficients can be employed to generate Fibonacci numbers, as the following theorem shows. This property, discovered in 1876 by Lucas, follows by the strong version of PMI.

Theorem 11.1 (Lucas' formula) Let $n \geq 1$. Then

$$F_n = \sum_{i=0}^{\lfloor (n-1)/2 \rfloor} \binom{n-i-1}{i} \qquad (11.1)$$ ■

Pascal Meets Fibonacci

It follows by the theorem that the sums along the northeast diagonals in Pascal's triangle yield Fibonacci numbers, as Figure 11.3 shows.

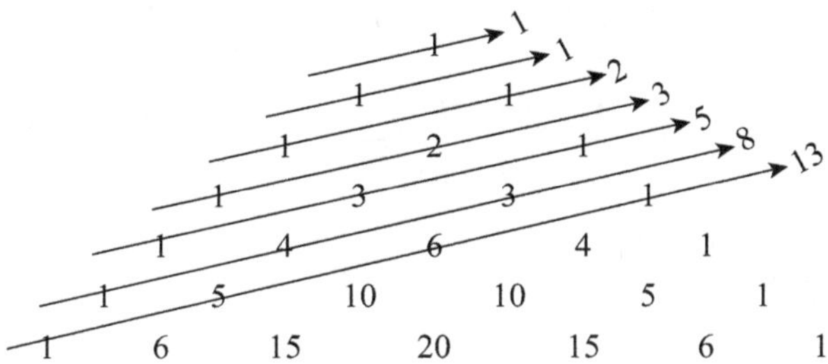

Figure 11.3

For example, $F_8 = \sum_{i=0}^{3} \binom{7-i}{i} = \binom{7}{0} + \binom{6}{1} + \binom{5}{2} + \binom{4}{3} = 21$, as expected.

Fibonacci, Lucas, and the Binomial Theorem

Using Binet's and Lucas' formulas, coupled with the binomial theorem, we can develop a host of identities. For example, notice that

$$\sum_{r=0}^{6} \binom{6}{r} F_r = \binom{6}{0} F_0 + \binom{6}{1} F_1 + \binom{6}{2} F_2 + \binom{6}{3} F_3 + \binom{6}{4} F_4 + \binom{6}{5} F_5 + \binom{6}{6} F_6 = 144 = F_{12}$$

More generally, we have the following result.

Theorem 11.2 (E. Lucas)

$$\sum_{r=0}^{n} \binom{n}{r} F_r = F_{2n} \tag{11.2}$$

Proof: Let α and β be the solutions of the equation $x^2 = x + 1$. Then $\alpha^2 = \alpha + 1$ and $\beta^2 = \beta + 1$. By Binet's formula, we have:

$$\begin{aligned}\sum_{r=0}^{n}\binom{n}{r}F_r &= \sum_{r=0}^{n}\binom{n}{r}\left(\frac{\alpha^r - \beta^r}{\alpha - \beta}\right)\\ &= \frac{1}{\alpha-\beta}\left[\sum_{r=0}^{n}\binom{n}{r}\alpha^r - \sum_{r=0}^{n}\binom{n}{r}\beta^r\right]\\ &= \frac{(1+\alpha)^n - (1+\beta)^n}{\alpha - \beta} = \frac{\alpha^{2n} - \beta^{2n}}{\alpha - \beta} = F_{2n}\end{aligned}$$ ■

Similarly, it can be shown that

$$\sum_{r=0}^{n}\binom{n}{r}L_r = L_{2n} \tag{11.3}$$

For example, $\binom{5}{0}L_0 + \binom{5}{1}L_1 + \binom{5}{2}L_2 + \binom{5}{3}L_3 + \binom{5}{4}L_4 + \binom{5}{5}L_5 = 2 + 5 + 30 + 40 + 35 + 11 = 123 = L_{10}$.

Fibonacci Again

Figure 11.4 shows a displaced Pascal's triangle, where each row, except the apex, is shifted to the right by one column from the previous row. It appears that the column sums are the Fibonacci numbers. This is indeed true, since the column sums satisfy the Fibonacci recurrence relation and the two initial conditions. Thus, the $(n+1)$st column sum is F_{n+1}, where $n \geq 0$.

n \ k	0	1	2	3	4	5	6	7	8	9	10	11	12	13	14	15	16
0	1																
1		1	1														
2			1	2	1												
3				1	3	3	1										
4					1	4	6	4	1								
5						1	5	10	10	5	1						
6							1	6	15	20	15	6	1				
7								1	7	21	35	35	21	7	1		
8									1	8	28	56	70	56	28	8	1
9										1	9	36	84	126	126	84	36
Column sums →	1	1	2	3	5	8	13	21	34	55	…						

Figure 11.4 A Displaced Pascal's Triangle

Pascal's Triangle Revisited

In Theorem 11.1, we found a relationship between the rising diagonals in Pascal's triangle and Fibonacci numbers. There is another way the two are related.

$n \backslash r$	0	1	2	3	4	5
0	1					
1	1	1				
2	1	2	1			
3	1	3	3	1		
4	1	4	6	4	1	
5	1	5	10	10	5	1

Figure 11.5

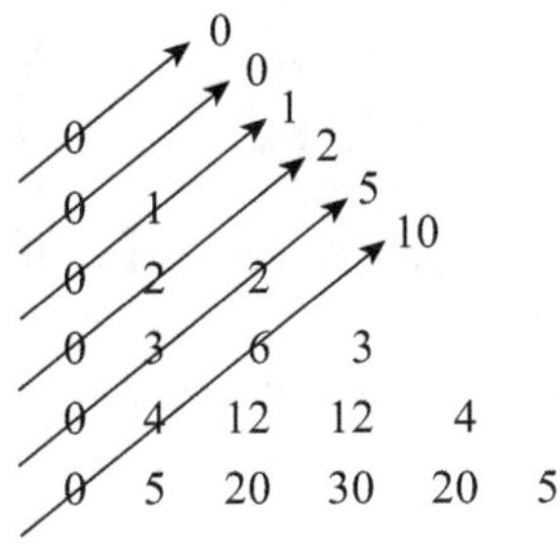

Figure 11.6

To see this, consider the left-justified Pascal's triangle in Figure 11.5. Multiply column r by r and then compute the diagonal sums D_n, where $n \geq 0$.

For example, $D_5 = 0+4+6 = 10 = 0\cdot 5+1\cdot 3+1\cdot 2+2\cdot 1+3\cdot 1+5\cdot 0 = \sum_{r=0}^{5} F_r F_{5-r}$.

More generally,

$$\sum_{r=0}^{\lfloor n/2 \rfloor} r\binom{n-r}{r} = \sum_{r=0}^{n} F_r F_{n-r}$$

This identity was developed in 1968 by Hoggatt. We will now establish this using Carlitz's proof, given in the following year, Maclaurin's series, and the generating function for F_n.

Proof: Let $D_n = \sum_{r=0}^{\lfloor n/2 \rfloor} r\binom{n-r}{r}$. Notice that $D_0 = 0 = D_1$; see Figure 11.6. Then:

$$\sum_{n=0}^{\infty} D_n x^n = \sum_{n=0}^{\infty}\left[\sum_{r=0}^{\lfloor n/2 \rfloor} r\binom{n-r}{r}\right] x^n$$

$$= \sum_{n=0}^{\infty} r x^{2r} \sum_{n=0}^{\infty}\binom{n+r}{r} x^n = \sum_{r=0}^{\infty} r x^{2r}(1-x)^{-r-1}$$

$$= \sum_{s=-1}^{\infty}(s+1)x^{2s+2}(1-x)^{-s-2} = \frac{x^2}{(1-x)^2}\sum_{s=0}^{\infty}(s+1)x^{2s}(1-x)^{-s}$$

$$= \frac{x^2}{(1-x)^2} \sum_{s=0}^{\infty} (s+1) \left(\frac{x^2}{1-x} \right)^s = \frac{x^2}{(1-x)^2} \left(1 - \frac{x^2}{1-x} \right)^{-2}$$

$$= \left[\frac{x}{1-x-x^2} \right]^2 = \left(\sum_{n=0}^{\infty} F_n x^n \right) \left(\sum_{n=0}^{\infty} F_n x^n \right)$$

$$= \sum_{n=0}^{\infty} \left(\sum_{r=0}^{n} F_r F_{n-r} \right) x^n$$

Equating the coefficients of x^n from both sides yields the desired result. ■

Finally, we add that Fibonacci and Lucas numbers belong to a large family of *generalized Fibonacci numbers* G_n; they are defined recursively as follows:

$$G_1 = a, \quad G_2 = b$$

$$G_n = \lambda G_{n-1} + G_{n-2}, \quad n \geq 3$$

where a, b, and λ are arbitrary integers.

12

Pascal's Triangle and Generating Functions

For over three centuries, Pascal's triangle has been investigated by people around the world. As we saw in the last four chapters, they succeeded in extracting numerous properties of the array. This chapter continues the exploration based on the investigation by Hoggatt using generating functions in 1968.

Column Generators

$n \backslash r$	0	1	2	3	4		...	n
0	1							
1	1	1						
2	1	2	1					
3	1	3	3	1				
4	1	4	6	4	1			
5	1	5	10	10	5	1		
6	1	6	15	20	15	6	1	
				⋮				
n	$\binom{n}{0}$	$\binom{n}{1}$	$\binom{n}{2}$	$\binom{n}{3}$		...		$\binom{n}{n}$
	$\frac{1}{1-x}$	$\frac{x}{(1-x)^2}$	$\frac{x^2}{(1-x)^3}$	$\frac{x^3}{(1-x)^4}$		...		$\frac{x^n}{(1-x)^{n+1}}$ ← Column generators

Figure 12.1 Left-justified Pascal's Triangle

To this end, consider the left-justified Pascal's triangle in Figure 12.1. As shown in the figure, column r is generated by the function

$$\begin{aligned} g_r(x) &= \frac{x^r}{(1-x)^{r+1}} \\ &= \sum_{n=0}^{\infty} \binom{n}{r} x^n = \sum_{n=r}^{\infty} \binom{n}{r} x^n \end{aligned}$$

where $r \geq 0$. Notice that the column generator $g_r(x)$ correctly and automatically aligns the binomial coefficients $\binom{n}{r}$ in column r for each value of r.

Using the generators $g_r(x)$, the generator function for the sum of the binomial coefficients in row n is given by

$$\begin{aligned} G(x) &= \sum_{n=0}^{\infty} g_r(x) = \frac{1}{1-x} \sum_{r=0}^{\infty} \left(\frac{x}{1-x}\right)^r \\ &= \frac{1}{1-x} \cdot \frac{1}{1 - \dfrac{x}{1-x}} = \frac{1}{1-2x} \end{aligned}$$

$$\sum_{r=0}^{\infty} \left[\sum_{n=r}^{\infty} \binom{n}{r} x^n\right] = \sum_{n=0}^{\infty} 2^n x^n$$

Equating the coefficients of x^n from both sides, yields Corollary 3.1: $\sum\limits_{r=0}^{n} \binom{n}{r} = 2^n$.

Rising Diagonal Sums

Recall from Theorem 11.1 that every rising diagonal sum in Pascal's triangle is a Fibonacci number. We will now establish this formula using generating functions.

The rising diagonals in Figure 12.1 can be transformed into rows by a suitable transformation: move each column down, except column 0, by one position. Figure 12.2 shows the resulting array. Consequently, the row sums in Figure 12.2 yield the rising diagonal sums in Pascal's triangle. (We will encounter this array in the next chapter.)

The columns of this array are generated by $g_r^*(x) = \frac{x^{2r}}{(1-x)^{r+1}}$, where $r \geq 0$; see Figure 12.2. Thus, the generating function for the sum of the binomial coefficients in row n is given by

$$G^*(x) = \sum_{r=0}^{\infty} g_r^*(x)$$

$$\sum_{r=0}^{\infty} \left[\sum_{n=0}^{\infty} \binom{n}{r} x^{n+r}\right] = \frac{1}{1-x} \sum_{r=0}^{\infty} \left(\frac{x^2}{1-x}\right)^r$$

					Row sums
1					1
1					1
1	1				2
1	2				3
1	3	1			5
1	4	3			8
1	5	6	1		13
1	6	10	4		21
		$\vdots$			$\uparrow$
$\binom{n}{0}$	$\binom{n-1}{1}$	$\binom{n-2}{2}$	$\binom{n-3}{3}$	$\cdots$	F_n
$\frac{1}{1-x}$	$\frac{x^2}{(1-x)^2}$	$\frac{x^4}{(1-x)^3}$	$\frac{x^6}{(1-x)^4}$	$\cdots$	$\leftarrow$ Column generators

Figure 12.2

$$\sum_{r=0}^{\infty}\left[\sum_{m=r}^{\infty}\binom{m-r}{r}x^m\right]=\frac{1}{1-x}\cdot\frac{1}{1-\dfrac{x^2}{1-x}}$$

$$\sum_{r=0}^{\infty}\left[\sum_{m=0}^{\infty}\binom{m-r}{r}x^m\right]=\frac{1}{1-x-x^2}$$

$$\sum_{r=0}^{\infty}\left[\sum_{n=0}^{\infty}\binom{n-r}{r}x^n\right]=\sum_{n=0}^{\infty}F_{n+1}x^n$$

Thus

$$F_{n+1}=\sum_{r=0}^{\infty}\binom{n-r}{r}=\sum_{r=0}^{\lfloor n/2\rfloor}\binom{n-r}{r},\qquad n\geq 0$$

Next, we turn to the even-numbered columns in Pascal's triangle.

Even-numbered Columns in Pascal's Triangle

In the following discussion of the even-numbered rows in Pascal's triangle, we will use the generating function

$$\frac{1-x}{1-3x+x^2}=\sum_{n=0}^{\infty}F_{2n+1}x^n$$

The even-numbered rows are generated by

$$g_{2r}(x) = \frac{x^{2r}}{(1-x)^{2r+1}} = \sum_{r=0}^{\infty} \binom{n}{2r} x^{n-r}$$

So

$$\frac{x^{2r}}{(1-x)^{2r+1}} = \sum_{n=0}^{\infty} \binom{n}{2r} x^{n-r}$$

Now, delete all odd-numbered columns in Figure 12.1. Figure 12.3 shows the resulting array. Now, move column $2r$ up by r spaces. Figure 12.4 shows the resulting array. Its rows are the falling southeast diagonals of the array in Figure 12.3.

n \ r	0	2	4	6	8
0	1				
1	1				
2	1	1			
3	1	3			
4	1	6	1		
5	1	10	5		
6	1	15	15	1	
7	1	21	35	7	

Figure 12.3

n \ r	0	2	4	6	8	10	12	14	Row sums
0	1								1
1	1	1							2
2	1	3	1						5
3	1	6	5	1					13
4	1	10	15	7	1				34
5	1	15	35	28	9	1			89
6	1	21	70	84	45	11	1		233
7	1	28	126	210	165	66	13	1	610

Figure 12.4

Its columns are generated by $\frac{x^r}{(1-x)^{2r+1}} = \sum_{n=0}^{\infty} \binom{n}{2r} x^{n-r}$ and the row sums by the function

$$\sum_{r=0}^{\infty} \frac{x^r}{(1-x)^{2r+1}} = \sum_{r=0}^{\infty} \left[\sum_{n=0}^{\infty} \binom{n}{2r} x^{n-r} \right]$$

$$\frac{1}{1-x} \sum_{r=0}^{\infty} \left[\frac{x}{(1-x)^2} \right]^r = \sum_{r=0}^{\infty} \left[\sum_{m=-r}^{\infty} \binom{m+r}{2r} x^m \right]$$

$$\frac{1}{1-x} \cdot \frac{1}{1 - \dfrac{x}{(1-x)^2}} = \sum_{r=0}^{\infty} \left[\sum_{m=0}^{\infty} \binom{m+r}{2r} x^m \right]$$

$$\frac{1-x}{1-3x+x^2} = \sum_{r=0}^{\infty}\left[\sum_{n=0}^{\infty}\binom{n+r}{2r}x^n\right]$$

$$\sum_{n=0}^{\infty} F_{2n+1}x^n = \sum_{r=0}^{\infty}\left[\sum_{n=0}^{\infty}\binom{n+r}{2r}x^n\right]$$

Thus, since $n + r \geq 2r$,

$$\begin{aligned} F_{2n+1} &= \sum_{r=0}^{\infty}\binom{n+r}{2r} = \sum_{r=0}^{n}\binom{n+r}{2r} \\ &= \binom{n}{0} + \binom{n+1}{2} + \binom{n+2}{4} + \cdots + \binom{2n}{2n} \qquad (12.1) \end{aligned}$$

In words, F_{2n+1} is the sum of the elements on the nth falling southeast diagonal, where $n \geq 0$. See Figure 12.5.

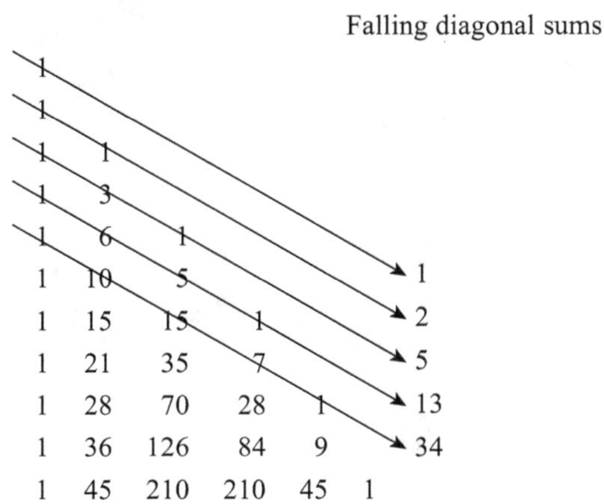

Figure 12.5

For example, $F_9 = \sum_{r=0}^{4}\binom{4+r}{2r} = \binom{4}{0} + \binom{5}{2} + \binom{6}{4} + \binom{7}{6} + \binom{8}{8} = 34$.

An Alternate Formula for F_{2n+1}

We will now derive an alternate combinatorial formula for F_{2n+1}.

Consider the generating function $a(x) = \sum_{n=0}^{\infty} a_n x^n$ and the column generators

$$g_i(x) = \frac{x^i}{(1-x)^{i+1}} = \sum_{n=0}^{\infty}\binom{n}{i}x^n$$

of the left-justified Pascal's triangle in Figure 12.1. Let $G(x) = \sum_{i=0}^{\infty} a_i g_i(x)$ and $a(x) = \frac{1}{1-x-x^2} = \sum_{n=0}^{\infty} F_{n+1}x^n$. Then:

$$G(x) = \sum_{i=0}^{\infty} F_{i+1} \frac{x^i}{(1-x)^{i+1}}$$

$$\sum_{i=0}^{\infty}\sum_{n=0}^{\infty} \binom{n}{i} F_{i+1}x^n = \frac{1}{1-x}\sum_{i=0}^{\infty} F_{i+1}\left(\frac{x}{1-x}\right)^i$$

$$= \frac{1}{1-x} \cdot \frac{1}{1 - \dfrac{x}{1-x} - \dfrac{x^2}{(1-x)^2}}$$

$$= \frac{1-x}{1-3x+x^2} = \sum_{n=0}^{\infty} F_{2n+1}x^n$$

This yields the following well-known result:[†]

$$\sum_{i=0}^{\infty} \binom{n}{i} F_{i+1} = F_{2n+1}$$

That is, $\sum_{i=0}^{n} \binom{n}{i} F_{i+1} = F_{2n+1}$.

For example, $\sum_{i=0}^{4} \binom{4}{i} F_{i+1} = \binom{4}{0}F_1 + \binom{4}{1}F_2 + \binom{4}{2}F_3 + \binom{4}{3}F_4 + \binom{4}{4}F_5 = 34 = F_9$.

On the other hand, suppose we use the column generator $g_i(x) = \frac{x^{2i}}{(1-x)^{i+1}}$. Then $a_i g_i(x) = F_{i+1} g_i(x)$ means the multiplication of F_{i+1} by the corresponding elements of the rising diagonals in Figure 12.2. Then:

$$G(x) = \sum_{i=0}^{\infty} a_i g_i(x) = \sum_{i=0}^{\infty} F_{i+1} \frac{x^{2i}}{(1-x)^{i+1}}$$

$$= \frac{1}{1-x}\sum_{i=0}^{\infty} F_{i+1}\left(\frac{x^2}{1-x}\right)^i = \frac{1}{1-x} \cdot \frac{1}{1 - \frac{x^2}{1-x} - \left(\frac{x^2}{1-x}\right)^2}$$

$$= \frac{1-x}{1-2x+x^3-x^4}$$

[†] See the author's *Fibonacci and Lucas Numbers with Applications* for an alternate proof.

Clearly, choosing different generating functions $a(x)$ will lead to different generating functions $G(x)$. As the next example, we let[†]

$$a(x) = \sum_{n=0}^{\infty} F_{n+1}^2 x^n = \frac{1-x}{1-2x-2x^2+x^3}$$

Then

$$\begin{aligned} G(x) &= \sum_{i=0}^{\infty} a_i g_i(x) \\ &= \sum_{i=0}^{\infty} F_{i+1}^2 \frac{x^{2i}}{(1-x)^{2i+1}} = \frac{1}{1-x}\sum_{i=0}^{\infty} F_{i+1}^2 \left(\frac{x}{1-x}\right)^{2i} \\ &= \frac{1}{1-x} \cdot \frac{1-\frac{1}{1-x}}{1-2\frac{x}{1-x}-2\frac{x^2}{(1-x)^2}+\frac{x^3}{(1-x)^3}} \\ &= \frac{(1-x)(1-2x)}{1-5x-5x^2} \end{aligned}$$

Finally, recall that column $r(\geq 0)$ in the left-justified Pascal's triangle in Figure 12.1 is generated by $g_r(x) = \frac{x^r}{(1-x)^{r+1}}$. Since $\frac{1}{(1-x)^2} = 1 + 2x + 3x^2 + 4x^3 + \cdots$, it follows that $\frac{x}{(1-x^2)^2} = x + 2x^3 + 3x^5 + 4x^7 + \cdots$. This generating function $h_1(x) = \frac{x}{(1-x^2)^2}$ introduces a zero between every two (non-zero) elements in column 1.

In fact, the generating function $h_r(x) = \frac{x^r}{(1-x^2)^{r+1}}$ generates the same left-justified Pascal's triangle, except that there is a zero between every two elements in each column, where $r \geq 0$. See Figure 12.6.

n \ r	0	1	2	3	4	5	6	7	8
0	1								
1	0	1							
2	1	0	1						
3	0	2	0	1					
4	1	0	3	0	1				
5	0	3	0	4	0	1			
6	1	0	6	0	5	0	1		
7	0	4	0	10	0	6	0	1	
8	1	0	10	0	15	0	7	0	1

Figure 12.6

[†] See the author's *Fibonacci and Lucas Numbers with Applications*.

$$\frac{1}{(1-x^2)^{r+1}} = \sum_{m=0}^{\infty} \binom{m}{r} x^{2m-2r}$$

$$\frac{x^r}{(1-x^2)^{r+1}} = \sum_{m=0}^{\infty} \binom{m}{r} x^{2m-r}$$

$$h_r(x) = \sum_{m=0}^{\infty} \binom{m}{r} x^{2m-r}$$

The row sums are generated by $G(x) = \sum_{r=0}^{\infty} h_r(x)$. That is,

$$\begin{aligned}\sum_{r=0}^{\infty}\sum_{m=0}^{\infty} \binom{m}{r} x^{2m-r} &= \frac{1}{1-x^2} \sum_{r=0}^{\infty} \left(\frac{x}{1-x^2}\right)^r \\ &= \frac{1}{1-x^2} \cdot \frac{1}{1-\frac{x}{1-x^2}} = \frac{1}{1-x-x^2} \\ &= \sum_{n=0}^{\infty} F_{n+1} x^n \end{aligned} \tag{12.2}$$

To equate the coefficients of x^{2n} from both sides, notice that r must be even. Let $r = 2k$. Then

$$\sum_{r=0}^{\infty}\sum_{m=0}^{\infty} \binom{m}{r} x^{2m-r} = \sum_{k=0}^{\infty}\sum_{m=0}^{\infty} \binom{m}{2k} x^{2m-2k} = \sum_{k=0}^{\infty}\sum_{n=0}^{\infty} \binom{n+k}{2k} x^{2n}$$

So the coefficient of x^{2n} on the LHS of (12.2) is

$$\sum_{k=0}^{\infty} \binom{n+k}{2k} = \sum_{k=0}^{n} \binom{n+k}{2k}$$

The coefficient of x^{2n} from the RHS of (12.2) is F_{2n+1}. Thus we have

$$\sum_{k=0}^{n} \binom{n+k}{2k} = F_{2n+1} \tag{12.3}$$

That is, the sum of the binomial coefficients on the nth falling southeast diagonal and the even columns in Pascal's triangle in Figure 12.1 is the Fibonacci number F_{2n+1}, as obtained earlier. See Figure 12.7.

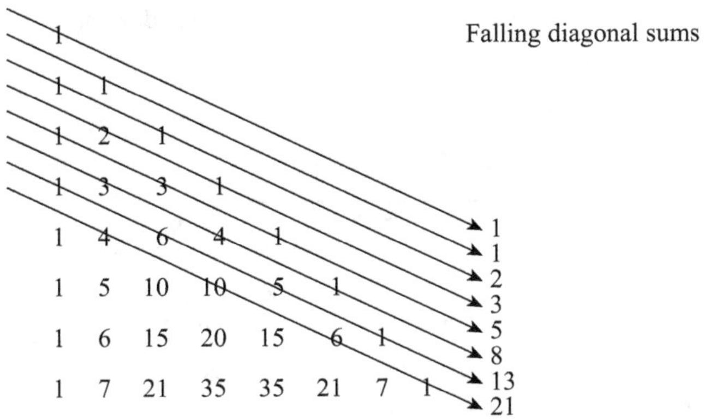

Figure 12.7 Left-justified Pascal's Triangle

For example, $\sum_{k=0}^{4}\binom{4+k}{2k}=\binom{4}{0}+\binom{5}{2}+\binom{6}{4}+\binom{7}{6}+\binom{8}{8}=1+10+15=34=F_9$.

On the other hand, suppose we would like to equate the coefficients of x^{2n-1} from both sides. For this, r must be odd, say, $r=2k+1$, where $k\geq 0$. Then

$$\sum_{r=0}^{\infty}\sum_{m=0}^{\infty}\binom{m}{r}x^{2m-r}=\sum_{k=0}^{\infty}\sum_{m=0}^{\infty}\binom{m}{2k+1}x^{2m-2k-1}=\sum_{k=0}^{\infty}\sum_{n=0}^{\infty}\binom{n+k}{2k+1}x^{2n-1}$$

So the coefficient of x^{2n-1} on the LHS of (12.2) is

$$\sum_{k=0}^{\infty}\binom{n+k}{2k+1}=\sum_{k=0}^{n-1}\binom{n+k}{2k+1}$$

The coefficient of x^{2n-1} on the RHS of (12.2) is F_{2n}. Thus

$$\sum_{k=0}^{n-1}\binom{n+k}{2k+1}=F_{2n} \tag{12.4}$$

For example, $\sum_{k=0}^{4}\binom{5+k}{2k+1}=\binom{5}{1}+\binom{6}{3}+\binom{7}{5}+\binom{8}{7}+\binom{9}{9}=5+20+21+8+1=55=F_{10}$; see Figure 12.7.

Returning to the above generating function $h_r(x)$, we note that it can be modified in an obvious way to construct a left-justified Pascal-like array with $s-1$ consecutive zeros between every two non-zero elements in the Pascal's triangle. It is given by $h_r(x)=\frac{x^r}{(1-x^s)^{r+1}}$, where $r\geq 0$. The non-zero coefficients in $h_r(x)$ are the same as those of $g_r(x)=\frac{x^r}{(1-x)^{r+1}}$, except that there are $s-1$ zeros between any two elements in the new array.

We can in fact modify this generating function $h_r(x)$ to duplicate each non-zero entry s times by multiplying it by $1 + x + x^2 + \cdots + x^{s-1}$:

$$h_r^*(x) = (1 + x + x^2 + \cdots + x^{s-1})h_r(x)$$

$$= \frac{1 + x + x^2 + \cdots + x^{s-1}}{(1 - x^s)^{r+1}}$$

$$= \frac{1 - x^s}{(1 - x)(1 - x^s)^{r+1}} = \frac{1}{(1 - x)(1 - x^s)^r}$$

To properly align the coefficients of x^n for various values of $n(\geq 0)$, we need the generating function $g_r^*(x) = \frac{x^{rs}}{(1-x)(1-x^s)^r}$.

For example, let $s = 2$. Then $g_r^*(x) = \frac{x^{2r}}{(1-x)(1-x^2)^r}$. Figure 12.8 shows the corresponding array, where each element appears twice in each column

1								
1	1							
1	1	1						
1	2	1	1					
1	2	3	1	1				
1	3	3	4	1	1			
1	3	6	4	5	1	1		
1	4	6	10	5	6	1	1	
1	4	10	10	15	6	7	1	1

Figure 12.8

13

Pascal-like Triangles

This chapter investigates two triangular arrays which occur in the study of races with ties.

Races with Ties

E. Mendelson of Queens College, Flushing, New York, in his study of races with ties, encountered two interesting Pascal-like triangles. Before we present them, let us consider a race with n runners. If no ties are allowed, then there are $n!$ possible outcomes. But this is unrealistic, so assume that the runners are allowed to tie for arbitrarily many positions. Let J_n denote the total number of such outcomes.

For example, $J_0 = 1 = J_1$. With two runners A and B, there are three possible outcomes: (A,B), (B,A), or AB (tie); with three runners A, B, and C, there are 13 possibilities: (A,B,C), (A,C,B), (B,A,C), (B,C,A), (C,A,B), (C,B,A), (AB,C), (AC,B), (BC,A) (two tied for the first place), (A,BC), (B,CA), (C,AB) (two tied for the second), and ABC (all tied for first). Thus $J_2 = 3$ and $J_3 = 13$.

More generally, suppose there are n runners and r of them do not finish first. Then those r runners can finish 2nd, 3rd, $\cdots$ places in J_r ways and they can be chosen from the n runners in $\binom{n}{r}$ different ways. So the total number of possible outcomes is $\binom{n}{r}J_r$. Since $0 \le r < n$, by the addition principle, $J_n = \sum_{r=0}^{n-1} \binom{n}{r}J_r$.

Thus, J_n can be defined recursively:

$$J_0 = 1$$

$$J_n = \sum_{r=0}^{n-1} \binom{n}{r} J_r, \quad n \geq 1 \tag{13.1}$$

For example, $J_4 = \sum_{r=0}^{4} \binom{n}{r} J_r = J_0 + 4J_1 + 6J_2 + 4J_3 = 1 + 4 + 18 + 52 = 75$.

Table 13.1 lists the values of J_n for $1 \leq n \leq 12$. The ones digits in the J_n's seem to follow the cyclic pattern 1–3–3–5. This was proved to be the case in 1962 by O. A. Gross of the Rand Corporation, Santa Monica, California, who established that $J_{n+4} \equiv J_n \pmod{10}$.

Table 13.1

n	J_n	n	J_n
1	1	7	47293
2	3	8	545835
3	13	9	7087261
4	75	10	102247563
5	541	11	1622632573
6	4683	12	28091567595

Arthur Cayley showed that J_n counts the number of trees of a special type and Gross showed that it occurs in the study of preferential arrangements.

Let $J(n, r)$ denote the number of possible outcomes of a race with n runners and r positions. Then it can be shown that $J(n, r) = r[J(n-1, r-1) + J(n-1, r)]$, where $J(n, 1) = 1$. Consequently, $J_n = \sum_{r=1}^{n} J(n, r)$.

The various values of $J(n, r)$ can be used to construct a Pascal-like triangle, as in Table 13.2. Each element $J(n, r)$ is obtained by adding $J(n-1, r-1)$ and $J(n-1, r)$ in the previous row, and then multiplying the sum by the column label r, where $n \geq 2$; see Table 13.2.

It follows from the recurrence relation and induction that $J(n, n) = n!$.

Mendelson also showed that

$$J_n = \sum_{r=0}^{n-1} (-1)^r C(n, r+1)(n-r)^n \tag{13.2}$$

where

$$C(n, j) = \sum_{k=0}^{j-1} (-1)^{k+j-1} \binom{k+n-j+1}{k}$$

Table 13.2

n \ r	1	2	3	4	5	6	7	J_n
1	1							1
2	1	2						3
3	1	6	6					13
4	1	14	36	24				75
5	1	30	150 —	240 ↓	120			541
6	1	62	540	1560	1800	720		4683
7	1	126	1806	8400	16800	15120	5040	47293

The values of $C(n, r)$ can be used to construct the Pascal-like array in Table 13.3. It satisfies Pascal's identity. The main diagonal of the array consists of pairs of the same positive integer in increasing order; namely, $C(n, n) = \lceil n/2 \rceil$.

Table 13.3

n \ j	1	2	3	4	5	6	7	8
1	1							
2	1	1						
3	1	2	2					
4	1	3	4	2				
5	1	4	7	6	3			
6	1	5	11 —	13 ↓	9	3		
7	1	6	16	24	22	12	4	
8	1	7	22	40	46	34	16	4

Using formula (13.2), $J_4 = \sum_{r=0}^{4}(-1)^r C(4, r+1)(4-r)^4 = 1 \cdot 4^4 - 3 \cdot 3^4 + 4 \cdot 2^4 - 2 \cdot 1^4 = 75$, as expected.

14

Fibonacci Triangles

Over the years, a number of Pascal-like triangular arrays have been developed that are related to Fibonacci numbers in a variety of ways. Many of them are creations of Vernon E. Hoggatt, Jr. (1921–1980) of then San Jose State College.

1						
1	1					
2	2	1				
5	4	3	1			
13	9	7	4	1		
34	22	16	11	5	1	
89	56	38	27	16	6	1

Figure 14.1

n \ j	0	1	2	3	4	5	6	Row sums
0	1							1
1	1	1						2
2	2	2	1					5
3	5	4	3	1				13
4	13	9	7 —	4 ↓	1			34
5	34	22	16	11	5	1		89
6	89	56	38	27	16	6	1	233

Figure 14.2

The Fibonacci array in Figure 14.1 was investigated by Hoggatt in 1971. It has three obvious properties:

- The apex element in row 0 is $1 = F_{-1}$.
- Every element in column 0 is an odd-numbered Fibonacci number.
- Every internal element, including the last entry in each row, can be obtained by adding its northern neighbor and this neighbor's western neighbor; see Figure 14.2. In other words, the array satisfies the Pascal-like relationship.

This Pascal-like behavior can be used to define the array H recursively:

$$H(n,0) = F_{2n-1}$$
$$H(n,j) = H(n-1,j-1) + H(n-1,j), \quad 0 \le j \le n$$

A quick inspection of the row sums reveals an interesting pattern: The nth row sum is F_{2n+1}; that is,

$$\sum_{j=0}^{n} H(n,j) = F_{2n+1} = H(n+1,0) \tag{14.1}$$

For example, $\sum_{j=0}^{5} H(5,j) = 34 + 22 + 16 + 11 + 5 + 1 = 89 = F_{11}$.

To establish formula (14.1), assume it is true for every nonnegative integer $\le n$. Then:

$$\sum_{j=0}^{n} H(n,j) = F_{2n+1}$$

$$H(n,0) + \sum_{j=1}^{n} H(n,j) = F_{2n-1} + F_{2n}$$

$$F_{2n-1} + \sum_{j=1}^{n} H(n,j) = F_{2n-1} + F_{2n}$$

That is, $\sum_{j=1}^{n} H(n,j) = F_{2n}$. Then:

$$\begin{aligned}
\sum_{j=0}^{n+1} H(n+1,j) &= H(n+1,0) + \sum_{j=1}^{n+1} H(n+1,j) \\
&= F_{2n+1} + \sum_{j=1}^{n+1} [H(n,j-1) + H(n,j)] \\
&= F_{2n+1} + \sum_{j=0}^{n} H(n,j) + \sum_{j=1}^{n} H(n,j) \\
&= F_{2n+1} + F_{2n+1} + F_{2n} \\
&= F_{2n+1} + F_{2n+2} = F_{2n+3}
\end{aligned}$$

Since $H(0,0) = 1 = F_1$, it follows by strong induction that formula (14.1) holds for every nonnegative integer n.

In 1971, Hoggatt established that $\frac{1-x}{1-3x+x^2} = \sum_{n=0}^{\infty} F_{2n+1}x^n$. Consequently, the various row sums are generated by the function $f(x) = \frac{1-x}{1-3x+x^2}$. He also showed that the various rising diagonal sums D_n can be generated by the function

$$\frac{(1-x)^3}{(1-x-x^2)(1-3x+x^2)} = \sum_{n=0}^{\infty} D_n x^n$$

Additional Fibonacci Triangles

The array in Figure 14.3 was investigated by Hoggatt in 1972 and studied two years later by P. S. Bruckman of the University of Illinois, Urbana, Illinois. It is interesting in its own right. The leftmost column (column 0) consists of the Fibonacci numbers F_{2n-2} and each of the remaining columns j is obtained by lowering the previous column $j-1$ by one level, where $n, j \geq 1$.

n \ j	0	1	2	3	4	5	6
0	1						
1	1	1					
2	3	1	1				
3	8	3	1	1			
4	21	8	3	1	1		
5	55	21	8	3	1	1	
6	144	55	21	8	3	1	1

Figure 14.3

n \ j	0	1	2	3	4	5	6	Row sums
0	1							1
1	1	1						2
2	3	1	1					5
3	8	3	1	1				13
4	21	8	3	1	1			34
5	55	21	8	3	1	1		89
6	144	55	21	8	3	1	1	233

Figure 14.4

This array satisfies three intriguing properties; they are about row sums, weighted row sums, and rising diagonal sums.

- First, notice that the row sums in Figure 14.4 follow an interesting pattern. It appears that the nth row sum is F_{2n+1}, where $n \geq 0$. This, in fact, can be established fairly easily.

  Let R_n denote the row sum of the Fibonacci numbers in row n, where $R_0 = 1$. Let $n \geq 1$. Then:

$$R_n = 1 + \sum_{j=0}^{n-1} F_{2n-2j} = 1 + \sum_{j=1}^{n} F_{2j}$$

$$= 1 + \sum_{j=1}^{n-1} \left(F_{2k+1} - F_{2k-1}\right) = 1 + (F_{2n+1} - F_1)$$

$$= 1 + (F_{2n+1} - 1) = F_{2n+1}$$

where we have used the fact[†] that

$$\sum_{j=1}^{n} F_{2j} = F_{2n+1} - 1 \tag{14.2}$$

Notice that the formula for R_n works even if $n = 0$.

For example, $R_5 = 55 + 21 + 8 + 3 + 1 + 1 = 89 = F_{11}$.

- Secondly, suppose we multiply column j by $j + 1$, where $j \geq 0$. Figure 14.5 shows the resulting array and the corresponding row sums. Interestingly, the computed row sums in the figure are Fibonacci numbers with even subscripts. So, we conjecture that the sum S_n in row n equals F_{2n+2}, where $n \geq 0$.

n \ j	0	1	2	3	4	5	6	Row sums
0	1							1
1	1	2						3
2	3	2	3					8
3	8	6	3	4				21
4	21	16	9	4	5			55
5	55	42	24	12	5	6		144
6	144	110	63	32	15	6	7	377

Figure 14.5

Using Lucas identity (14.2) and the fact that $\sum_{j=1}^{n} jF_{2j} = nF_{2n+1} - F_{2n}$, this can be confirmed as follows:

$$S_n = \sum_{j=1}^{n} jF_{2n+2-2j} + (n+1) = \sum_{k=1}^{n} (n-k+1)F_{2k} + (n+1)$$

$$= (n+1)\sum_{k=1}^{n} F_{2k} - \sum_{k=1}^{n} kF_{2k} + (n+1)$$

$$= (n+1)(F_{2n+1} - 1) - (nF_{2n+1} - F_{2n}) + (n+1)$$

$$= (n+1)F_{2n+1} - (n+1) - nF_{2n+1} + F_{2n} + (n+1)$$

$$= F_{2n+1} + F_{2n} = F_{2n+2}, \quad \text{as desired.}$$

For example, $S_5 = 55 + 42 + 24 + 12 + 5 + 6 = 144 = F_{12}$.

[†]This formula was established in 1876 by Lucas. See the author's *Fibonacci and Lucas Numbers with Applications*.

- Finally, we investigate the various diagonal sums of the array in Figure 14.3. Using Figure 14.6, we conjecture that each rising diagonal sum is a Fibonacci square. More specifically, we conjecture that the nth rising diagonal sum is F_{n+1}^2, where $n \geq 0$.

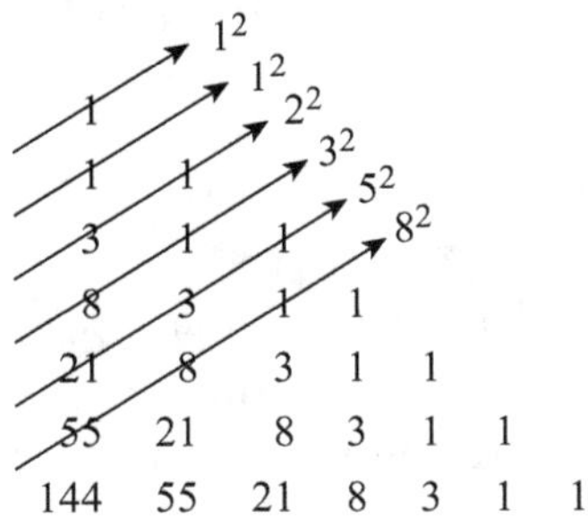

Figure 14.6

To establish this formula, we will employ the identity $F_n^2 + F_{n+1}^2 = F_{2n+1}$, where $n \geq 0$.

Let D_n denote the nth rising diagonal sum, where $n \geq 0$. Clearly, $D_0 = 1 = F_{0+1}^2$ and $D_1 = 1 = F_{1+1}^2$.

Let $n \geq 2$. Then

$$D_n = \begin{cases} \sum_{j=1}^{n/2} F_{4j} + 1 & \text{if } n \text{ is even} \\ \sum_{j=1}^{(n+1)/2} F_{4j-2} & \text{otherwise} \end{cases}$$

Then

$$D_{2m} = \sum_{j=1}^{m} F_{4j} + 1$$

$$= F_1 + \sum_{j=1}^{m} \left(F_{4j+1} - F_{4j-1}\right) = \sum_{i=0}^{2m} (-1)^i F_{2i+1}$$

and

$$D_{2m+1} = \sum_{j=1}^{m+1} F_{4j-2}$$

$$= \sum_{j=1}^{m+1} \left(F_{4j-1} - F_{4j-3}\right) = \sum_{i=0}^{2m+1} (-1)^{i+1} F_{2i+1}$$

Combining these two cases, we have:

$$D_n = \sum_{i=0}^{n}(-1)^{n-i}F_{2i+1} = \sum_{i=0}^{n}(-1)^{n-i}(F_{i+1}^2 + F_i^2)$$

$$= \sum_{i=0}^{n}\left[(-1)^{n-i}F_{i+1}^2 - (-1)^{n-i+1}F_i^2\right] = (-1)^{n-n}F_{n+1}^2 - (-1)^{n-0+1}F_0^2$$

$$= F_{n+1}^2 - 0 = F_{n+1}^2$$

This formula holds for $n = 0$ and $n = 1$ also.

For example, $S_6 = 144 + 21 + 3 + 1 = 169 = F_7^2$.

The array in Figure 14.7 was created by Hoggatt in 1977. Each column consists of the odd-numbered Fibonacci numbers F_{2n+1} in a staggered format, where $n \geq 0$.

1						
2	1					
5	2	1				
13	5	2	1			
34	13	5	2	1		
89	34	13	5	2	1	
233	89	34	13	5	2	1

Figure 14.7

n \ j	0	1	2	3	4	5	6	Row sums
0	1							1
1	2	1						3
2	5	2	1					8
3	13	5	2	1				21
4	34	13	5	2	1			55
5	89	34	13	5	2	1		144
6	233	89	34	13	5	2	1	377

Figure 14.8

Here also the row sums follow a nice pattern; see Figure 14.8. They are the Fibonacci numbers F_{2n+2}, where $n \geq 0$. This follows from the formula $\sum_{j=1}^{n} F_{2j-1} = F_{2n}$, established in 1876 by Lucas.

Using formula (14.2), it follows that the cumulative sum S of rows 0 through n is given by

$$S = \sum_{i=0}^{n} F_{2i+2} = \sum_{i=1}^{n} F_{2i} = F_{2n+3} - 1$$

For example, the cumulative sum of the sums in rows 0 through 5 is given by

$$S = F_{2\cdot 5+3} - 1 = F_{13} - 1 = 232 = 1 + 3 + 8 + 21 + 55 + 144$$

The sums along the rising diagonals also follow a pattern; see Figure 14.9. They are the products $F_{n+1}F_{n+2}$ of consecutive Fibonacci numbers, where $n \geq 0$.

To establish this, let $n = 2k + 1$. Then the rising diagonal sum is $\sum_{i=0}^{k} F_{4i+3}$. Using PMI, this sum can be shown to be $F_{2k+2}F_{2k+3} = F_{n+1}F_{n+2}$.

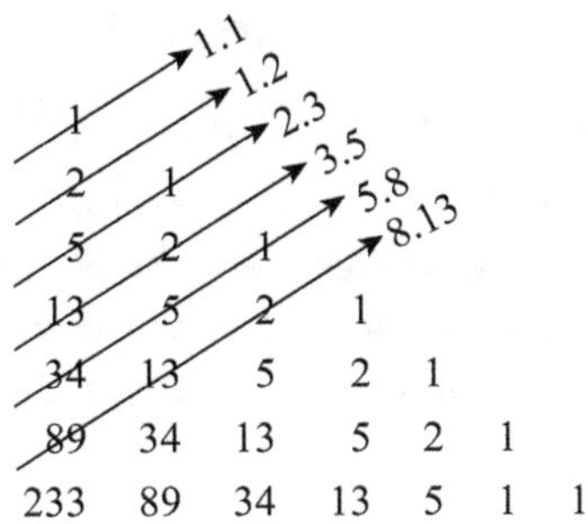

Figure 14.9

On the other hand, let $n = 2k$. Then the diagonal sum equals $\sum_{i=0}^{k} F_{4i+1} = F_{2k+1}F_{2k+2} = F_{n+1}F_{n+2}$.

Thus, in both cases, the nth rising diagonal sum equals $F_{n+1}F_{n+2}$, where $n \geq 0$.

Next, we multiply the columns in Figure 14.7 by $1, 2, 3, \cdots$ to the right. What can we say about the resulting row sums? It appears from Figure 14.10 that each row sum is one less than a Fibonacci number. So, we conjecture that the nth row sum is $F_{2n+3} - 1$, where $n \geq 0$.

n \ j	0	1	2	3	4	5	6	Row sums
0	1							1 = 2 – 1
1	2	2						4 = 5 – 1
2	5	4	3					12 = 13 – 1
3	13	10	6	4				33 = 34 – 1
4	34	26	15	8	5			88 = 89 – 1
5	89	68	39	20	10	6		232 = 233 – 1
6	233	178	102	52	25	12	7	609 = 610 – 1

Figure 14.10

To confirm this, it suffices to show that $\sum_{j=0}^{n}(n+1-j)F_{2j+1} = F_{2n+3} - 1$, as G. Brezsenyi of Lamar University, Beaumont, Texas, did in 1977.

When $n = 0$, the result is true. Now assume it is true for $n = k$. Then:

$$\sum_{j=0}^{k+1}[(k+1)+1-j]F_{2j+1} = \sum_{j=0}^{k}(k+2-j)F_{2j+1} + F_{2k+3}$$

$$= \sum_{j=0}^{k}(k+1-j)F_{2j+1} + \sum_{j=0}^{k}F_{2j+1} + F_{2k+3}$$

$$= (F_{2k+3} - 1) + (F_{2k+2} + F_{2k+3}) = F_{2k+3} - 1 + F_{2k+4}$$

$$= F_{2k+5} - 1 = (F_{2(k+1)+3} - 1)$$

Thus, by PMI, the formula works for every integer $n \geq 0$.

The array in Figure 14.11 was investigated by Hoggatt in 1972. The various row sums manifest an interesting pattern: Every row sum is a Fibonacci number. Except for row 1, the nth row sum is F_{2n}, where $n \geq 2$.

n \ j	1	2	3	4	5	6	7	8	Row sums
1	1	1							2
2	1	2							3
3	2	4	1	1					8
4	5	9	3	4					21
5	13	22	7	11	1	1			55
6	34	56	16	27	5	6			144
7	89	145	38	65	16	22	1	1	377

Figure 14.11

To establish this, we will need the generating function for Fibonacci numbers with odd subscripts:[†]

$$\frac{x - x^2}{1 - 3x + x^2} = \sum_{n=1}^{\infty} F_{2n-1} x^n$$

and follow the technique employed by Carlitz in 1973. We omit the proof in the interest of brevity.

[†] See the author's *Fibonacci and Lucas Numbers with Applications.*

15

Josef's Triangle

In 1981, S. Josef of Technical University at Brno, Czech Republic, developed a triangular array based on the same technique used in the construction of Hosoya's triangle. Although he called it a *Lucas triangle,* we will call it *Josef's triangle* for obvious reasons. Unlike Hosoya's triangle, which is generated by Fibonacci numbers, Josef's triangle is generated by using both Fibonacci and Lucas numbers. The top northeast and southeast diagonals consist of Lucas numbers, and the next northeast and southeast diagonals consist of Fibonacci numbers. Since both Fibonacci and Lucas numbers satisfy the same recurrence relation, each remaining element in the array is obtained by adding the two preceding elements in the same northeast (or southeast) diagonal. See Figure 15.1.

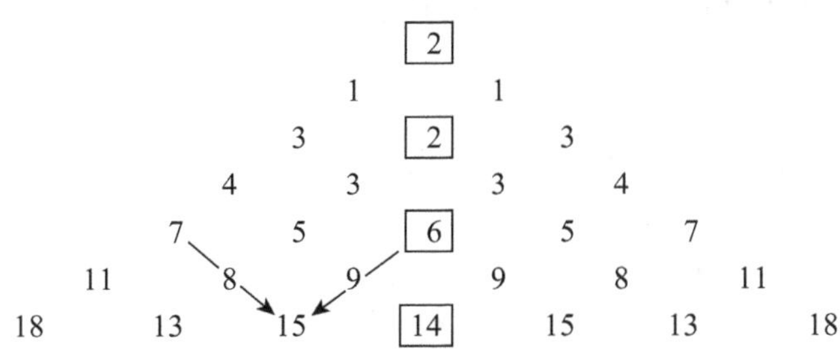

Figure 15.1 Josef's Triangle

Accordingly, Josef's array can be defined recursively, as follows:

$$g(0,0) = 2, \qquad g(1,0) = 1 = g(1,1), \qquad g(2,1) = 2$$

$$g(n,j) = g(n-1,j) + g(n-2,j), \qquad n \geq 2, \ \ n \geq j \geq 0 \tag{15.1}$$

$$= g(n-1,j-1) + g(n-2,j-2), \qquad n \geq j \geq 2 \tag{15.2}$$

For example, $g(8,4) = g(7,4) + g(6,4) = 23 + 15 = 38 = g(7,3) + g(6,2)$.

Since $g(1,0) = 1, g(2,0) = g(1,0) + g(0,0) = 1 + 2 = 3$, and $g(n,0) = g(n-1,0) + g(n-2,0)$, it follows that $g(n,0) = L_n$, where $n \geq 0$. Likewise, $g(n,n) = L_n, g(n,1) = F_{n+1} = g(n,n-1)$ for every $n \geq 1$.

Clearly, the array is symmetric about the vertical line through the apex; that is, $g(n,j) = g(n,n-j)$. Also, row n contains $n+1$ elements. If n is even, then row n contains exactly one central element; otherwise, there are two middle elements. The various central elements are boxed in Figure 15.1.

There is an interesting relationship between the numbers $g(n,j)$ and Fibonacci numbers:

$$g(n,j) = F_{k+1} \cdot g(n-k,j) + F_k \cdot g(n-k-1,j) \tag{15.3}$$

where $1 \leq k \leq n-j-1$. This can indeed be confirmed by induction.

Likewise, we also have:

$$g(n,j) = F_{k+1} \cdot g(n-k,j-k) + F_k \cdot g(n-k-1,j-k-1) \tag{15.4}$$

where $1 \leq k \leq n-1$.

It follows by properties (15.3) and (15.4) that each element in the array is a linear combination of two consecutive elements in its northeast (or southeast) diagonal.

Suppose we let $k = n-j-1$ in formula (15.3). Then it becomes:

$$\begin{aligned} g(n,j) &= F_{n-j} \cdot g(j+1,j) + F_{n-j-1} \cdot g(j,j) \\ &= F_{n-j}F_{j+2} + F_{n-j-1}L_j \qquad (15.5) \end{aligned}$$

Central Elements J_n in Josef's Triangle

Using formula (15.5) and the fact that $L_m = F_{m-1} + F_{m+1}$, we have:

$$\begin{aligned} J_n &= g(2n,n) \\ &= F_nF_{n+2} + F_{n-1}L_n \qquad (15.6) \\ &= F_n(F_n + F_{n+1}) + F_{n-1}(F_{n+1} + F_{n-1}) \\ &= F_n^2 + F_{n+1}(F_n + F_{n-1}) + F_{n-1}^2 \\ &= F_{n-1}^2 + F_n^2 + F_{n+1}^2 \qquad (15.7) \end{aligned}$$

Thus, every central element in Josef's triangle is the sum of the squares of three consecutive Fibonacci numbers; see the boxed numbers in Figure 15.1.

As a byproduct, since $F_{n+1}^2 \equiv F_n^2 + F_{n-1}^2 \pmod 2$, it follows that J_n is even for every integer $n \geq 0$. It follows from equation (15.7) that no central element is a square. This is so since the sequence $F_{n-1}^2 + F_n^2 + F_{n+1}^2 \pmod 9$ is periodic with period 12, and the only distinct elements are 2, 5, 6, and 8. But the square of every integer is congruent to 0, 1, 4, or 7 modulo 9.

In fact, formula (15.6) can be generalized to include the two middle elements in row n, where n is odd. To this end, let $j = \lfloor n/2 \rfloor$ in formula (15.5). Then it yields:

$$g(n, \lfloor n/2 \rfloor) = F_{n-\lfloor n/2 \rfloor}F_{\lfloor n/2 \rfloor+2} + F_{n-\lfloor n/2 \rfloor-1}L_{\lfloor n/2 \rfloor} \tag{15.8}$$

Notice that formula (15.7) is a special case of (15.8). So the two middle elements in row n, where n is odd, are given by

$$\begin{aligned} g(n, \lfloor n/2 \rfloor) &= g(n, n - \lfloor n/2 \rfloor) \\ &= F_{\lfloor n/2 \rfloor}F_{n-\lfloor n/2 \rfloor+2} + F_{\lfloor n/2 \rfloor-1}L_{n-\lfloor n/2 \rfloor} \end{aligned} \tag{15.9}$$

Formula (15.9) gives an alternate way of computing the middle elements when n is odd.

Formula (15.5), coupled with Binet's formulas for Fibonacci and Lucas numbers, can be used to develop yet another explicit formula for $g(n,j)$. We have:

$$\begin{aligned} g(n,j) &= F_{n-j}F_{j+2} + F_{n-j-1}L_j \\ &= \frac{\alpha^{n-j} - \beta^{n-j}}{\sqrt{5}} \cdot \frac{\alpha^{j+2} - \beta^{j+2}}{\sqrt{5}} + \frac{\alpha^{n-j-1} - \beta^{n-j-1}}{\sqrt{5}} \cdot (\alpha^j + \beta^j) \\ &= \frac{(\alpha^{n+2} + \beta^{n+2}) - (\alpha\beta)^{j+2}(\alpha^{n-2j-2} + \beta^{n-2j-2})}{5} + \\ &\quad \frac{(\alpha^{n-1} - \beta^{n-1}) + (\alpha\beta)^j(\alpha^{n-2j-1} - \beta^{n-2j-1})}{\sqrt{5}} \\ &= \frac{L_{n+2} - (-1)^j L_{n-2j-2}}{5} + F_{n-1} + (-1)^j F_{n-2j-1} \end{aligned} \tag{15.10}$$

As a bonus, since $g(n,j)$ is an integer, it follows from formula (15.10) that $L_{n+2} \equiv (-1)^j L_{n-2j-2} \pmod{5}$; that is, $L_n \equiv (-1)^j L_{n-2j-4} \pmod{5}$. Consequently, $L_n \equiv (-1)^{\lfloor n/2 \rfloor} L_{n-2\lfloor n/2 \rfloor} \pmod{5}$.

In particular, $L_{2n} \equiv 2(-1)^n \pmod{5}$ and $L_{2n+1} \equiv (-1)^n \pmod{5}$. It follows from these two congruences that every Lucas number is congruent to 1, 2, 3, or 4 modulo 5. Consequently, *no* Lucas number is divisible by 5.

In fact, since $L_1 = 1, L_2 = 3, L_3 = 4$, and $L_4 = 7 \equiv 2 \pmod{5}$, it follows by the Lucas recurrence relation that the least positive residues of Lucas numbers modulo 5 follow the cyclic pattern 1–3–4–2.

An Alternate Formula for J_n

We can use formula (15.10) to develop an alternate formula for the central element J_n. Since $L_{-2} = 3$ and $F_{-1} = 1$, we have:

$$\begin{aligned} J_n &= g(2n, n) \\ &= \frac{L_{2n+2} - (-1)^n L_{-2}}{5} + F_{2n-1} + (-1)^n F_{-1} \\ &= \frac{L_{2n+2} - 3(-1)^n}{5} + F_{2n-1} + (-1)^n \end{aligned} \tag{15.11}$$

Since $L_{m-1} + L_{m+1} = 5F_m$, it follows from properties (15.7) and (15.11) that

$$\frac{L_{2n+2} - 3(-1)^n}{5} + F_{2n-1} + (-1)^n = F_{n-1}^2 + F_n^2 + F_{n+1}^2$$

$$L_{2n+2} + 5F_{2n-1} + 2(-1)^n = 5(F_{n-1}^2 + F_n^2 + F_{n+1}^2)$$

$$L_{2n+2} + (L_{2n-2} + L_{2n}) + 2(-1)^n = 5(F_{n-1}^2 + F_n^2 + F_{n+1}^2)$$

That is,

$$L_{2n-2} + L_{2n} + L_{2n+2} = 5(F_{n-1}^2 + F_n^2 + F_{n+1}^2) - 2(-1)^n \tag{15.12}$$

Consequently, $L_{2n-2} + L_{2n} + L_{2n+2} \equiv \pm 2 \pmod 5$.

A Recursive Formula for J_n

Interestingly, formulas (15.1) and (15.5), and that $F_{2n} = F_n L_n$ and $F_{n+1} + F_{n-1} = L_n$ can be used to generate a recurrence relation for J_n; see Figure 15.2.

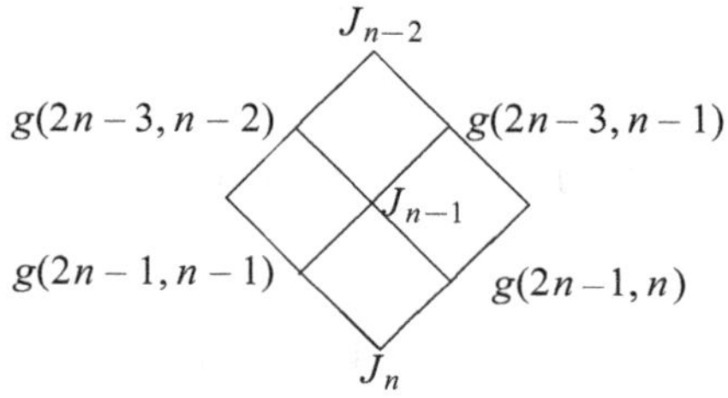

Figure 15.2

$$\begin{aligned}
J_n &= g(2n, n) \\
&= g(2n-1, n-1) + g(2n-2, n-2) \\
&= g(2n-1, n-1) + g(2n-3, n-2) + g(2n-4, n-2) \\
&= g(2n-1, n-1) + g(2n-3, n-2) + J_{n-2} \\
&= (F_nF_{n+1} + F_{n-1}L_{n-1}) + (F_{n-1}F_n + F_{n-2}L_{n-2}) + J_{n-2} \\
&= F_n(F_{n+1} + F_{n-1}) + F_{2n-2} + F_{2n-4} + J_{n-2} \\
&= F_nL_n + F_{2n-2} + F_{2n-4} + J_{n-2} \\
&= F_{2n} + F_{2n-2} + F_{2n-4} + J_{n-2}
\end{aligned} \tag{15.13}$$

Consequently, the sum of the southwest and northwest neighbors of J_{n-1} is $F_{2n} + F_{2n-2} + F_{2n-4}$, where $n \geq 2$.

It follows from equations (15.7) and (15.13) that

$$\begin{aligned}
F_{2n} - F_{2n-2} - F_{2n-4} + J_{n-2} &= F_{n-1}^2 + F_n^2 + F_{n+1}^2 \\
J_{n-2} &= F_{n-1}^2 + F_n^2 + F_{n+1}^2 - F_{2n} - F_{2n-2} - F_{2n-4}
\end{aligned}$$

This yields

$$J_n = F_{n+1}^2 + F_{n+2}^2 + F_{n+3}^2 - F_{2n+4} - F_{2n+2} - F_{2n} \tag{15.14}$$

For example, $J_4 = F_5^2 + F_6^2 + F_7^2 - F_{12} - F_{10} - F_8 = 5^2 + 8^2 + 13^2 - 144 - 55 - 21 = 38$.

Next, we turn to four addition formulas satisfied by $g(n, j)$.

Addition Formulas

Using the recurrence relations (15.1) and (15.2), we have:

$$\begin{aligned}
g(n+2, j+1) &= g(n+1, j) + g(n, j) \\
&= [g(n, j) + g(n-1, j)] + [g(n-1, j-1) + g(n-2, j-1)] \\
&= g(n, j) + g(n-1, j) + g(n-1, j-1) + g(n-2, j-1)
\end{aligned} \tag{15.15}$$

That is, every element $g(n+2, j+1)$ in the array can be obtained by adding the vertices of the rhombus with vertices $g(n, j)$, $g(n-1, j)$, $g(n-1, j-1)$, and $g(n-2, j-1)$; see Figures 15.3 and 15.4.

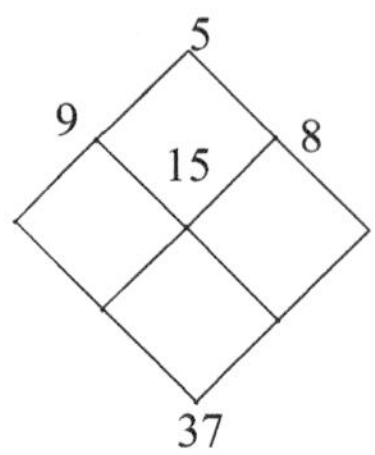

Figure 15.3

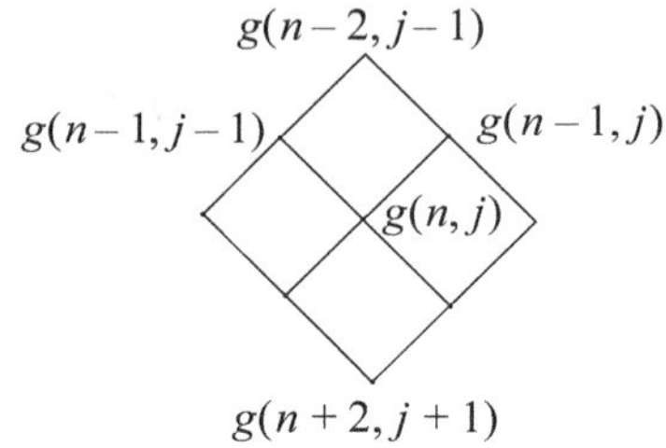

Figure 15.4

Similarly, we have:

$$g(n-4,j-2) = g(n,j) - g(n-1,j) - g(n-1,j-1) + g(n-2,j-1)$$

$$g(n-1,j-2) = g(n,j) + g(n-1,j) - g(n-1,j-1) - g(n-2,j-1)$$

$$g(n-1,j+1) = g(n,j) - g(n-1,j) + g(n-1,j-1) - g(n-2,j-1)$$

Returning to the rhombus in Figure 15.4, consider the triangle with vertices $g(n,j), g(n-1,j)$, and $g(n-1,j-1)$. Keeping n fixed, the sum of the three numbers remains the same for every value of j; that is,

$$g(n,j) + g(n-1,j) + g(n-1,j-1) = g(n,k) + g(n-1,k) + g(n-1,k-1) \tag{15.16}$$

for every $k \geq 1$, where $1 \leq j \leq n-1$. See Figure 15.5.

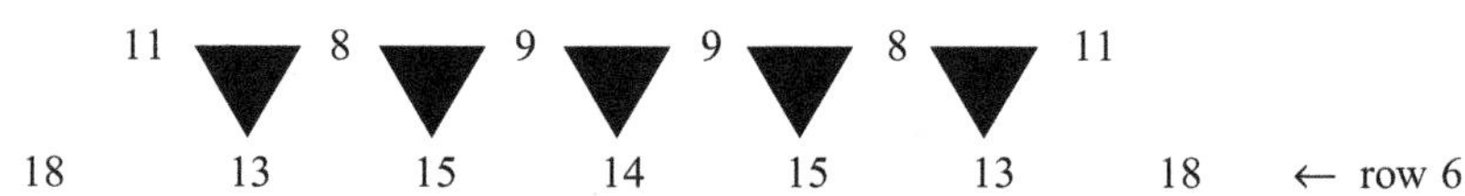

Figure 15.5 Each Triangular Sum Equals 32

In Figure 15.5, notice that $32 = 11 + 21 = L_5 + F_8$. More generally, the triangular sum (15.16) is always the sum of a Lucas number and a Fibonacci number, namely, L_{n-1} and F_{n+2}:

$$g(n,j) + g(n-1,j) + g(n-1,j-1) = L_{n-1} + F_{n+2} \tag{15.17}$$

where $1 \leq j \leq n-1$. See Figure 15.6.

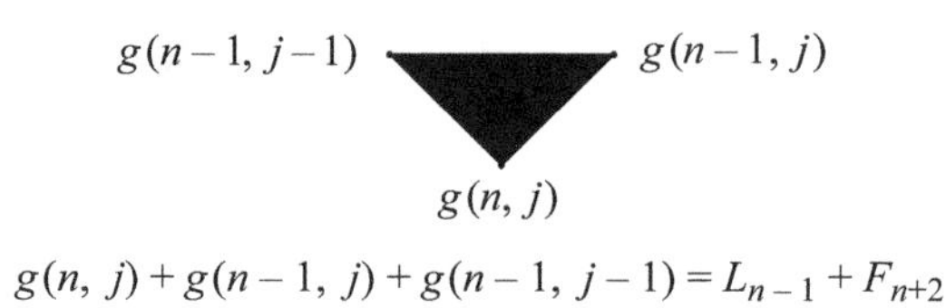

Figure 15.6

Formula (15.17) can be established using strong induction. When $n = 2$:

$$\text{LHS} = g(2, 1) + g(1, 1) + g(1, 0) = 2 + 1 + 1 = 1 + 3 = L_1 + F_4 = \text{RHS}$$

Now, assume that the formula holds for all positive integers $k \geq 2$:

$$g(k, j) + g(k-1, j) + g(k-1, j-1) = L_{k-1} + F_{k+2}$$

where $1 \leq j \leq k - 1$. Then

$$\begin{aligned} g(k+1,j)+g(k,j)+g(k,j-1) &= [g(k,j)+g(k-1,j)]+[g(k-1,j)+g(k-2,j)] \\ &\quad +[g(k-1,j-1)+g(k-2,j-1)] \\ &= [g(k,j)+g(k-1,j)+g(k-1,j-1)] \\ &\quad +[g(k-1,j)+g(k-2,j)+g(k-2,j-1)] \\ &= (L_{k-1}+F_{k+2})+(L_{k-2}+F_{k+1}) = L_k+F_{k+3} \end{aligned}$$

Thus, by strong induction, formula (15.17) is true for every $n \geq 2$, where $1 \leq j \leq n - 1$.

Finally, consider Figure 15.7. The sum of the numbers on the base minus the one at the apex is four times a Fibonacci number: $g(n, j) + g(n, j-1) - g(n-1, j-1) = 4F_{n-1}$, where $1 \leq j \leq n - 1$. This also can be confirmed using strong induction.

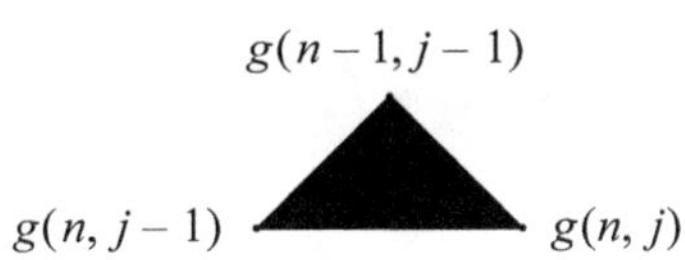

$$g(n, j) + g(n, j-1) - g(n-1, j-1) = 4F_{n-1}$$

Figure 15.7

For example, $g(7, 3) + g(7, 2) - g(6, 2) = 23 + 24 - 15 = 32 = 4 \cdot 8 = 4F_6$, as expected.

16

Lucas Triangles

A number of triangular arrays are made up of Lucas numbers or lead to Lucas numbers. Like the Fibonacci arrays in Chapter 14, many of them are the brain children of Hoggatt, either directly or indirectly. We call them *Lucas triangles* for obvious reasons.

Recall from Chapter 12 that Fibonacci numbers can be generated from the rising diagonals of Pascal's triangle. This motivated Mark Feinberg, then a student at the University of Pennsylvania, to develop a similar triangular array that would generate the various Lucas numbers from its rising diagonals. In 1967, four years after publishing his work on tribonacci numbers, Feinberg developed such an invaluable treasure by investigating the coefficients in the expansion of the polynomial $f_n(x, y) = (x + y)^{n-1}(x + 2y)$.

n \ j	0	1	2	3	4	5	6	7
1	1	2						
2	1	3	2					
3	1	4	5	2				
4	1	5	9	7	2			
5	1	6	14	16	9	2		
6	1	7	20	30	25	11	2	
7	1	8	27	50	55	36	13	2

Figure 16.1

The first five such polynomials are:

$f_1(x,y)=x+2y$ $\qquad$ $f_4(x,y)=x^4+5x^3y+9x^2y^2+7xy^3+2y^4$

$f_2(x,y)=x^2+3xy+2y^2$ $\qquad$ $f_5(x,y)=x^5+6x^4y+14x^3y^2+16x^2y^3+9xy^4+2y^5$

$f_3(x,y)=x^3+4x^2y+5xy^2+2y^3$ $\qquad$ $f_6(x,y)=x^6+7x^5y+20x^4y^2+30x^3y^3+25x^2y^4+11xy^5+2y^6$

The various coefficients in these polynomials can be used to construct the triangular array in Figure 16.1, which Feinberg called a *Lucas triangle*.

Feinberg's Lucas triangle manifests a number of interesting properties.[†] Each element $A(n,j)$ of this array is given by $A(n,j) = \binom{n}{j} + \binom{n-1}{j-1}$, where $A(n, 0) = 1$, $A(1, 1) = 2$, and $n \geq j \geq 0$; $A(n,j)$ satisfies the same recurrence relation as Pascal's identity. For example, $A(8, 5) = 91 = 55 + 36 = A(7, 4) + A(7, 5)$.

DiDomenico's Triangle

n \ j	0	1	2	3	4	5	
0	1						
1	2	1					
2	2	3	1				
3	2	5	4	1			
4	2	7	9	5	1		
5	2	9	16	14	6	1	
6	2	11	25	30	20	7	1

Figure 16.2 DiDomenico's Triangle

In the early 1990s, independent of Feinberg's work, Angelo Salvatore DiDomenico (1932–), when he was a teacher at Framingham High School, Framingham, Massachusetts, developed the triangular array T in Figure 16.2. It is the same as Feinberg's Lucas triangle in Figure 16.1, except that it has an apex and that each row is the reverse of the corresponding row in Figure 16.1; that is, $T(n,j) = A(n, n-j) = \binom{n}{n-j} + \binom{n-1}{n-j-1}$. Its rising diagonal sums yield the Fibonacci numbers F_{n+1}; see Figure 16.3; and ignoring the apex, the falling diagonal sums yield the Lucas numbers L_n, where $n \geq 1$; see Figure 16.4.

DiDomenico's triangle can be rearranged to construct the rectangular array in Figure 16.5. Each northeast diagonal in Figure 16.2 becomes a column in Figure 16.5 and each southeast diagonal becomes a row.

[†] See the author's *Fibonacci and Lucas Numbers with Applications*.

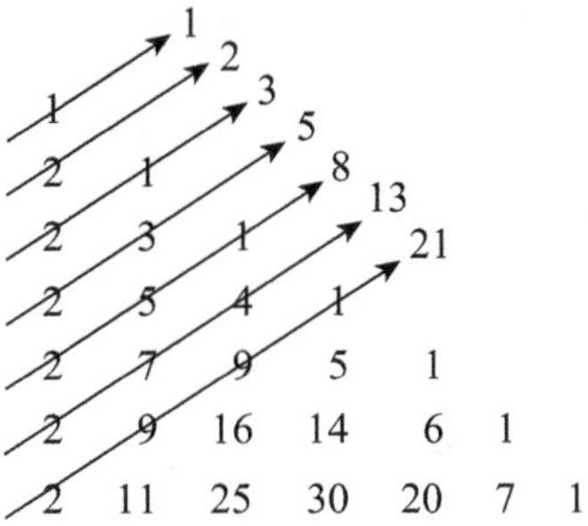

Figure 16.3

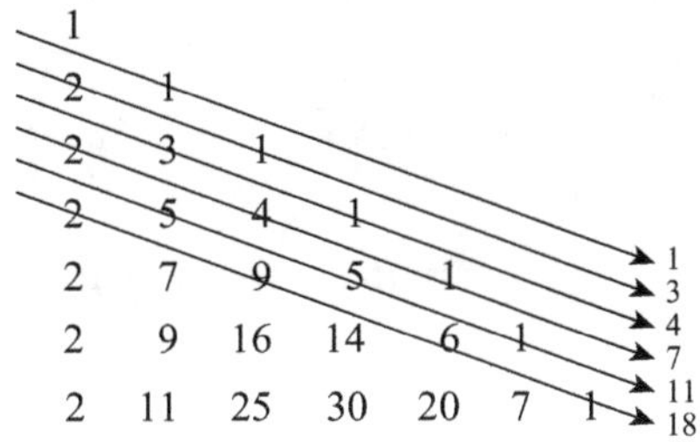

Figure 16.4

$n \backslash j$	0	1	2	3	4
0	1	1	1	1	1
1	2	3	4	5	6
2	2	5	9	14	20
3	2	7	16	30	50
4	2	9	25	55	105

Figure 16.5

$n \backslash j$	1	2	3	4	5	6	Row sums
1	1						1 = 3 − 2
2	4	1					5 = 7 − 2
3	11	4	1				16 = 18 − 2
4	29	11	4	1			45 = 47 − 2
5	76	29	11	4	1		121 = 123 − 2

Figure 16.6 Brady's Triangle

Brady's Array

The next triangular array B we will pursue was developed in 1977 by W. G. Brady of Slippery Rock State College, Slippery Rock, Pennsylvania. It is patterned after Hoggatt's array in Figure 14.3. The triangular array consists of staggered columns of the various Lucas numbers L_{2n-1}, where $n \geq 1$; see Figure 16.6.

The row sums, the rising diagonal sums, and the row sums when the columns j are weighted j times provide intriguing results, as shown in 1978 by A. G. Shannon of The New South Wales Institute of Technology, Australia.

Row Sums

It appears from Figure 16.6 that the nth row sum is $L_{2n} - 2$, where $n \geq 1$. In fact, we can establish this to be true:

$$\sum_{j=1}^{n} L_{2n-2j+1} = \sum_{j=1}^{n} \left[L_{2n-2(j-1)} - L_{2n-2j}\right]$$

$$= \sum_{j=1}^{n} L_{2n-2(j-1)} - \sum_{j=2}^{n+1} L_{2n-2(j-1)}$$

$$= L_{2n} + \sum_{j=2}^{n} L_{2n-2(j-1)} - \sum_{j=2}^{n} L_{2n-2(j-1)} - L_0$$

$$= L_{2n} - L_0 = L_{2n} - 2$$

Next, we turn to the rising diagonal sums.

Rising Diagonal Sums

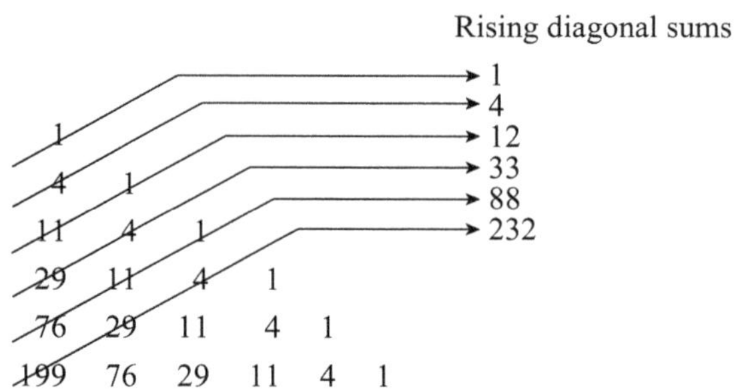

Figure 16.7

Figure 16.7 shows the various rising diagonal sums. Each sum is one less than a Fibonacci number. More generally, we shall prove that the nth diagonal sum is $F_{2n+1} - 1$, using the fact that $L_m = F_{m+2} - F_{m-2}$, where $m \geq 1$.

The nth diagonal sum is given by:

$$\sum_{j=0}^{\lfloor (n-1)/2 \rfloor} L_{2n-4j-1} = \sum_{j=0}^{\lfloor (n-1)/2 \rfloor} \left(F_{2n-4j+1} - F_{2n-4j-3}\right)$$

$$= \sum_{j=0}^{\lfloor (n-1)/2 \rfloor} F_{2n-4j+1} - \sum_{j=1}^{\lfloor (n+1)/2 \rfloor} F_{2n-4j+1}$$

$$= F_{2n+1} + \sum_{j=1}^{\lfloor (n-1)/2 \rfloor} F_{2n-4j+1} - \sum_{j=1}^{\lfloor (n-1)/2 \rfloor} F_{2n-4j+1} - F_{2n-4\lfloor \frac{n+1}{2} \rfloor + 1}$$

$$= F_{2n+1} - F_{2n-4\lfloor \frac{n+1}{2} \rfloor + 1}$$

$$= \begin{cases} F_{2n+1} - F_1 & \text{if } n \text{ is even} \\ F_{2n+1} - F_{-1} & \text{otherwise} \end{cases}$$

$$= F_{2n+1} - 1, \text{ as desired}$$

Weighted Row Sums

Suppose we multiply the elements in column j by j, where $j \geq 1$. Using Figure 16.8, we conjecture that the resulting row sum is $L_{2n+1} - (2n+1)$, where $n \geq 1$.

$n \backslash j$	1	2	3	4	5	6	Weighted row sums
1	1						$1 = 4 - (2\cdot 1+1)$
2	4	1					$6 = 11 - (2\cdot 2+1)$
3	11	4	1				$22 = 29 - (2\cdot 3+1)$
4	29	11	4	1			$67 = 76 - (2\cdot 4+1)$
5	76	29	11	4	1		$188 = 199 - (2\cdot 5+1)$
6	199	76	29	11	4	1	$508 = 521 - (2\cdot 6+1)$

Figure 16.8

To confirm this, we will use the following facts:

$$L_m = F_{m+1} + F_{m-1} \qquad \sum_{i=1}^{m} F_{2i} = F_{2m+1} - 1 \qquad \sum_{i=1}^{m} iF_{2i} = mF_{2m+1} - F_{2m}$$

The weighted row sum in row n is given by

$$\begin{aligned}
\sum_{j=1}^{n} jL_{2n-2j+1} &= \sum_{j=1}^{n} j(F_{2n-2j+2} + F_{2n-2j}) \\
&= \sum_{k=0}^{n-1}(n-k)(F_{2k+2} + F_{2k}) = n + \sum_{k=1}^{n-1}(n-k)(F_{2k} + F_{2k+2}) \\
&= n + n\sum_{k=1}^{n-1} F_{2k} + n\sum_{k=1}^{n-1} F_{2k+2} - \sum_{k=1}^{n-1} kF_{2k} - \sum_{k=1}^{n}(k-1)F_{2k} \\
&= n + n(F_{2n-1} - 1) + n(F_{2n+1} - 2) - \left[(n-1)F_{2n-1} - F_{2n-2}\right] - \\
&\quad \left[(nF_{2n+1} - F_{2n}) - (F_{2n+1} - 1)\right] \\
&= (F_{2n+1} + F_{2n-1}) + (F_{2n} + F_{2n-2}) - (2n+1) \\
&= (L_{2n} + L_{2n-1}) - (2n+1) = L_{2n+1} - (2n+1), \text{ as conjectured}
\end{aligned}$$

Another DiDomenico Triangle

Next, we present a Pascal-like array D, developed in 1995 by DiDomenico. It is defined recursively as follows:

$$D(0,0) = 1, \quad D(1,0) = 3$$

$$D(n,j) = D(n-1,j) + D(n-1,j-1), \quad n \geq j \geq 0 \tag{16.1}$$

n \ j	0	1	2	3	4	5	6
0	1						
1	3	1					
2	3	4	1				
3	3	7	5	1			
4	3	10	12	6	1		
5	3	13	22	18	7	1	
6	3	16	35	40	25	8	1

Falling diagonal 1

Figure 16.9 DiDomenico's Triangle

DiDomenico's triangle (Figure 16.9) also satisfies a number of properties:

- It satisfies the same recurrence relation as Pascal's.
- Let $n \geq 2$. Suppose $D(n-1,0) = 3$. Then, by the recurrence relation (16.1), $D(n,0) = D(n-1,0) + D(n-1,-1) = 3 + 0 = 3$. Since $D(1,0) = 3$, it follows by PMI that $D(n,0) = 3$ for every $n \geq 1$.
- Using the recurrence relation (16.1), $D(n,n) = D(n-1,n) + D(n-1, n-1) = 0 + D(n-1,n-1) = D(n-1,n-1)$. Since $D(0,0) = 1$, again it follows by PMI that $D(n,n) = 1$ for every $n \geq 0$.
- Again, by the recurrence relation, $D(n,n-1) = D(n-1,n-1) + D(n-1,n-2) = 1 + D(n-1,n-2)$. Since $D(1,0) = 3$, once again it follows by PMI that $D(n,n-1) = n+2$ for every $n \geq 1$.

Row Sums

Next, we compute the row sums in DiDomenico's triangle. It appears from Figure 16.10 that each row sum is a power of 2. In fact, we conjecture that the nth row sum R_n is 2^{n+1}, where $n \geq 1$.

n \ j	0	1	2	3	4	5	6	Row sums
0	1							1
1	3	1						4
2	3	4	1					8
3	3	7	5	1				16
4	3	10	12	6	1			32
5	3	13	22	18	7	1		64
6	3	16	35	40	25	8	1	128

Figure 16.10

To establish this, notice that the conjecture is true when $n = 1$: $R_1 = 2^2$. Now, assume that it is true for an arbitrary integer $n \geq 1$. Then, by the recursive definition of D, we have:

$$\begin{aligned} R_{n+1} &= \sum_{j=0}^{n+1} D(n+1, j) = \sum_{j=0}^{n+1} [D(n,j) + D(n,j-1)] \\ &= \sum_{j=0}^{n+1} D(n,j) + \sum_{j=0}^{n+1} D(n,j-1) = \sum_{j=0}^{n} D(n,j) + \sum_{j=0}^{n} D(n,j) \\ &= R_n + R_n = 2R_n = 2^{n+2} \end{aligned}$$

Consequently, by PMI, $R_n = 2^{n+1}$ for every $n \geq 1$, as desired.

Rising Diagonal Sums

To reveal a hidden treasure in the array, we compute the various rising diagonal sums; see Figure 16.11. Each appears to be a Lucas number. So, we conjecture that the nth rising diagonal sum D_n is L_{n+1}.

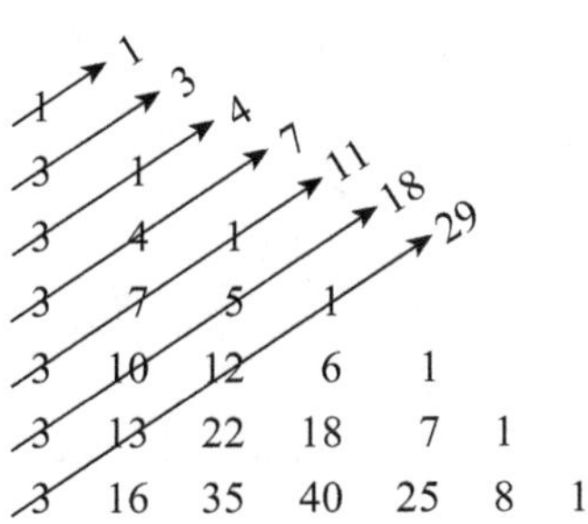

Figure 16.11

To confirm that $D_n = L_{n+1}$, it suffices to show that D_n satisfies the recursive definition of L_n. To this end, first notice that

$$D_n = \sum_{j=0}^{\lfloor n/2 \rfloor} D(n-j,j) \tag{16.2}$$

Then $D_0 = \sum_{j=0}^{0} D(0-j,j) = D(0,0) = 1$, and $D_1 = \sum_{j=0}^{0} D(1-j,j) = D(1,0) = 3$. Thus, D_n satisfies both initial conditions.

Now, assume that formula (16.2) works for every nonnegative integer $< n$. Then:

$$\begin{aligned}
D_{n-1} + D_{n-2} &= \sum_{j=0}^{\lfloor \frac{n-1}{2} \rfloor} D(n-1-j,j) + \sum_{j=0}^{\lfloor \frac{n-2}{2} \rfloor} D(n-2-j,j) \\
&= \sum_{j=0}^{\lfloor \frac{n-1}{2} \rfloor} D(n-1-j,j) + \sum_{r=1}^{\lfloor n/2 \rfloor} D(n-1-r,r-1) \\
&= \sum_{j=0}^{\lfloor \frac{n-1}{2} \rfloor} D(n-1-j,j) + \sum_{j=0}^{\lfloor n/2 \rfloor} D(n-1-j,j-1)
\end{aligned}$$

Suppose n is even, say, $n = 2k$. Then:

$$\begin{aligned}
D_{n-1} + D_{n-2} &= \sum_{j=0}^{k-1} D(2k-1-j,j) + \sum_{j=0}^{k} D(2k-1-j,j-1) \\
&= \sum_{j=0}^{k} D(2k-1-j,j) + \sum_{j=0}^{k} D(2k-1-j,j-1) \\
&= \sum_{j=0}^{k} [D(2k-1-j,j) + D(2k-1-j,j-1)] \\
&= \sum_{j=0}^{k} D(2k-j,j) = \sum_{j=0}^{\lfloor n/2 \rfloor} D(n-j,j) \\
&= D_n
\end{aligned}$$

Similarly, it can be shown that D_n satisfies the Lucas recurrence relation when n is odd.

Thus, by the strong version of PMI, D_n satisfies the recurrence definition of L_n, so $D_n = L_{n+1}$, as desired.

17

Catalan Numbers and Pascal's Triangle

In the preceding chapters, we saw a number of quite unexpected occurrences of Catalan numbers.† Catalan found them in 1838 in his study of well-formed sequences of parentheses. Although they are named after Catalan, he did not discover them. About 1751, Euler found them while investigating the triangulations of convex polygons. In fact, the Chinese mathematician Antu Ming (1692?–1763?) discovered them about 1730, through his geometric models.

Recall from Chapter 2 that *Catalan numbers* C_n are given by

$$C_n = \frac{(2n)!}{(n+1)!n!} = \frac{1}{n+1}\binom{2n}{n}, \qquad n \geq 0 \tag{17.1}$$

Since $(n+1) \mid \binom{2n}{n}$, it follows that every Catalan number is an integer. The various Catalan numbers are

$$1, 1, 2, 5, 14, 42, 132, 429, 1430, 4862, 16796, \cdots$$

Catalan numbers also provide an array of opportunities for experimentation, exploration, and conjecturing. To our delight and at the same time total surprise, they pop up in numerous seemingly unrelated places and situations, including combinatorics, abstract algebra, linear algebra, algebraic geometry, and sports.

There are a number of ways of characterizing Catalan numbers. First, we turn to Euler's triangulation problem. Over the centuries, a number of outstanding mathematicians have worked on the polygonal dissection problem.

† For an extensive discussion of Catalan numbers, see the author's *Catalan Numbers with Applications*.

Example 17.1 Find the number of ways T_n (the interior of) a convex n-gon† can be partitioned into triangles by drawing non-intersecting diagonals, where $n \geq 3$.

Solution There is exactly one way of triangulating a triangle, two different ways of triangulating a square, five different ways of triangulating a pentagon, and 14 different ways of triangulating a hexagon, as shown in Figure 17.1. Thus, we have the Catalan numbers 1, 2, 5, and 14.

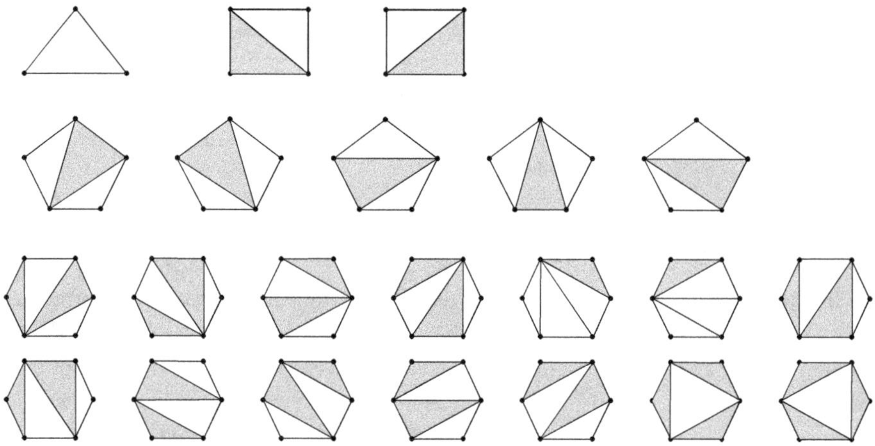

Figure 17.1 Triangulations of an n-gon, where $3 \leq n \leq 6$

Using induction, Euler showed in 1761 that

$$T_n = \frac{2 \cdot 6 \cdot 10 \cdots (4n-10)}{(n-1)!}, \qquad n \geq 3$$

Letting $C_n = T_{n+2}$, this yields

$$C_n = \frac{4n-2}{n+1} \cdot \frac{2 \cdot 6 \cdot 10 \cdots (4n-6)}{n!} = \frac{4n-2}{n+1} C_{n-1}$$

When $n = 1$, this yields $C_0 = C_1 = 1$.

Thus, C_n can be defined recursively, as follows:

$$C_0 = 1$$

$$C_n = \frac{4n-2}{n+1} C_{n-1}, \qquad n \geq 1 \tag{17.2}$$

†A *convex n-gon* is a polygon with n sides such that every diagonal lies in its interior.

An Explicit Formula for C_n

The explicit formula (17.1) can be recovered from the recursive formula (17.2) using iteration:

$$\begin{aligned}
C_n &= \frac{4n-2}{n+1}C_{n-1} \\
&= \frac{(4n-2)(4n-6)}{(n+1)n}C_{n-2} \\
&= \frac{(4n-2)(4n-6)(4n-10)}{(n+1)n(n-1)}C_{n-3} \\
&\vdots \\
&= \frac{(4n-2)(4n-6)(4n-10)\cdots 6\cdot 2}{(n+1)n\cdots 3\cdot 2}C_0 \\
&= \frac{(2n-1)(2n-3)(2n-5)\cdots 3\cdot 1}{(n+1)!}\cdot 2^n \\
&= \frac{(2n)!2^n}{(n+1)!2^n n!} = \frac{1}{n+1}\binom{2n}{n}
\end{aligned}$$

Segner's Recursive Formula

In 1761, using the addition and multiplication principles, the Hungarian mathematician, physicist, and physician Johann von Segner (1704–1777) discovered a second-order recurrence formula for T_n:

$$T_n = T_2T_{n-1} + T_3T_{n-2} + \cdots + T_{n-1}T_2$$

where $T_2 = 1$ and $n \geq 3$.

Since $C_n = T_{n+2}$, this yields *Segner's recurrence relation* for C_n:

$$\begin{aligned}
C_n &= C_0C_{n-1} + C_1C_{n-2} + \cdots + C_{n-1}C_0 \qquad (17.3)\\
&= (C_0, C_1, \ldots, C_{n-1})\cdot(C_{n-1}, C_{n-2}, \ldots, C_0)
\end{aligned}$$

For example, $C_4 = C_0C_3 + C_1C_2 + C_2C_1 + C_3C_0 = 1\cdot 5 + 1\cdot 2 + 2\cdot 1 + 5\cdot 1 = 14$.

An Alternate Derivation of the Explicit Formula

Segner's recurrence relation provides a delightful way of deriving the explicit formula (17.1). To this end, consider the generating function

$$C(x) = C_0 + C_1x + C_2x^2 + \cdots + C_nx^n + \cdots \qquad (17.4)$$

Then:

$$C^2(x) = \sum_{n=0}^{\infty}(C_0C_n + C_1C_{n-1} + \cdots + C_nC_0)x^n$$
$$= \sum_{n=1}^{\infty} C_nx^n = \frac{C(x) - C_0}{x}$$

This implies, $xC^2(x) - C(x) + 1 = 0$. Consequently, $C(x) = \frac{1\pm\sqrt{1-4x}}{2x}$. Since C_n is always a positive integer, we choose the minus sign. So $C(x) = \frac{1-\sqrt{1-4x}}{2x}$. But, from Chapter 10,

$$\sqrt{1-4x} = 1 - 2x\sum_{n=1}^{\infty}\frac{1}{n}\binom{2n-2}{n-1}x^{n-1}$$

$$\frac{1-\sqrt{1-4x}}{2x} = \sum_{n=0}^{\infty}\frac{1}{n+1}\binom{2n}{n}x^n$$

That is,

$$C(x) = \sum_{n=0}^{\infty}\frac{1}{n+1}\binom{2n}{n}x^n \tag{17.5}$$

Equating the coefficients of x^n from (17.4) and (17.5), we get the desired explicit formula.

Next, we study Catalan's parenthesization problem. First notice that the expressions $(w(x(yz)))$ and $((wx)(yz))$ are correctly parenthesized, whereas $((wx)($ and $)(wx)yz($ are not; thus ((())) and (()()) are *well-formed* sequences, but (()(and)()(are not.

Example 17.2 Find the number P_n of *correctly parenthesized sequences* of left and right parentheses that can be formed with n pairs; that is, find the number of different ways $n + 1$ operands can be multiplied, where $n \geq 0$.

Solution Table 17.1 lists the various possible correctly parenthesized expressions for $0 \leq n \leq 4$ and the corresponding values of P_n; λ denotes the *null word* consisting of no characters.

Notice, for example, that $P_3 = 5$. Consequently, there are five different ways the product $abcd$ can be evaluated. They are $((ab)(cd))$, $(((ab)c)d)$, $((a(bc))d)$, $(a(b(cd)))$, and $(a((bc)d))$.

It appears from the table that $P_n = C_n$. To confirm this, first, notice that $P_0 = 1 = C_0$ and $P_1 = 1 = C_1$. We will now develop a recurrence relation for P_n.

Now, let $n \geq 2$. Let $0 \leq r \leq n-1$. The first r pairs can be correctly parenthesized in P_r ways and the remaining $n - 1 - r = n - r - 1$ pairs in P_{n-r-1} ways. Using the multiplication principle, these two subexpressions can be

Table 17.1 Well-formed Sequences with *n* Pairs of Left and Right Parentheses

n	Correctly parenthesized sequences					P_n
0	λ					1
1	()					1
2	()()	(())				2
3	()()()	(()())	(())()	()(())	((()))	5
	()()()()	((()))()	(())(())	()((()))	(())()()	
4	()(())()	()()(())	(()())()	()(()())	(()()())	14
	((()))()	(()(()))	((())())	(((())))		

parenthesized in P_rP_{n-r-1} different ways. This is true for each value of r; so by the addition principle, $P_n = \sum_{r=0}^{n-1} P_rP_{n-r-1}$.

Since P_n satisfies the same recursive formula and the initial condition as C_n, $P_n = C_n$ for every $n \geq 0$. ■

In 1961, H. G. Forder of the University of Auckland, New Zealand showed that there is a bijection between the set of triangulations of an $(n+2)$-gon and the set of correctly parenthesized sequences with n pairs of parentheses. Figures 17.2 and 17.3 illustrate the underlying technique for the correspondence when $n = 3$.

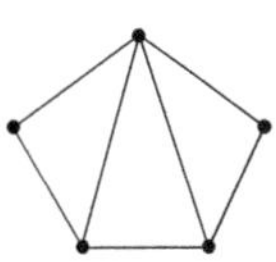

Figure 17.2

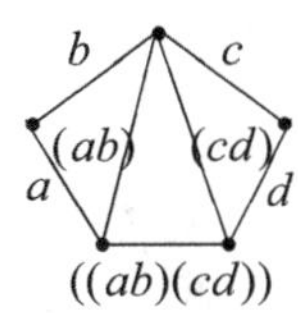

Figure 17.3

More generally, every triangulation of an $(n+2)$-gon yields a unique correctly parenthesized expression consisting of $n+1$ operands and n binary operators, and vice versa.

In 1961, Forder also investigated the number of paths a rook can take from the upper left-hand corner to the lower right-hand corner on an $n \times n$ chessboard, without crossing over the main southeast diagonal. When $n = 4$, there are five such paths, as Figure 17.4 shows.

In general, it can be shown that there is a bijection between the set of possible paths of the rook from the northwest corner to the southeast corner on an $n \times n$ chessboard and the set of well-formed sequences with n pairs of parentheses.

An added dividend, it would be a rewarding exercise to construct a triangular array of the number of paths from the upper left-hand corner to any square on or

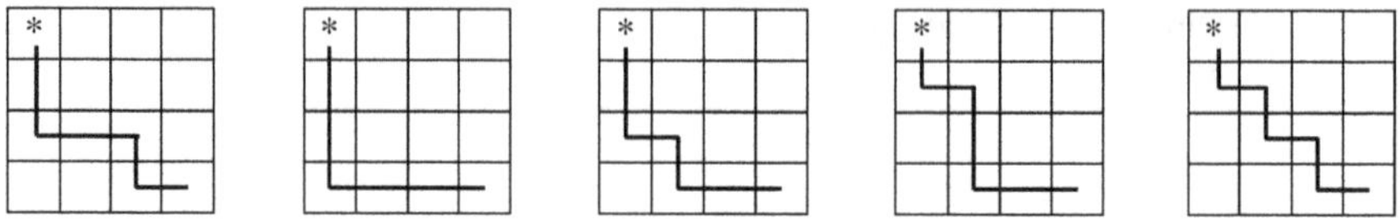

$n = 4$

Figure 17.4

below the main diagonal on an $n \times n$ chessboard, and investigate its properties, as Forder did.

There are a number of ways Catalan numbers can be read from Pascal's triangle. They are among the many treasures of Pascal's triangle.

Catalan Numbers and Pascal's Triangle

The most obvious way is by using the explicit formula $C_n = \frac{1}{n+1}\binom{2n}{n}$. So C_n can be computed by dividing the CBC $\binom{2n}{n}$ by $n+1$. For example, $C_3 = \frac{1}{4}\binom{6}{3} = \frac{20}{4} = 5$. See Figure 17.5, where the CBCs are circled for easy identification.

```
                        (1)
                     1       1
                  1     (2)     1
               1     3       3     1
            1     4     (6)     4     1
         1     5    10      10     5     1
      1     6    15    (20)    15     6     1
   1     7    21    35      35    21     7     1
```

Figure 17.5 Pascal's Triangle

There are several additional ways of extracting Catalan numbers from Pascal's triangle. In the interest of brevity, we will simply list some of them, which can be justified algebraically without much effort.

1) $C_n = \frac{1}{n}\binom{2n}{n-1}$

2) $C_n = \binom{2n}{n} - \binom{2n}{n-1}$

3) $C_{n+1} = \binom{2n}{n} - \binom{2n}{n-2}$

4) $C_n = \frac{1}{2n+1}\binom{2n+1}{n}$

5) $C_n = \binom{2n-1}{n-1} - \binom{2n-1}{n-2}$

6) $C_n = 2\binom{2n}{n} - \binom{2n+1}{n}$

7) $C_n = \binom{2n+1}{n+1} - 2\binom{2n}{n+1}$ (R. K. Guy, 1988)

18

Catalan's Parenthesization Problem Revisited

Catalan's parenthesization problem can be restated in a number of different ways. For example, it can be worded as follows: Find the number of arrangements of n positive ones and n negative ones such that every partial sum is nonnegative.

Lobb's Generalization of the Parenthesization Problem

Table 18.1

n \ m	0	1	2	3
0	λ			
1	+−			
2	++−− +−+−	+++− ++−+ +−++	++++	
3	+++−−− ++−+−− ++−−+− +−++−− +−+−+−	++++−− +++−+− ++−++− +−+++− +++−−+ ++−+−+ ++−−++ +−++−+ +−+−++	+++++− ++++−+ +++−++ ++−+++ +−++++	++++++

In 1999, A. Lobb of Merton College, Oxford, England, investigated an interesting generalization of the parenthesization problem: Find the number $L_{n,m}$ of arrangements of $n+m$ positive ones and $n-m$ negative ones such that every partial sum is nonnegative, where $0 \le m \le n$; that is, find the number of arrangements of $n+m$ pluses and $n-m$ minuses such that the number of pluses is greater than or equal to the number of minuses in each substring when scanned from left to right. Table 18.1 gives the possible arrangements for $0 \le m \le n \le 3$, where λ denotes the null arrangement.

Notice that every arrangement must end in $++$, $+-$, $-+$, or $--$. So $L_{n,m} = L_{n-1,m-1} + 2L_{n-1,m} + L_{n-1,m+1}$, where $m \ge 1$. When $m = 0$, every arrangement must end in $++$ or $--$; so $L_{n,0} = L_{n-1,0} + L_{n-1,1}$. Clearly, $L_{0,0} = 1$. Thus, the *Lobb numbers* $L_{n,m}$ can be defined recursively as follows:

$$
\begin{aligned}
L_{0,0} &= 1 \\
L_{n,0} &= L_{n-1,0} + L_{n-1,1}, \quad n \ge 1 \\
L_{n,m} &= L_{n-1,m-1} + 2L_{n-1,m} + L_{n-1,m+1} \quad 1 \le m \le n
\end{aligned}
$$

When $m > n$, $L_{n,m} = 0$.

Using PMI, we now establish that

$$
L_{n,m} = \frac{2m+1}{n+m+1}\binom{2n}{n+m} \tag{18.1}
$$

as Lobb did. It follows that $L_{n,0} = C_n$, $L_{n,n} = 1$, and $L_{n,n-1} = 2n-1$; see Table 18.3.

Case 1 Let $m = 0$. Clearly, the formula works when $n = 0$. Now assume that it works for an arbitrary integer $n \ge 0$. Then:

$$
\begin{aligned}
L_{n+1,0} &= L_{n,0} + L_{n,1} \\
&= \frac{1}{n+1}\binom{2n}{n} + \frac{3}{n+2}\binom{2n}{n+1} \\
&= \frac{(2n)!}{(n+1)!(n+2)!}[(n+1)(n+2) + 3n(n+1)] \\
&= \frac{(2n)!}{(n+1)!(n+2)!}(2n+1)(2n+2) \\
&= \frac{1}{n+2}\binom{2n+2}{n+1}
\end{aligned}
$$

Case 2 Let $m \geq 1$. Notice that the formula holds when $n = 1$. Assume that it is true for any integer $n \geq 1$. Then:

$$
\begin{aligned}
L_{n,m} &= L_{n-1,m-1} + 2L_{n-1,m} + L_{n-1,m+1} \\
&= \frac{2m-1}{n+m}\binom{2n}{n+m-1} + \frac{2(2m+1)}{n+m+1}\binom{2n}{n+m} + \frac{2m+3}{n+m+2}\binom{2n}{n+m+1} \\
&= \frac{(2n)!}{(n+m+2)!(n-m+1)!}[(2m-1)(n+m+1)(n+m+2) \\
&\quad + (4m+2)(n-m+1)(n+m+2)(n-m+1) \\
&\quad + (2m+3)(n-m)(n-m+1)] \\
&= \frac{(2n)!}{(n+m+2)!(n-m+1)!}(2m+1)(2n+1)(2n+2) \\
&= \frac{2m+1}{n+m+2}\binom{2n+2}{n+m+1}
\end{aligned}
$$

Thus, the result is true for every $m \geq 0$. So, by PMI, the formula works for every $n \geq 0$. ■

We will shortly explore an alternate and simpler way of computing the Lobb numbers. As byproducts of this pursuit, we will develop several interesting properties of $L_{n,m}$, and explore the various ways Lobb and Catalan numbers are bonded together.

To begin with, it follows from (18.1) that the values of $L_{n,m}$ can be read from the even-numbered rows of Pascal's triangle, beginning at the middle column, by assigning weights to the elements; see Table 18.2.

Table 18.2 Values of $L_{n,m}$

$n \backslash m$	0 0	1	2	3	4
0	$\frac{1}{1}\binom{0}{0}$				
1	$\frac{1}{2}\binom{2}{1}$	$\frac{3}{3}\binom{2}{2}$			
2	$\frac{1}{3}\binom{4}{2}$	$\frac{3}{4}\binom{4}{3}$	$\frac{5}{5}\binom{4}{4}$		
3	$\frac{1}{4}\binom{6}{3}$	$\frac{3}{5}\binom{6}{4}$	$\frac{5}{6}\binom{6}{5}$	$\frac{7}{7}\binom{6}{6}$	
4	$\frac{1}{5}\binom{8}{4}$	$\frac{3}{6}\binom{8}{5}$	$\frac{5}{7}\binom{8}{6}$	$\frac{7}{8}\binom{8}{7}$	$\frac{9}{9}\binom{8}{7}$

Parity of $L_{n,m}$

It is well known that C_n is odd if and only if either $n = 0$ or n is a Mersenne number $M_k = 2^k - 1$, where $k \geq 1$.† Using this fundamental property of Catalan numbers, we now establish a similar property for $L_{n,m}$.

Theorem 18.1 $L_{n,m}$ is odd for every m if and only if either $n = 0$ or n is a Mersenne number.

Proof: Clearly, $L_{0,0}$ is odd. So, assume that n is a Mersenne number. Since $L_{n,n} = 1$, assume that $0 < m < 2^k - 1$. Then

$$L_{n,m} = \frac{2m+1}{2^k+m}\binom{2^{k+1}-2}{2^k+m-1}$$

But

$$\begin{aligned}\binom{2^{k+1}-2}{2^k+m-1} &= \frac{(2^{k+1}-2)!}{(2^k+m-1)!(2^k-m-1)!}\\ &= \frac{(2^{k+1}-2)(2^{k+1}-3)\cdots[2^{k+1}-(2^k-m-1)][2^{k+1}-(2^k-m)]}{(2^k-m-1)!}\\ &= \left(\prod_{i=1}^{2^k-m-1}\frac{2^{k+1}-i}{i}\right)\frac{2^k+m}{2^{k+1}-1}\end{aligned}$$

is an integer.

Let $i = 2^a b$, where $0 \leq a \leq k$ and b is odd. Then:

$$\begin{aligned}\frac{2^{k+1}-i}{i} &= \frac{2^{k+1}-2^a b}{2^a b}\\ &= \frac{2^{k-a+1}-b}{b}\end{aligned}$$

This is the ratio of two odd integers, so $\prod_{i=1}^{2^k-m-1}\frac{2^{k+1}-i}{i}$ is the ratio of products of odd integers. But, since $\prod_{i=1}^{2^k-m-1}(2^{k+1}-i)$ is the product of $2^k - m - 1$ consecutive integers, it is divisible by $(2^k - m - 1)!$. So, $\prod_{i=1}^{2^k-m-1}\frac{2^{k+1}-i}{i}$ is an odd integer.

† See the author's *Catalan Numbers with Applications*.

Since $\binom{2^{k+1}-2}{2^k+m-1}$ is an integer, it follows that

$$\frac{2m+1}{2^{k+1}-1}\left(\prod_{i=1}^{2^k-m-1}\frac{2^{k+1}-i}{i}\right)=\frac{2m+1}{2^k+m}\binom{2^{k+1}-2}{2^k+m-1}$$

is an odd integer. That is, $L_{n,m}$ is an odd integer.

Thus, $L_{n,m}$ is odd for $0 \le m \le n$.

Conversely, suppose $L_{n,m}$ is odd for every m, where $0 \le m \le n$. Then $L_{n,0} = C_n$ is odd. Therefore, either $n = 0$ or n is a Mersenne number. ■

Thus, we have three strikingly similar properties:

- $\binom{n}{r}$ is odd for every r if and only if either $n = 0$ or n is a Mersenne number.
- C_n is odd if and only if either $n = 0$ or n is a Mersenne number.
- $L_{n,m}$ is odd for every m if and only if either $n = 0$ or n is a Mersenne number.

It now follows that $L_{n,m}$ is odd for every m if and only if C_n is odd.

An Alternate Formula for $L_{n,m}$

Next, we derive an alternate and simple explicit formula for $L_{n,m}$ using elementary algebra:

$$\begin{aligned}
L_{n,m} &= \frac{2m+1}{n+m+1}\binom{2n}{n+m}\\
&= \frac{2m+1}{n+m+1}\cdot\frac{(2n)!}{(n-m)!(n+m)!}\\
&= \frac{(2n)!}{(n-m)!(n+m)!}\left[1-\frac{n-m}{n+m+1}\right]\\
&= \frac{(2n)!}{(n-m)!(n+m)!}-\frac{(2n)!}{(n-m-1)!(n+m+1)!}\\
&= \binom{2n}{n-m}-\binom{2n}{n-m-1} \qquad (18.2)
\end{aligned}$$

A Combinatorial Interpretation

Interestingly, formula (18.2) has a simple combinatorial interpretation. Since the total number of arrangements with exactly $n-m$ negative ones equals $\binom{2n}{n-m}$ and

$L_{m,m}$ counts the number of acceptable arrangements, it follows that $\binom{2n}{n-m-1}$ counts the number of unacceptable arrangements.

Lobb Numbers and Pascal's Triangle

By virtue of formula (18.2), the values of $L_{n,m}$ can be read fairly easily from Pascal's triangle; beginning with the middle column consisting of the CBCs in row $2n$, subtract the adjacent elements successively to the left (or to the right by symmetry).

For example, the various values of $L_{4,m}$ can be computed using row 8 in the reverse order: $70-56=14$, $56-28=28$, $28-8=20$, $8-1=7$, and $1-0=1$; see the loops in Figure 18.1.

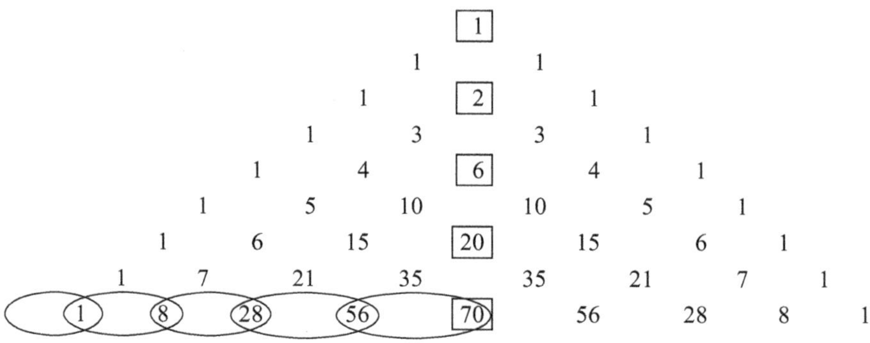

Figure 18.1

It follows from $L_{n,0}=L_{n-1,0}+L_{n-1,1}$ and (18.2) that:

$$\begin{aligned} C_n &= C_{n-1} + \binom{2n}{n-2} - \binom{2n}{n-3} \\ &= \frac{4n-2}{n+1} C_{n-1} \end{aligned}$$

This recurrence relation was conjectured by H. Urban in 1941 after noticing an interesting pattern in the values of $\frac{C_1}{C_0}, \frac{C_2}{C_1}, \cdots,$ and $\frac{C_5}{C_4}$. The explicit formula for C_n can be recovered from this recursive formula, as we did in Chapter 17.

The various values of $L_{n,m}$ can be arranged in a triangular array, as in Table 18.3. Figure 18.2 shows a *Sierpiński-like* geometric representation of its binary version, where a black triangle indicates that the corresponding entry is odd; a lighter triangle indicates that the corresponding entry is even; and $0 \le n \le 63$. Notice that the rows corresponding to Mersenne numbers consist of only black triangles.

Table 18.3

$n \backslash m$	0	1	2	3	4	5	Row sums
0	1						1
1	1	1					2
2	2	3	1				6
3	5	9	5	1			20
4	14	28	20	7	1		70
5	42	90	75	35	9	1	252
	C_n					$2n-1$	CBCs

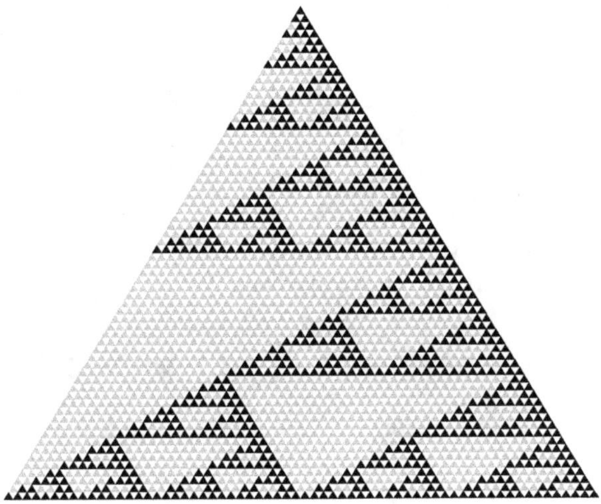

Figure 18.2 Lobb's Binary Triangle

An Alternate Proof of Theorem 18.1

Using Pascal's identity and formula (18.2), we have:

$$\begin{aligned}\binom{2n+1}{n-m}-\binom{2n+1}{n-m-1} &= \left[\binom{2n}{n-m}+\binom{2n}{n-m-1}\right]-\left[\binom{2n}{n-m-1}+\binom{2n}{n-m-2}\right]\\ &= \binom{2n}{n-m}-\binom{2n}{n-m-2}\\ &= \left[\binom{2n}{n-m}-\binom{2n}{n-m-1}\right]+\left[\binom{2n}{n-m-1}-\binom{2n}{n-m-2}\right]\\ &= L_{n,m}+L_{n,m+1}\end{aligned}$$

Let $n = 0$ or M_k. Then $2n + 1 = 2^{k+1} - 1 = M_{k+1}$. Then $\binom{2n+1}{n-m}$ and $\binom{2n+1}{n-m-1}$ are odd, so $L_{n,m} + L_{n,m+1}$ is even. Consequently, $L_{n,m}$ and $L_{n,m+1}$ have the same parity for every m. Since $L_{n,0} = C_n$ is odd, it follows that $L_{n,m}$ is odd for every m.

The converse follows as before. ■

Interestingly, using a repeated application of Pascal's identity, we can rewrite formula (18.2):

$$L_{n,m} = \binom{2n}{n-m} - \binom{2n-1}{n-m-1} - \binom{2n-2}{n-m-2} - \cdots - \binom{n+m}{0} \tag{18.3}$$

$$= \binom{2n}{n+m} - \binom{2n-1}{n+m} - \binom{2n-2}{n+m} - \cdots - \binom{n+m}{n+m} \tag{18.4}$$

In particular,

$$C_n = \binom{2n}{n} - \binom{2n-1}{n-1} - \binom{2n-2}{n-2} - \cdots - \binom{n}{0} \tag{18.5}$$

$$= \binom{2n}{n} - \binom{2n-1}{n} - \binom{2n-2}{n} - \cdots - \binom{n}{n} \tag{18.6}$$

Thus, we can extract the various Catalan numbers from Pascal's triangle by subtracting the elements on each rising (or falling) diagonal from the corresponding CBC.

For example, $C_4 = \binom{8}{4} - \binom{7}{3} - \binom{6}{2} - \binom{5}{1} - \binom{4}{0} = 14$; see the loops in Figure 18.1.

Next, we pursue several hidden treasures of the Lobb numbers.

Properties of Array *L*

1) The nth row sum is the nth CBC:

$$\sum_{m=0}^{n} L_{n,m} = \binom{2n}{n} = (n+1)C_n \tag{18.7}$$

This follows by (18.2).

2)
$$\sum_{m=0}^{n} (-1)^m L_{n,m} = 0 \tag{18.8}$$

This can be established using (18.2) and the fact that

$$\sum_{s \geq 0} (-1)^s \binom{r}{s} = 0 \tag{18.9}$$

3) The sum of every other term in a row is one-half of a CBC:

$$\sum_{m\geq 0} L_{n,2m} = \frac{1}{2}\binom{2n}{n} = \sum_{m\geq 0} L_{n,2m+1} \tag{18.10}$$

This follows by (18.7) and (18.8).

4) The sum of the squares of the numbers in row n is C_{2n}:

$$\sum_{m=0}^{n} L_{n,m}^2 = C_{2n} \tag{18.11}$$

To establish this result, we need the following formulas:[†]

$$C_k = \binom{2k}{k} - \binom{2k}{k-1}$$

$$\sum_{r=0}^{n} \binom{n}{r}^2 = \binom{2n}{n}$$

$$\sum_{\text{integers } k} \binom{r}{a+k}\binom{s}{b-k} = \binom{r+s}{a+b}$$

This formula implies that:

$$\begin{aligned} 2\sum_{m=0}^{n}\binom{2n}{n-m}\binom{2n}{n-m-1} &= 2\sum_{m=0}^{n}\binom{2n}{n-m}\binom{2n}{(n-1)-m} \\ &= \binom{4n}{n+(n-1)} \\ &= \binom{4n}{2n-1} \end{aligned} \tag{18.12}$$

Using formulas (18.2) and (18.12), and Lagrange's identity, we have:

$$\begin{aligned} \sum_{m=0}^{n} L_{n,m}^2 &= \sum_{m=0}^{n}\binom{2n}{n+m}^2 + \sum_{m=0}^{n}\binom{2n}{n-m-1}^2 - 2\sum_{m=0}^{n}\binom{2n}{n-m}\binom{2n}{n-m-1} \\ &= \sum_{r=0}^{n}\binom{2n}{r}^2 + \sum_{r=0}^{n-1}\binom{2n}{n+r+1}^2 - \binom{4n}{2n-1} \\ &= \sum_{r=0}^{2n}\binom{2n}{r}^2 - \binom{4n}{2n-1} \\ &= \binom{4n}{2n} - \binom{4n}{2n-1} = C_{2n} \end{aligned}$$

[†]See R. L. Graham et al, *Concrete Mathematics,* 2nd edition, Addison-Wesley, Reading, Massachusetts, 1994.

Thus, we have the following interesting result.

Theorem 18.2 Every Catalan number C_{2n} is the sum of $n+1$ squares. ■

For example, $\sum_{m=0}^{4} L_{4,m}^2 = 14^2 + 28^2 + 20^2 + 7^2 + 1^2 = 1{,}430 = C_8$.

Formulas (18.7), (18.10), and (18.11), coupled with (18.1), yield the following combinatorial identities:

$$\sum_{m=0}^{n} \frac{2m+1}{n+m+1}\binom{2n}{n+m} = (n+1)C_n$$

$$2\sum_{m\geq 0} \frac{4m+1}{n+2m+1}\binom{2n}{n+2m} = (n+1)C_n$$

$$2\sum_{m\geq 0} \frac{4m+3}{n+2m+2}\binom{2n}{n+2m+1} = (n+1)C_n$$

$$(2n+1)\sum_{m=0}^{n} \left[\frac{2m+1}{n+m+1}\binom{2n}{n+m}\right]^2 = (2n+1)C_{2n}$$

Recall from the previous chapter that there are C_n ordered rooted trees with n edges each in a forest and it contains a total of $(n+1)C_n$ vertices; exactly one-half of the vertices are leaves.

5) Finally, let $p_n^{(r)}$ denote the nth r-gonal number.[†] Then

$$2\sum_{m=0}^{n} p_{L_{n,m}}^{(r)} = (r-2)C_{2n} + (n+1)(4-r)C_n \tag{18.13}$$

This follows by (18.7) and (18.11), and the fact that $2p_n^{(r)} = r[(r-1)n - 2(r-2)]$.

In particular, for example,

$$2\sum_{m=0}^{n} t_{L_{n,m}} = C_{2n} + (n+1)C_n \tag{18.14}$$

where t_k is the kth triangular number.

For example,

$$2\sum_{m=0}^{4} t_{L_{4,m}} = 2\,(t_{14} + t_{28} + t_{20} + t_7 + t_1) = 2(105 + 406 + 210 + 28 + 1)$$

$$= 1{,}430 + 70 = C_8 + 5C_4$$

[†] A. H. Beiler, *Recreations in the Theory of Numbers*, 2nd edition, Dover, New York, 1966.

19

Leibniz's Harmonic Triangle

In 1665, the renowned Dutch physicist, mathematician, and astronomer, Christiaan Huygens (1629–1695) was offered a senior position by Louis XIV at the newly founded Academy of Sciences, along with a substantial pension and free housing. Huygens' 15-year stay in Paris had a profound influence on the mathematical development of the young Gottfried Wilhelm Leibniz (1646–1716), an outstanding German mathematician, philosopher, physicist, diplomat, and linguist.

While in Paris, Huygens proposed the following problem to Leibniz:

Find the sum S of the reciprocals of the triangular numbers t_n: $S = \sum_{n=0}^{\infty} \frac{1}{t_n}$

Leibniz recognized that each term of the series is twice the difference between the successive terms of the **harmonic sequence** $1, 1/2, 1/3, 1/4, \cdots$:

Harmonic sequence:	1		$\frac{1}{2}$		$\frac{1}{3}$		$\frac{1}{4}$		$\frac{1}{5}$	$\cdots$
Differences:		$\frac{1}{2}$		$\frac{1}{6}$		$\frac{1}{12}$		$\frac{1}{20}$		$\cdots$

Let

$$\begin{aligned} S_n &= \frac{1}{2} + \frac{1}{6} + \frac{1}{12} + \frac{1}{20} + \cdots + \frac{1}{2t_n} \\ &= \left(1 - \frac{1}{2}\right) + \left(\frac{1}{2} - \frac{1}{3}\right) + \left(\frac{1}{2} - \frac{1}{3}\right) + \cdots + \left(\frac{1}{n} - \frac{1}{n+1}\right) \\ &= 1 - \frac{1}{n+1} \end{aligned}$$

So $S = \lim_{n\to\infty} 2S_n = 2$.

Using the harmonic sequence and its successive sequences of differences, Leibniz constructed a delightful triangular array of numbers, which he called the *harmonic triangle*. But unlike all arrays we have encountered thus far, the harmonic triangle is made up of reciprocals of positive integers; they are called *Leibniz numbers* $L(n, r)$. They are defined by

$$L(n, r) = \frac{1}{(n+1)\binom{n}{r}}, \quad 0 \le r \le n \tag{19.1}$$

For example, $L(3, 2) = \frac{1}{4\binom{3}{2}} = \frac{1}{4 \cdot 3} = \frac{1}{12}$ and $L(5, 3) = \frac{1}{6\binom{5}{3}} = \frac{1}{6 \cdot 10} = \frac{1}{60}$.

Notice that:

$$L(n, 0) = \frac{1}{(n+1)\binom{n}{0}} = \frac{1}{n+1} \tag{19.2}$$

$$L(n, n) = \frac{1}{(n+1)\binom{n}{n}} = \frac{1}{n+1} = L(n, 0) \tag{19.3}$$

and

$$\begin{aligned}
L(n, r) + L(n, r-1) &= \frac{1}{(n+1)\binom{n}{r}} + \frac{1}{(n+1)\binom{n}{r-1}} \\
&= \frac{1}{n+1}\left[\frac{r!(n-r)!}{n!} + \frac{(r-1)!(n-r+1)!}{n!}\right] \\
&= \frac{r!(n-r)!}{(n+1)n!}\left[1 + \frac{n-r+1}{r}\right] \\
&= \frac{r!(n-r)!}{(n+1)n!} \cdot \frac{n+1}{r} = \frac{(r-1)!(n-r)!}{n!} \\
&= \frac{1}{n\binom{n-1}{r-1}} = L(n-1, r-1)
\end{aligned}$$

That is,

$$L(n, r) = L(n-1, r-1) - L(n, r-1) \tag{19.4}$$

where $r \ge 1$.

Using this recurrence relation coupled with the boundary conditions (19.2) and (19.3), we can construct Leibniz's harmonic triangle, as in Figure 19.1.

As in the case of Pascal's triangle, the harmonic triangle also possesses several remarkable properties:

1) By virtue of the boundary conditions, each row begins with and ends in the same number $\frac{1}{n+1}$, where $n \ge 0$.
2) The leftmost northeast diagonal and the rightmost southwest diagonal consists of the numbers $\frac{1}{n+1}$, where $n \ge 0$.

$$
\begin{array}{ccccccccccccc}
 & & & & & & \frac{1}{1} & & & & & & \leftarrow \text{row } 0 \\
 & & & & & \frac{1}{2} & & \frac{1}{2} & & & & & \\
 & & & & \frac{1}{3} & & \frac{1}{6} & & \frac{1}{3} & & & & \\
 & & & \frac{1}{4} & & \frac{1}{12} & & \frac{1}{12} & & \frac{1}{4} & & & \\
 & & \frac{1}{5} & & \frac{1}{20} & & \frac{1}{30} & & \frac{1}{20} & & \frac{1}{5} & & \\
 & \frac{1}{6} & & \frac{1}{30} & & \frac{1}{60} & & \frac{1}{60} & & \frac{1}{30} & & \frac{1}{6} & \\
\frac{1}{7} & & \frac{1}{42} & & \frac{1}{105} & & \frac{1}{140} & & \frac{1}{105} & & \frac{1}{42} & & \frac{1}{7}
\end{array}
$$

Figure 19.1 Leibniz's Harmonic Triangle

3) Since $L(n, r) = \frac{1}{(n+1)\binom{n}{r}} = \frac{1}{(n+1)\binom{n}{n-r}} = L(n, n-r)$, the array is symmetric about the vertical axis through the apex.
4) In Pascal's triangle, each entry is the sum of its northwest and northeast neighbors. But in Leibniz's triangle, each entry is the sum of its southwest and southeast neighbors; in other words, each internal entry is obtained by subtracting its left neighbor from its northwest neighbor. For example, $\frac{1}{20} = \frac{1}{12} - \frac{1}{30}$; see Figure 19.1.
5) Since $L(n, n-1) = \frac{1}{(n+1)\binom{n}{n-1}} = \frac{1}{n(n+1)} = \frac{1}{2} \cdot \frac{1}{\binom{n+1}{2}} = \frac{1}{2t_n}$, the second rightmost southeast diagonal consists of one-half of the reciprocals of the triangular numbers, where $n \geq 1$; see Figure 19.2.
6) Since

$$
\begin{aligned}
L(n, n-2) &= \frac{1}{(n+1)\binom{n}{n-2}} = \frac{1}{(n+1)\binom{n}{2}} \\
&= \frac{2}{(n+1)n(n-1)} = \frac{1}{3} \cdot \frac{1}{\binom{n+1}{3}} \\
&= \frac{1}{3T_{n-1}},
\end{aligned}
$$

the third southeast diagonal consists of one-third of the reciprocals of the tetrahedral numbers, where $n \geq 2$; see Figure 19.2.
7) Likewise, the fourth southeast diagonal from the right consists of one-fourths of the reciprocals of the figurate numbers of dimension four, namely, $\frac{1}{4}\binom{n+1}{4}$, where $n \geq 3$.
8) $\sum_{k=m}^{n} L(k, r) = L(m-1, r-1) - L(n, r-1)$

$$
\begin{array}{ccccccccccccc}
 & & & & & & \frac{1}{1} & & & & & & \\
 & & & & & \frac{1}{2} & & \frac{1}{2} & & & & & \\
 & & & & \frac{1}{3} & & \frac{1}{6} & & \frac{1}{3} & & & & \\
 & & & \frac{1}{4} & & \frac{1}{12} & & \frac{1}{12} & & \frac{1}{4} & & & \\
 & & \frac{1}{5} & & \frac{1}{20} & & \frac{1}{30} & & \frac{1}{20} & & \frac{1}{5} & & \\
 & \frac{1}{6} & & \frac{1}{30} & & \frac{1}{60} & & \frac{1}{60} & & \frac{1}{30} & & \frac{1}{6} & \\
\frac{1}{7} & & \frac{1}{42} & & \frac{1}{105} & & \frac{1}{140} & & \frac{1}{105} & & \frac{1}{42} & & \frac{1}{7}
\end{array}
$$

$$\nwarrow \quad \frac{1}{3T_{n-1}} \qquad \nwarrow \quad \frac{1}{2t_n} \qquad \nwarrow \quad \frac{1}{n+1}$$

Figure 19.2 Leibniz's Harmonic Triangle

To establish this, we return to the recurrence relation (19.4) and express it as a telescoping sum:

$$
\begin{aligned}
\sum_{k=m}^{n} L(k,r) &= \sum_{k=m}^{n} [L(k-1,r-1) - L(k,r-1)] \\
&= [L(m-1,r-1) - L(m,r-1)] \\
&\quad + [L(m,r-1) - L(m+1,r-1)] + \cdots + \\
&\quad [L(n-1,r-1) - L(n,r-1)] \\
&= [L(m-1,r-1) - L(n,r-1)], \quad \text{as desired}
\end{aligned}
$$

9) Although the harmonic series $\sum_{n=0}^{\infty} L(n,0) = \sum_{n=0}^{\infty} \frac{1}{n+1}$ is divergent, the sum of the numbers along each of the remaining southwest diagonal is convergent. More generally, $\sum_{n=0}^{\infty} L(n,r) = L(m-1,r-1)$. This follows from the above sum.

10) To derive the next property, consider the sum $\sum_{r=0}^{k} (-1)^r L(n,r)$. Using the recurrence relation, we have:

$$
\begin{aligned}
\sum_{r=0}^{k} (-1)^r L(n,r) &= L(n,0) - L(n,1) + L(n,2) - L(n,3) + \cdots + (-1)^k L(n,k) \\
&= [L(n+1,0) + L(n+1,1)] - [L(n+1,1) + L(n+1,2)]
\end{aligned}
$$

$$+[L(n+1,2)+L(n+1,3)]-\cdots+(-1)^k$$

$$[L(n+1,k)+L(n+1,k+1)]$$

$$=L(n+1,0)+(-1)^k L(n+1,k+1)$$

For example, $\sum_{r=0}^{3}(-1)^r L(4,r) = \frac{1}{5}-\frac{1}{20}-\frac{1}{30}-\frac{1}{20}=\frac{2}{15}=\frac{1}{6}-\frac{1}{30}=L(5,0)+(-1)^3L(5,4)$.

Harmonic Numbers

We close this chapter with a brief introduction to *harmonic numbers* H_n, which are partial sums of numbers on the leftmost southwest diagonal in Figure 19.1:

$$H_n=\sum_{i=1}^{n}\frac{1}{i},\quad \text{so } H_n = H_{n-1}+\frac{1}{n}$$

where $H_0 = 0$ and $n \geq 1$. The first five harmonic numbers are $H_1 = 1$, $H_2 = \frac{3}{2}, H_3 = \frac{11}{6}, H_4 = \frac{25}{12}$, and $H_5 = \frac{137}{60}$. The name harmonic number for H_n comes from the fact that "a tone of wavelength $\frac{1}{n}$ is the nth harmonic of a tone of wavelength 1".*

Harmonic numbers occur in real-world situations. For example, suppose we stack up n playing cards overhanging an edge of a table. Then the maximum overhang is H_n. This example was studied by R. T. Sharp of McGill University, Montreal, Canada in 1954. Harmonic numbers also occur in combinatorics (see Chapter 20) and in the analysis of algorithms in computer science.†

The following are a few interesting properties of harmonic numbers:

1) $\sum_{i=0}^{n} H_i = (n+1)H_n - n$

2) $H_{2^m} \geq 1+\frac{m}{2}, \quad m \geq 0$

3) $H_n \leq \frac{n+1}{2}$

4) $H_n \leq \ln n$

5) $\sum_{k=1}^{n} kH_k = \left[\frac{n(n+1)}{2}\right]H_{n+1} - \frac{n(n+1)}{4}$

6) $\sum_{i=0}^{n}\frac{1}{2i+1} = H_{2n+1} - \frac{1}{2}H_n$

7) $\sum_{k=m}^{n-1}\binom{k}{m}H_k = \binom{n}{m+1}\left(H_n - \frac{1}{m+1}\right)$

*See R. L. Graham et al., *Concrete Mathematics*.

†See R. L. Kruse, *Data Structures and Program Design*, 3rd edition, Prentice-Hall, Englewood Cliffs, NJ.

Growth of Harmonic Numbers

Finally, we add that the growth of harmonic numbers can be investigated by properly grouping the terms:*

$$H_n = 1 + \left(\frac{1}{2} + \frac{1}{3}\right) + \left(\frac{1}{4} + \frac{1}{5} + \frac{1}{6} + \frac{1}{7}\right) + \left(\frac{1}{8} + \frac{1}{9} + \cdots + \frac{1}{15}\right) + \cdots$$

Let s_k denote the kth group:

$$s_k = \frac{1}{2^{k-1}} + \frac{1}{2^{k-1}+1} + \cdots + \frac{1}{2^k - 1}$$

Since s_k contains 2^{k-1} terms and $\frac{1}{2^k} <$ each term $\leq \frac{1}{2^{k-1}}$, it follows that $\frac{2^{k-1}}{2^k} < s_k \leq \frac{2^{k-1}}{2^{k-1}}$; that is, $\frac{1}{2} < s_k \leq 1$.

Suppose $\frac{1}{n}$ falls in the kth group. Then, by PMI, $\frac{k}{2} < H_n \leq k$. Since $\frac{1}{2^k} < \frac{1}{n} \leq \frac{1}{2^{k-1}}$, it follows that $k - 1 \leq \lg n < k$, so $k = \lfloor \lg n \rfloor + 1$, where $\lg n = \log_2 n$. Thus

$$\frac{\lfloor \lg n \rfloor + 1}{2} < H_n \leq \lfloor \lg n \rfloor + 1$$

Consequently, $H_n \to \infty$ as $n \to \infty$; in other words, H_n grows larger and larger logarithmically as $n \to \infty$; that is, $H_n = O(\lg n)$.

A good approximation of H_n can be found by computing the area under the curve $y = \frac{1}{x}$ between $x = \frac{1}{2}$ and $x = n + \frac{1}{2}$; see Figure 19.3:

$$\int_{1/2}^{n+1/2} \frac{dx}{x} = \ln\left(n + \frac{1}{2}\right) - \ln\frac{1}{2} = \ln\left(n + \frac{1}{2}\right) + \ln 2$$

Thus, when n is sufficiently large, $H_n \approx \ln n$.

A better approximation for H_n is given by*

$$H_n \approx \ln n + \gamma + \frac{1}{2n} - \frac{1}{12n^2} + \frac{1}{120n^4} - \frac{1}{2560n^6} + \frac{1}{240n^8} - \frac{1}{132n^{10}}$$

where γ is the *Euler's constant* ≈ 0.57721664901533, which Euler discovered in 1734.

In closing, we add that the harmonic number H_n can be expressed in terms of binomial coefficients: $H_n = \sum_{r=1}^{n} \frac{(-1)^{r-1}}{n}\binom{n}{r}$. This follows by evaluating the integrals in the equation

$$\int_0^1 \frac{1-x^n}{1-x} dx = \int_0^1 [1-(1-t)^n]\frac{dx}{t}$$

*R. L. Graham et al., *Concrete Mathematics*.

*http://numbers.computations.free.fr/Constants/Gamma/gamma.html

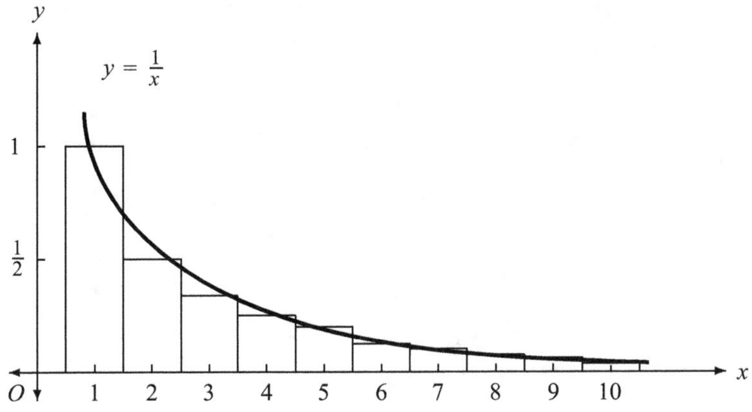

Figure 19.3

as I. W. Busbridge did in 1949.† Using integral calculus, she also showed that

$$\sum_{r=1}^{n} \frac{(-1)^{r-1}}{n} \binom{n}{r} H_r = \sum_{r=1}^{n} \frac{1}{r^2}$$

† See *Mathematical Gazette* 33 (1949), 294–295.

20

Stirling's Triangles

There are two kinds of Stirling numbers: Stirling numbers of the first kind and those of the second kind. They are named after the Scottish mathematician James Stirling (1692–1770), who published them in 1730.

We first investigate Stirling numbers of the first kind.

Stirling Numbers of the First Kind

Recall from Chapter 1 that there are $P(m, n) = m(m-1)(m-2)\cdots(m-n+1)$ permutations of m distinct objects taken n at a time; that is, there are $m(m-1)\cdot(m-2)\cdots(m-n+1)$ different ways of distributing m distinct objects into n distinct boxes, where $n \le m$. Letting $m = x$, we get the product

$$\begin{aligned}[x]_n &= x(x-1)(x-2)\cdots(x-n+1) \qquad (20.1)\\ &= (x-0)(x-1)(x-2)\cdots[x-(n-1)]\end{aligned}$$

which is often known as a *descending factorial.*

For example,

$$\begin{aligned}[x]_1 &= x &&= x\\ [x]_2 &= x(x-1) &&= x^2 - x\\ [x]_3 &= x(x-1)(x-2) &&= x^3 - 3x^2 + 2x\\ [x]_4 &= x(x-1)(x-2)(x-3) &&= x^4 - 6x^3 + 11x^2 - 6x\end{aligned}$$

The descending factorial $[x]_n$ can be expressed in terms of a binomial coefficient:

$$[x]_n = \binom{x}{n} n! = P(x, n)$$

For example, $[x]_3 = x(x-1)(x-2) = \binom{x}{3}3!$.

Besides, $[x]_n$ is a linear combination of the monomials $1, x, x^2 \cdots x^n$; that is, $[x]_n$ lies in the vector space spanned by the basis vectors $1, x, x^2 \cdots x^n$ of the vector space of polynomials of degree $\leq n$ over the field of real numbers.

As in the case of the factorial $n!$, $[x]_n$ also can be defined recursively:

$$[x]_0 = 1$$

$$[x]_n = (x - n + 1)[x]_{n-1}, \quad n \geq 1$$

For example,

$$\begin{aligned}
[x]_4 &= (x-3)[x]_3 = (x-3)(x-2)[x]_2 \\
&= (x-3)(x-2)(x-1)[x]_1 = (x-3)(x-2)(x-1)(x-0)[x]_0 \\
&= (x-3)(x-2)(x-1)x
\end{aligned}$$

Expanding $[x]_n$ yields a polynomial of degree n in x, where the coefficient of x^n is 1 and the remaining coefficients alternate in sign; also the constant term is zero. Let $(-1)^r s(n,r)$ denote the coefficient of x^r, where $1 \leq r \leq n$. Thus

$$[x]_n = \sum_{r=0}^{n} (-1)^{n-r} s(n,r)x^r \tag{20.2}$$

$$= s(n,n)x^n - s(n,1)x^{n-1} + \cdots + (-1)^n s(n,0) \cdot 1$$

In particular, let $x = n$. Then $[x]_n = [n]_n = n!$. So $n!$ can be expressed as a sum of multiples of powers of n:

$$n! = s(n,n)n^n - s(n,1)n^{n-1} + \cdots + (-1)^n s(n,0)$$

Two Interesting Observations

- Since the constant term is 0, it follows that $s(n,0) = 0$.
- $s(n,r)$ is the absolute value of the coefficient of x^r.

For example, considering the product $[x]_4$, $s(4,3) = |-6| = 6$ and $s(4,2) = |11| = 11$.

Clearly, the numbers $s(n,r)$ can also be realized as the coefficients in the expansion of the *ascending factorial*:

$$\langle x\rangle_n = x(x+1)(x+2)\cdots(x+n-1) = \sum_{r=0}^{n} s(n,r)x^r$$

where $\langle x\rangle_0 = 1$.

The numbers $s(n, r)$ are *Stirling numbers of the first kind.* The notation $s(n, r)$ was popularized in 1958 by J. Riordan of then Bell Telephone Laboratories, Inc., Murray Hill, New Jersey.

We will investigate a few fundamental properties of the numbers $s(n, r)$ in the next two theorems. One of them is a Pascal-like identity, a recurrence relation we can employ to generate the Stirling numbers $s(n, r)$.

Theorem 20.1 Let n be a positive integer. Then:

1) $s(n, 0) = 0$ 2) $s(n, n) = 1$
3) $s(n, 1) = (n-1)!$ 4) $s(n, n-1) = \binom{n}{r}$

Proof:

1) Since the constant term in the product $[x]_n$ is zero, $s(n, 0) = 0$.
2) The coefficient of x^n in the product $[x]_n$ is 1; so $s(n, n) = 1$.
3) The coefficient of x in $[x]_n$ is the constant term in the product $(x-1)\cdot(x-2)\cdots[x-(n-1)]$; namely, $(-1)(-2)\cdots[-(n-1)] = (-1)^{n-1}\cdot(n-1)!$. So $s(n, 1) = |(-1)^{n-1}(n-1)!| = (n-1)!$.
4) The coefficient of x^{n-1} in $[x]_n$ equals that of x^{n-2} in $(x-1)(x-2)\cdots[x-(n-1)]$. It is obtained by choosing the constant term from one of the factors in $(x-1)(x-2)\cdots[x-(n-1)]$ and equals

$$(-1)+(-2)+\cdots+[-(n-1)] = -\frac{n(n-1)}{2} = -\binom{n}{2}$$

Thus $s(n, n-1) = \binom{n}{2}$. ■

For example, $s(4, 0) = 0$, $s(4, 4) = 1$, $s(4, 1) = 3! = 6!$, and $s(4, 3) = \binom{4}{2} = 6$. The next theorem gives us a Pascal-like identity for $s(n, r)$.

Theorem 20.2 Let n be a positive integer and $1 < r < n$. Then

$$s(n, r) = s(n-1, r-1) + (n-1)s(n-1, r)$$

Proof: First notice that $[x]_n = [x-(n-1)][x]_{n-1}$. Using formula (20.2), we have $[x]_n = \sum_{r=0}^{n}(-1)^{n-r}s(n, r)x^r$. Then $[x]_{n-1} = \sum_{r=0}^{n-1}(-1)^{n-r-1}s(n-1, r)x^r$. Now, multiply both sides by $(x-n+1)$:

$$[x]_n = [x-(n-1)]\sum_{r=0}^{n-1}(-1)^{n-r-1}s(n-1, r)x^r$$

$$= \sum_{r=0}^{n-1}(-1)^{n-r-1}s(n-1, r)x^{r+1} - (n-1)\sum_{r=0}^{n-1}(-1)^{n-r-1}s(n-1, r)x^r$$

That is,

$$\sum_{r=0}^{n}(-1)^{n-r}s(n,r)x^r = \sum_{r=0}^{n-1}(-1)^{n-r-1}s(n-1,r)x^{r+1}$$
$$- (n-1)\sum_{r=0}^{n-1}(-1)^{n-r-1}s(n-1,r)x^r$$

Equating the coefficients of x^r from both sides we get:

$$(-)^{n-r}s(n,r) = (-1)^{n-r-2}s(n-1,r-1) - (n-1)(-1)^{n-r-1}s(n-1,r)$$

$$s(n,r) = s(n-1,r-1) + (n-1)s(n-1,r), \quad \text{as desired.}$$ ■

For example, $s(5,3) = s(4,2) + (5-1)s(4,3) = 11 + 4 \cdot 6 = 35$.

The recurrence relation in Theorem 20.2 together with the initial conditions in Theorem 20.1 yields the following recursive definition of $s(n, r)$.

A Recursive Definition of $s(n, r)$

$$\begin{aligned} &s(n,1) = (n-1)! \\ &s(n,n) = 1 \\ &s(n,r) = s(n-1,r-1) + (n-1)s(n-1,r), \quad 1 < r < n \end{aligned} \tag{20.3}$$

This recursive definition enables us to construct the Pascal-like triangle in Figure 20.1. Row n begins with $(n-1)!$ and each row ends in a 1. By virtue of the recurrence relation, each element $s(n, r)$ can be obtained by adding $n - 1$ times the element $s(n-1, r)$ just above it to its left neighbor $s(n-1, r-1)$.

n \ r	1	2	3	4	5	6	7	8
1	1							
2	1	1						
3	2	3	1					
4	6	11	6	1				
5	24	50	35	10	1			
6	120	274	225	85	15	1		
7	720	1764	1624	735	175	21	1	
8	5040	13068	13132	6769	1960	322	28	1

Figure 20.1 Stirling's Triangle for Cyclic Permutations

Since $s(n, n-1) = \binom{n}{2}$, the second diagonal from above consists of triangular numbers; see Figure 20.1.

Using this triangular array, the polynomial $[x]_n$ can be expanded conveniently and quickly. For example:

$$\begin{aligned}[x]_5 &= x(x-1)(x-2)(x-3)(x-4) \\ &= s(5,5)x^5 - s(5,4)x^4 + s(5,3)x^3 - s(5,3)x^2 + s(5,1)x \\ &= x^5 - 10x^4 + 35x^3 - 50x^2 + 24x\end{aligned}$$

Since polynomials are often written in increasing powers of x, the coefficients must be read from right to left in the array; always remember to insert the correct sign for each coefficient.

Theorems 20.1 and 20.2 can be used in tandem to derive the following byproduct.

Corollary 20.1 Let $n \geq 2$. Then $s(n, 2) = (n-1)!H_{n-1}$

Proof: By the recurrence relation, we have:

$$\begin{aligned}s(n,2) &= s(n-1,1) + (n-1)s(n-1,2) \\ &= (n-2)! + (n-1)s(n-1,2) \\ \frac{s(n,2)}{(n-1)!} &= s(n-1,2) + \frac{1}{n-1}\end{aligned}$$

Since $s(2, 2) = 1$, it follows that $\frac{s(n,2)}{(n-1)!}$ satisfies the same recursive definition as H_{n-1}. So $\frac{s(n,2)}{(n-1)!} = H_{n-1}$, as desired. ■

For example, $s(5, 2) = 4!H - 4 = 24 \cdot \frac{25}{12} = 50$. See the triangular array in Figure 20.1.

Harmonic Numbers and Stirling Numbers $s(n, r)$

Corollary 20.1 can be rewritten as $s(n+1, 2) = n!H_n$, where $n \geq 1$ and H_n denotes the harmonic number $H_n = \sum_{i=1}^{n} \frac{1}{i}$. In addition, there are two other relationships linking Stirling numbers $s(n, r)$ with harmonic numbers:

- $s(n+1, 3) = \frac{n!}{2}\left(H_n^2 - \sum_{i=1}^{n} \frac{1}{i^2}\right)$
- $s(n+1, 4) = \frac{n!}{6}\left(H_n^3 - 3H_n \sum_{i=1}^{n} \frac{1}{i^2} + 2\sum_{i=1}^{n} \frac{1}{i^3}\right)$

A Combinatorial Interpretation

The Stirling number $s(n, r)$ has an interesting combinatorial interpretation, as the next theorem demonstrates.

Theorem 20.3 Suppose n people are to be seated around r indistinguishable round tables with at least one person at each table, where $n \geq r$. Two seating arrangements at a table are considered the same if one is a cyclic permutation of the other. The number of such arrangements is given by $s(n, r)$.

Before we present a proof, we would like to clarify the phrases "indistinguishable round tables" and "cyclic permutation" by examples. The arrangements in Figure 20.2*a* and 20.2*b* are considered the same; so are the arrangements in Figures 20.2*c* and 20.2*d*. But the arrangements in Figure 20.2*a* and 20.2*e* are distinct.

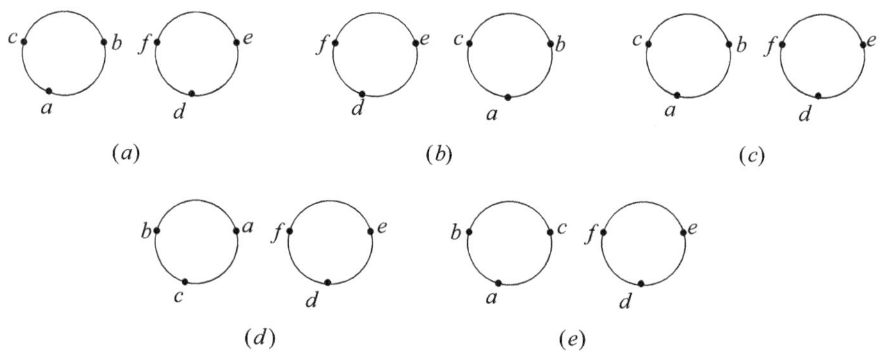

Figure 20.2

Proof: Let $A(n, r)$ denote the number of seating arrangements that satisfy the given conditions. We will show that $A(n, r)$ satisfies the recursive definition for $s(n, r)$.

First, suppose $r = n$. Since there are n people and n round tables, and no tables can be empty, there must be exactly one person at each table. But the tables are not distinguishable. There is exactly one such arrangement, so $A(n, n) = 1 = s(n, n)$.

Suppose $r = 1$. So there are n people and one round table. There are $\frac{n!}{n} = (n-1)!$ such cyclic permutations. So, $A(n, 1) = (n-1)! = s(n, 1)$.

Thus, $A(n, r)$ satisfies both initial conditions in (20.3).

It remains to show that $A(n, r)$ satisfies the same recurrence relation as $s(n, r)$. To this end, suppose we label the n people 1 through n.

Case 1 Suppose person n is sitting alone at a table.

This leaves $n-1$ people and $r-1$ tables. By definition, they can be seated in $A(n-1, r-1)$ ways.

Case 2 Suppose person n is sitting at a table with at least one other person.

Consider the persons 1 through $n-1$ and the r round tables. By definition, there are $A(n-1, r)$ ways they can be seated around the r round tables. Let $1 \leq i \leq n-1$. Now adding person n to the left of person i yields a seating arrangement of the n people around the r tables for each value of i. Since $1 \leq i \leq n-1$, the number of such arrangements equals $(n-1)A(n-1, r)$.

Thus, by the addition principle, the total number of seating arrangements is given by $A(n, r) = A(n-1, r-1) + (n-1)A(n-1, r)$. This is exactly the same as the recurrence relation satisfied by $s(n, r)$.

Thus, $A(n, r) = s(n, r)$. ■

Cycles of Permutations

Theorem 20.3 can be restated in terms of cycles of permutations, but first a definition. Consider the permutation σ of the elements of the set $\{1, 2, 3, 4, 5\}$, given by the 2×5 array

$$\sigma = \begin{pmatrix} 1 & 2 & 3 & 4 & 5 \\ 3 & 5 & 2 & 1 & 4 \end{pmatrix}$$

where $\sigma(i)$ appears below the element i and $1 \leq i \leq 5$. The permutation σ can be displayed pictorially in a cyclic fashion, as in Figure 20.3; so σ is often written as $\sigma = (13254)$, where each element is the image of the preceding element in a cyclic order. Notice that σ consists of exactly one *cycle*.

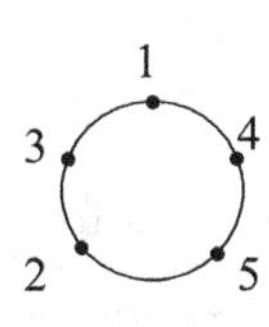

Figure 20.3

More generally, any permutation of n elements can be written as a product (that is, composition) of cycles. For example, the permutation

$$\mu = \begin{pmatrix} 1 & 2 & 3 & 4 & 5 \\ 3 & 1 & 2 & 5 & 4 \end{pmatrix}$$

can be written as $\mu = (132)(45)$ and

$$\begin{pmatrix} 1 & 2 & 3 & 4 & 5 \\ 1 & 3 & 2 & 5 & 4 \end{pmatrix} = (1)(23)(45)$$

It now follows from Theorem 20.3 that the Stirling numbers $s(n, r)$ represent the number of permutations of n items with r cycles.

For example, there are $s(3, 1) = 2$ permutations of the elements 1, 2, and 3 with $r = 1$ cycle: (123) and (132); there are $s(3, 2) = 3$ permutations with 2 cycles: (1)(23), (2)(13), and (3)(12); and there is exactly $s(3, 3) = 1$ permutation containing $r = 3$ cycles: (1)(2)(3). Thus the total number of permutations of three elements is given by $s(3, 1) + s(3, 2) + s(3, 3) = 2 + 3 + 1 = 6 = 3!$, as expected.

Table 20.1 lists the various permutations of the elements $1, 2, 3, \cdots, n$ as products of r cycles, where $1 \leq n \leq 4$.

Table 20.1 Permutations of the Set $\{1, 2, 3, \cdots, n\}$ with r Cycles

n \ r	1	2	3	4	Total
1	(1)				1 = 1!
2	(12)	(1)(2)			2 = 2!
3	(123) (132)	(1)(23) (2)(13) (3)(12)	(1)(2)(3)		6 = 3!
4	(1234) (1243) (1324) (1342) (1423) (1432)	(1)(234) (1)(243) (2)(134) (2)(143) (3)(124) (3)(142) (4)(123) (4)(132) (12)(34) (13)(24) (14)(23)	(1)(2)(34) (1)(3)(24) (1)(4)(23) (2)(3)(14) (2)(4)(13) (3)(4)(12)	(1)(2)(3)(4)	24 = 4!

It follows by Corollary 20.1 that the number of permutations of n items with exactly two cycles is also given by $(n-1)!H_{n-1}$; see Table 20.1 also.

Another interesting property of the Stirling numbers $s(n, r)$ is the fact that $\sum_{r=1}^{n} rs(n, r) = s(n+1, 2)$. This can be established by exhibiting a bijection between the set of permutations of n elements with an arbitrary number of cycles and the set of permutations of the set $\{0, 1, 2, \cdots, n\}$ with exactly two cycles.[†] By Corollary 20.1, this implies that $\frac{\sum_{r=1}^{n} rs(n,r)}{n!} = H_n$. That is, the average number of cycles in a permutation of n elements equals H_n.

Stirling's triangle in Figure 20.1 manifests yet another interesting property: The sum of the numbers in row n equals $n!$; that is, $\sum_{r=1}^{n} s(n, r) = n!$; see Figure 20.4. For example, $6 + 11 + 6 + 1 = 24 = 4!$.

[†]See A. Benjamin and J. J. Quinn, *Proofs That Really Count*, MAA, Washington, D.C., 2003; L. Comtet, *Advanced Combinatorics*, Dreidel, Boston, Massachusetts, 1974.

n \ r	1	2	3	4	5	6	7	8	Row sums
1	1								1 = 1!
2	1	1							2 = 2!
3	2	3	1						6 = 3!
4	6	11	6	1					24 = 4!
5	24	50	35	10	1				120 = 5!
6	120	274	225	85	15	1			720 = 6!
7	720	1764	1624	735	175	21	1		5040 = 7!
8	5040	13068	13132	6769	1960	322	28	1	40320 = 8!

Figure 20.4 Stirling's Triangle

We will now prove by induction that this observation was not just a coincidence, but is always the case.

Theorem 20.4 $\sum_{r=1}^{n} s(n, r) = n!$.

Proof (by PMI): Since $s(1, 1) = 1 = 1!$, the result is true when $n = 1$.

Now, assume that the result is true for an arbitrary positive integer m: $\sum_{r=1}^{m} s(m, r) = m!$. Then:

$$\sum_{r=1}^{m+1} s(m+1, r) = \sum_{r=1}^{m+1} [s(m, r-1) + ms(m, r)]$$

$$= \sum_{r=1}^{m+1} s(m, r-1) + m \sum_{r=1}^{m} s(m, r)$$

$$= \sum_{r=2}^{m+1} s(m, r-1) + m \cdot m! = \sum_{r=1}^{m} s(m, r) + m \cdot m!$$

$$= m! + m \cdot m! = (m+1)!$$

Thus, by induction, the result is true for every integer $n \geq 1$; that is, $\sum_{r=1}^{n} s(n, r) = n!$, as desired. ■

Just as the sum of the coefficients in $(1 + x)^n$ equals 2^n, this theorem implies that the sum of the absolute values of the coefficients in $[x]_n$ equals $n!$.

The following theorem[†] gives a generating function for the Stirling numbers $s(n, r)$.

Theorem 20.5

$$\sum_{n=r}^{\infty} s(n, r)\frac{x^n}{n!} = \frac{1}{r!}[\log(1+x)]^r$$ ■

We now turn to Stirling numbers of the second kind. (They occur in the study of Hanoi graphs.[†])

Stirling Numbers of the Second Kind

Stirling numbers of the second kind are denoted by $S(n, r)$, where $S(n, r)$ denotes the number of ways of partitioning an n-element set into r nonempty subsets. That is, $S(n, r)$ denotes the number of ways of distributing n distinct objects into r identical boxes, with no box left empty.

Although it is customary to use the set-notation using braces to list the various subsets in a partitioning, in the interest of brevity, we shall drop the braces and use vertical bars to separate one subset from another. For example, we will denote the partition $\{\{a, c\}, \{b, d, e\}\}$ of the set $\{a, b, c, d, e\}$ as $ac \mid bde$, which is obviously clearer and shorter.

The next example illustrates how to compute $S(n, r)$, when n is fairly small.

Example 20.1 Compute $S(4, 3)$.

Solution Consider the set $\{a, b, c, d\}$. There are exactly six ways of partitioning this set into three elements each. They are:

$ab \mid c \mid d$ $\quad ac \mid b \mid d$ $\quad ad \mid b \mid c$ $\quad bc \mid a \mid d$ $\quad bd \mid a \mid c$ $\quad cd \mid a \mid b$

Thus, $S(4, 3) = 6$. See Figure 20.5.

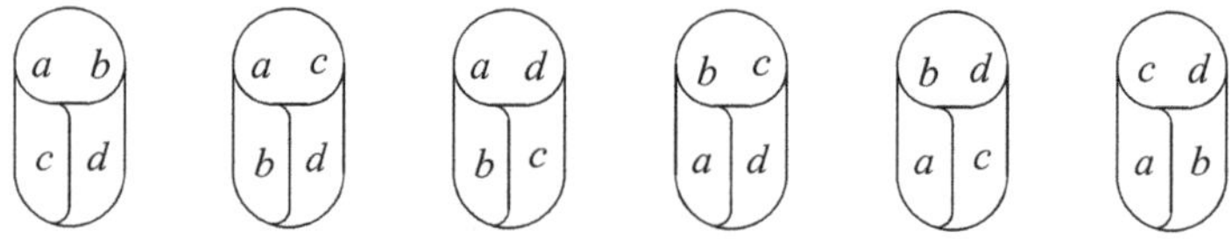

Figure 20.5

■

[†]See J. H. van Lint and R. M. Wilson, *A Course in Combinatorics*, Cambridge University Press, New York, 1992.

[†]See the article by S. Klavžar et al., 2001.

The next theorem provides us with a few fundamental properties of Stirling numbers $s(n, r)$. Some of them remind us of the basic properties of $s(n, r)$. But, first, let us study the following examples.

Example 20.2 Let $S = \{a, b, c, d\}$. Suppose the element d belongs to a subset of a partition of S by itself, so the partition looks as follows: $abc \mid d$. The set $\{a, b, c\}$ has $2^3 - 1 = 7$ nonempty subsets: $\{a\}$, $\{b\}$, $\{c\}$, $\{a, b\}$, $\{a, c\}$, $\{b, c\}$, and $\{a, b, c\}$. Now pair each subset A with the difference $S - A$:

$$\begin{array}{ccl} & \text{yields} & \\ a & \rightarrow & a \mid bcd \\ b & \rightarrow & b \mid acd \\ c & \rightarrow & c \mid abd \\ a, b, c & \rightarrow & abc \mid d \end{array} \qquad \begin{array}{ccl} & \text{yields} & \\ a, b & \rightarrow & ab \mid cd \\ a, c & \rightarrow & ac \mid bd \\ b, c & \rightarrow & bc \mid ad \end{array}$$

All partitions of S into two subsets

(Again, for simplicity, we have dropped the braces for subsets.) Each pairing results in a partitioning of S. Thus the number of nonempty subsets of $\{a, b, c\}$ equals the number of partitions of S into two subsets. ∎

Example 20.3 Let $S = \{a, b, c, d\}$. Suppose d belongs to a subset with at least one more element in a partition of S, say, $A = \{c, d\}$: $ab \mid cd$. The set $\{a, b\}$ has three nonempty subsets: $\{a\}$, $\{b\}$, and $\{a, b\}$. Now, pair each subset A with $S - A$:

$$\begin{array}{ccl} & \text{yields} & \\ a & \rightarrow & a \mid bcd \\ b & \rightarrow & b \mid acd \\ a, b & \rightarrow & ab \mid cd \end{array}$$

Notice that the resulting partitions are already accounted for in Example 20.2. In other words, we get no new partitions. ∎

We are now ready to present the fundamental results.

Theorem 20.6 Let $n \geq 1$. Then:

1) $S(n, 0) = 0$ 2) $S(n, 1) = 1$ 3) $S(n, 2) = 2^{n-1} - 1$
4) $S(n, n-1) = \binom{n}{2}$ 5) $S(n, n) = 1$

Proof: Let $S = \{a_1, a_2 \cdots a_n\}$.

1) Since it is impossible to partition a set with n elements into no subsets, $S(n, 0) = 0$.
2) There is only one way all the n elements can be put into one subset: namely, the set itself. So $S(n, 1) = n$.
3) **Case 1** Suppose a_n belongs to a subset by itself in a partition: $a_1, a_2 \cdots a_{n-1} \mid a_n$. The set $\{a_1, a_2 \cdots a_{n-1}\}$ has 2^{n-1} subsets, one of which is the null set.

So it has $2^{n-1} - 1$ nonempty subsets. Pairing each subset A with the difference $S - A$ gives a partition of S (see Example 20.2).

Case 2 Suppose a_n belongs to a subset with at least one other element. This case produces no new partitions of S, as we saw in Example 20.3.

Thus, the total number of partitions of S into two nonempty subsets equals $2^{n-1} - 1$; that is, $S(n, 2) = 2^{n-1} - 1$.

4) To produce a partition of S into $n - 1$ subsets, we must have one subset of two elements and $n - 2$ subsets of one element each: $a_1 \mid a_2 \mid \cdots \mid a_i a_j \mid \cdots \mid a_n$. Each two-element subset of S determines such a partition and vice versa. So the number of such partitions equals the number of two-element subsets of S: namely, $\binom{n}{2}$. Thus $S(n, n-1) = \binom{n}{2}$.
5) A set of n elements can be partitioned into n subsets in exactly one way, so $S(n, n) = 1$. ■

For example, $S(3, 0) = 0$, $S(3, 1) = 1$, $S(3, 2) = 3$, and $S(3, 3) = 1$. You can confirm this using the set $S = \{a, b, c\}$.

The next theorem provides us with a recurrence relation for $S(n, r)$; it looks analogous to the one for $s(n, r)$. But before we present it, let us study the next two examples in the hope that they will help us follow the proof of the theorem easily.

Example 20.4 Consider the following partition of the set $\{a, b, c, d, e\}$: $abcd \mid e$. Now partition the set $A = \{a, b, c, d\}$ into two subsets. There are seven such partitions:

$a \mid bcd$	$ab \mid cd$
$b \mid acd$	$ac \mid bd$
$c \mid abd$	$ad \mid bc$
$d \mid abc$	

Now, insert $\{e\}$ as a new subset in each partition of A:

	yields			yields	
$a \mid bcd$	$\rightarrow$	$a \mid bcd \mid e$	$ab \mid cd$	$\rightarrow$	$ab \mid cd \mid e$
$b \mid acd$	$\rightarrow$	$b \mid acd \mid e$	$ac \mid bd$	$\rightarrow$	$ac \mid bd \mid e$
$c \mid abd$	$\rightarrow$	$c \mid abd \mid e$	$ad \mid bc$	$\rightarrow$	$ad \mid bc \mid e$
$d \mid abc$	$\rightarrow$	$d \mid abc \mid e$			

All partitions of S

Each insertion results in a partition of S and there are seven such partitions of S. ■

This example will come in handy when we pursue Case 1 of the proof of Theorem 20.7.

Example 20.5 Consider the following partitions of the set $S = \{a, b, c, d\}$:

$$\begin{array}{ll} a \mid bcd & ab \mid cd \\ b \mid acd & ac \mid bd \\ c \mid abd & bc \mid ad \end{array}$$

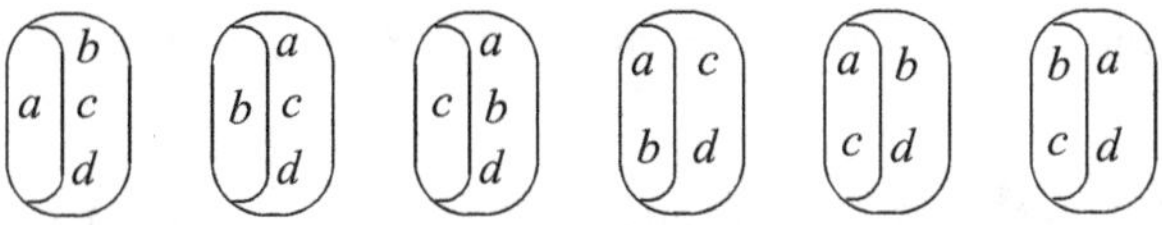

Figure 20.6 Subsets Containing d and at Least One More Element

See Figure 20.6. In each case, d belongs to a subset that contains at least one other element. Deleting d from each yields the following partitions of the set $\{a, b, c\}$ (see Figure 20.7 also):

$$\begin{array}{l} a \mid bc \\ b \mid ac \\ c \mid ab \end{array}$$

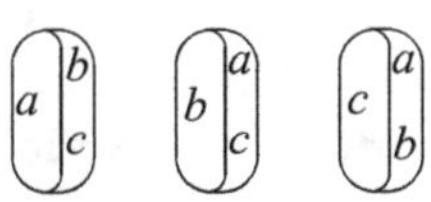

Figure 20.7 Three Partitions of $\{a, b, c\}$

Interestingly, these are the various partitions of the set $\{a, b, c\}$ into two subsets.

Now, reinsert d in each subset; since each of these partitions is made up of two subsets, d can be inserted in the subset on the left of the vertical bar or in the subset on its right, as in Figure 20.8.

	yields	
$a \mid bc$	$\rightarrow$	$ad \mid bc$ or $a \mid bcd$
$b \mid ac$	$\rightarrow$	$bd \mid ac$ or $b \mid acd$
$c \mid ab$	$\rightarrow$	$cd \mid ab$ or $c \mid abd$

Figure 20.8 Six Partitions of S

Notice that the set $\{a, b, c\}$ has three partitions into two subsets each. By inserting d in a subset of each partition, we were able to produce $6 = 2 \cdot 3$ partitions of S. See Figure 20.9.

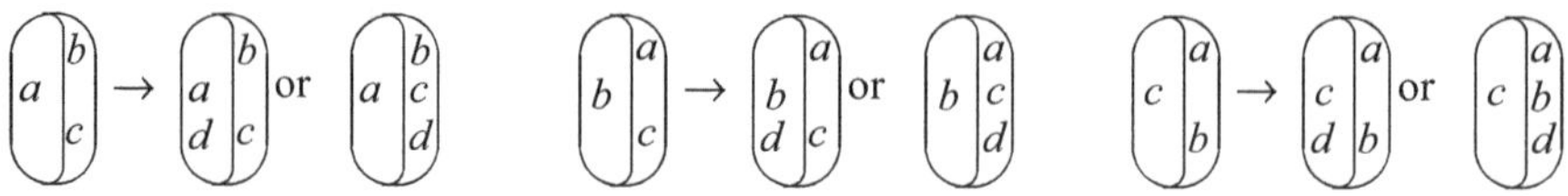

Figure 20.9

We will employ this technique in the development of Case 2 in the following theorem.

Theorem 20.7 Let $n \geq r \geq 1$. Then $S(n, r) = S(n-1, r-1) + rS(n-1, r)$.

Proof: Consider a partition of the set $A = \{a_1, a_2, \cdots, a_n\}$ into r subsets.

Case 1 Suppose $\{a_n\}$ is a subset of the partition.

By definition, the remaining elements $a_1, a_2, \cdots, a_{n-1}$ can be partitioned into $r-1$ subsets in $S(n-1, r-1)$ ways. Adding $\{a_n\}$ to each such partition of $\{a_1, a_2, \cdots, a_{n-1}\}$ results in a partition of S (see Example 20.4). Thus there are $S(n-1, r-1)$ such partitions of A.

Case 2 Suppose $\{a_n\}$ is not a subset of the partition.

Then a_n belongs to a subset that contains at least one other element. Deleting a_n from all such subsets yields a partition of the set $B = \{a_1, a_2, \cdots, a_{n-1}\}$ into r subsets. There are $S(n-1, r)$ such partitions of B. Inserting a_n in each subset of every partition of B produces a partition of A. Since each partition of B consists of r subsets, each such partition contributes r partitions of A. But there are $S(n-1, r)$ partitions of B. So the total number of such partitions of A equals $rS(n-1, r)$ (see Example 20.5).

Thus, by the addition principle, the total number of partitions of A into r subsets equals $S(n-1, r-1) + rS(n-1, r)$. That is, $S(n, r) = S(n-1, r-1) + rS(n-1, r)$. ■

For example, $S(4, 2) = S(3, 1) + 2S(3, 2) = 1 + 2 \cdot 3 = 7$. You may confirm this using the set $A = \{a, b, c, d\}$.

Theorem 20.7 coupled with the initial conditions in Theorem 20.6 yields the following recursive definition of Stirling numbers of the second kind.

A Recursive Definition of $S(n, r)$

$$
\begin{aligned}
&S(n, 1) = 1 \\
&S(n, n) = 1 \\
&S(n, r) = S(n-1, r-1) + rS(n-1, r), \qquad 1 < r < n
\end{aligned}
\tag{20.4}
$$

This recursive definition can be used to construct the triangular array of Stirling numbers $S(n, r)$, as in Figure 20.10. By virtue of the boundary conditions, as in Pascal's triangle, each row begins with and ends in a 1. Every interior entry $S(n, r)$ can be obtained by multiplying the element $S(n-1, r)$ just above it by the column number r and then adding its left neighbor $S(n-1, r-1)$ to the product. For example, $90 = 15 + 3 \cdot 25$ (see Figure 20.10).

$n \backslash r$	1	2	3	4	5	6	7	8
1	1							
2	1	1						
3	1	3	1					
4	1	7	6	1				
5	1	15	25	10	1			
6	1	31	90	65	15	1		
7	1	63	301	350	140	21	1	
8	1	127	966	1701	1050	266	28	1

Figure 20.10 Stirling's Triangle for Partitions

Two Observations About The Array

- Since $S(n, 2) = 2^{n-1} - 1$, column 2 consists of Mersenne numbers.
- Since $S(n, n-1) = \binom{n}{2}$, the next-to-the last southeast diagonal consists of triangular numbers.

Stirling Numbers $S(n, r)$, e^x, and Differentiation

In 1928, J. Ginsburg of Columbia University made a fascinating observation about Stirling numbers $S(n, r)$. They can be obtained as coefficients in $e^{-x}\phi_n(x)$ in the reverse order, where $\phi_n(x)$ is defined recursively as follows:

$$\phi_0(x) = e^x$$

$$\phi_n(x) = \left[x\phi_{n-1}(x)\right]', \qquad n \geq 1$$

where the prime denotes differentiation with respect to x.

For example:

$\phi_0(x)=e^x$

$\phi_1(x)=[x\phi_0(x)]'=(x+1)e^x$

$\phi_2(x)=[x\phi_1(x)]'=(x^2+3x+1)e^x$

$\phi_3(x)=[x\phi_2(x)]'=(x^3+6x^2+7x+1)e^x$

$\phi_4(x)=[x\phi_3(x)]'=(x^4+10x^3+25x^2+15x+1)e^x$

The coefficients in the polynomials $e^{-x}\phi_n(x)$ can be used to construct the triangular array in Figure 20.11. It is a reflection of the Stirling's triangle in Figure 20.10 about a vertical line on the far right.

n \ r	1	2	3	4	5	6	7	8
1	1							
2	1	1						
3	1	3	1					
4	1	6	7	1				
5	1	10	25	15	1			
6	1	15	65	90	31	1		
7	1	21	140	350	301	63	1	
8	1	28	966	1050	1701	960	127	1

Figure 20.11 Stirling's Triangle for Partitions Reflected About a Vertical Line

Earlier we found that

$$[x]_n = \sum_{r=0}^{n}(-1)^{n-r}s(n, r)x^r \tag{20.5}$$

The expansion is in fact "reversible"; known as the *Newton–Gregory expansion* of x^n, it is given by the following theorem.

Theorem 20.8 Let $n \geq 0$. Then

$$x^n = \sum_{r=0}^{n} S(n, r)[x]_r \tag{20.6}$$

Proof (by PMI): Since $S(0, 0) = 1$, $[x]_0 = 1 \cdot 1 = 1 = x^0$. When $n = 1$,

$$\sum_{r=0}^{n} S(n, r)[x]_r = S(1, 0)[x]_0 + S(1, 1)[x]_1 = 0 \cdot 1 + 1 \cdot x = x = x^1$$

Now, assume that the theorem is true for exponents $\leq n$; that is,

$$x^n = \sum_{r=0}^{n} S(n,r)[x]_r = \sum_{r=0}^{n} \binom{x}{r} r! S(n,r) \qquad \geq 0$$

Then

$$x^{n+1} = x \cdot x^n = \sum_{r=0}^{n} x \binom{x}{r} r! S(n,r)$$

By Pascal's identity, $\binom{x}{r} = \binom{x-1}{r} + \binom{x-1}{r-1}$, so $x\binom{x}{r} = (r+1)\binom{x}{r+1} + r\binom{x}{r}$. Therefore,

$$\begin{aligned}
x^{n+1} &= \sum_{r=0}^{n} \left[(r+1)\binom{x}{r+1} + r\binom{x}{r} \right] r! S(n,r) \\
&= \sum_{r=0}^{n} (r+1)! \binom{x}{r+1} S(n,r) + \sum_{r=0}^{n} r \cdot r! \binom{x}{r} S(n,r) \\
&= \sum_{r=1}^{n+1} r! \binom{x}{r} S(n,r-1) + \sum_{r=0}^{n+1} r \cdot r! \binom{x}{r} S(n,r) \\
&= \sum_{r=0}^{n+1} r! \binom{x}{r} \left[S(n,r-1) + rS(n,r) \right] \\
&= \sum_{r=0}^{n+1} \binom{x}{r} r! S(n+1,r)
\end{aligned}$$

Thus, by induction, the formula works for every integer $n \geq 0$. ■

For example,

$$\begin{aligned}
\sum_{r=0}^{3} S(3,r)[x]_r &= S(3,0)[x]_0 + S(3,1)[x]_1 + S(3,2)[x]_2 + S(3,3)[x]_3 \\
&= 0 \cdot 1 + 1 \cdot x + 3 \cdot x(x-1) + 1 \cdot x(x-1)(x-2) \\
&= x + 3x(x-1) + x(x-1)(x-2) = x^3
\end{aligned}$$

Notice that the Newton–Gregory expansion of x^n in Theorem 20.8 can also be written as[†]

$$x^n = \sum_{r=0}^{n} r! S(n,r) \binom{x}{r}$$

[†] See C. Jordan, *Calculus of Finite Differences*, Chelsea, New York, 1947.

Stirling Meets Fibonacci

In 1967, D. Lind of the University of Virginia, Charlotteville, Virginia, proposed a fascinating problem linking Fibonacci numbers F_n and Stirling numbers of both kinds. The solution given here is based on his 1969 solution and is a delightful application of Theorem 20.8 and Binet's formula for F_n.

Example 20.6 Show that $F_n = \sum_{r=0}^{n} \sum_{j=0}^{r} (-1)^{r-j} S(n,r)\, s(r,j) F_j$.

Proof: By Theorem 20.8,

$$x^n = \sum_{r=0}^{n} S(n,r)[x]_r$$

$$= \sum_{r=0}^{n} S(n,r) \sum_{j=0}^{r} (-1)^{r-j} s(r,j) x^j$$

$$= \sum_{r=0}^{n} \sum_{j=0}^{r} (-1)^{r-j} S(n,r) s(r,j) x^j$$

So

$$\alpha^n - \beta^n = \sum_{r=0}^{n} \sum_{j=0}^{r} (-1)^{r-j} S(n,r) s(r,j)(\alpha^j - \beta^j)$$

That is,

$$F_n = \sum_{r=0}^{n} \sum_{j=0}^{r} (-1)^{r-j} S(n,r) s(r,j) F_j$$

as desired. ■

In particular, let $n = 5$. Then $\sum_{r=0}^{5} \sum_{j=0}^{r} (-1)^{r-j} S(5,r) s(r,j) F_j = 1 + 0 + 25 - 40 + 19 = 5 = F_5$, as expected.

Two Interesting Observations

- It follows by Theorem 20.8 that all basis elements x^k in the vector space of polynomials in x of degree $\le n$ over the field of real numbers can be expressed in terms of the descending factorials $[x]_r$, where $0 \le r \le n$. Consequently, $\{[x]_1, [x]_2, [x]_3, \cdots, [x]_n\}$ is a basis of the vector space.

For example:

$$\begin{array}{ll} x^0 = 1 & x^1 = [x]_1 \\ x^2 = [x]_2 + [x]_1 & x^3 = [x]_3 + 3[x]_2 + [x]_1 \\ x^4 = [x]_4 + 6[x]_3 + 7[x]_2 + [x]_1 & x^5 = [x]_5 + 10[x]_4 + 25[x]_3 + 15[x]_2 + [x]_1 \end{array}$$

You may confirm these expansions. This is exactly how Stirling discovered the numbers $S(n, r)$, as a tool to express x^n as a linear combination of the descending factorials $[x]_r$, where $1 \leq r \leq n$.

- Substituting for $[x]_r = \sum_{j=0}^{r}(-1)^{r-j}s(r,j)x^j$ in formula (20.6), we get:

$$\begin{aligned} x^n &= \sum_{r=0}^{n} S(n, r) \sum_{j=0}^{r}(-1)^{r-j}s(r,j)x^j \\ &= \sum_{r=0}^{n}\sum_{j=0}^{r}(-1)^{r-j}S(n, r)s(r,j)x^j \end{aligned}$$

This is true if and only if

$$\sum_{r=j}^{n}(-1)^{r-j}S(n, r)s(r,j) = \delta_{nj} \tag{20.7}$$

where δ_{nj} denotes the Kronecker delta.
Property (20.7) implies that matrices of the Stirling numbers $S(n, r)$ and those of the numbers $s(n, r)$ with proper signs are multiplicative inverses of each other.

For example,

$$\begin{bmatrix} 1 & 0 & 0 \\ 1 & 1 & 0 \\ 1 & 3 & 1 \end{bmatrix}\begin{bmatrix} 1 & 0 & 0 \\ -1 & 1 & 0 \\ 2 & -3 & 1 \end{bmatrix} = \begin{bmatrix} 1 & 0 & 0 \\ 0 & 1 & 0 \\ 0 & 0 & 1 \end{bmatrix} = \begin{bmatrix} 1 & 0 & 0 \\ -1 & 1 & 0 \\ 2 & -3 & 1 \end{bmatrix}\begin{bmatrix} 1 & 0 & 0 \\ 1 & 1 & 0 \\ 1 & 3 & 1 \end{bmatrix}$$

The following theorem† can be used to generate the Stirling numbers $S(n, r)$.

Theorem 20.9

$$\sum_{n=r}^{\infty} S(n, r)\frac{x^n}{n!} = \frac{1}{r!}(e^x - 1)^r \qquad \blacksquare$$

†See J. H. van Lint and R. M. Wilson, *A Course in Combinatorics*, Cambridge University Press, New York, 1992.

Stirling Numbers $S(m, n)$ and Surjections

Interestingly, there is a close relationship between $S(m, n)$ and the number of surjections from a finite set with m elements to a finite set with n elements. To see this, we state the next two results, without proofs in the interest of brevity.

Theorem 20.10*

$$S(m, n) = \frac{1}{n!}\sum_{r=0}^{n}(-1)^r\binom{n}{r}(n-r)^m \quad \blacksquare$$

Theorem 20.11† Let A and B be finite sets with $|A| = m$ and $|B| = n$, where $|S|$ denotes the cardinality of the set S and $m \geq n$. The number of surjections N from A to B is given by

$$N = \sum_{r=0}^{n}(-1)^r\binom{n}{r}(n-r)^m \quad \blacksquare$$

These two theorems yield the following result between $S(m, n)$ and the number of surjections.

Corollary 20.2 There are $n!S(m, n)$ surjections from a finite set with m elements to a finite set with n elements, where $m \geq n$. ■

For example, let $m = 3$ and $n = 2$. Then there are $n!S(m, n) = 2!S(3, 2) = 2 \cdot 3 = 6$ surjections from $\{a, b, c\}$ to $\{x, y\}$. They are displayed in Figure 20.12.

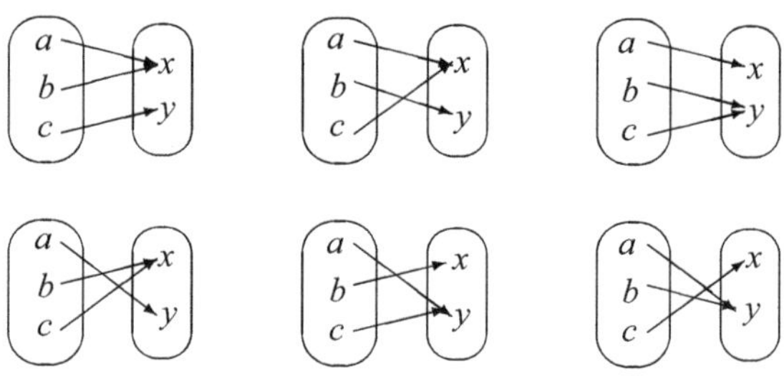

Figure 20.12 Surjections from $\{a, b, c\}$ to $\{x, y\}$

*See R. A. Brauldi, *Introductory Combinatorics*, 3rd edition, Prentice-Hall, Upper Saddle River, New Jersey, 1999.

†For a proof, see the author's *Discrete Mathematics with Applications*, Elsevier, Burlington, Massachusetts, 2004.

Non-surjections

Once again, let $|A| = m$ and $|B| = n$. Then n^m functions can be defined from A to B. Since $n!S(m, n)$ of them are surjections, it follows that $n^m - n!S(m, n)$ of them are non-surjections.

Corollary 20.3 Let $|A| = m$ and $|B| = n$, where $m \geq n$. Then $n^m - n!S(m, n)$ non-surjections can be defined from A to B. ∎

For example, let $m = 3$ and $n = 2$. Then $n^m - n!S(m, n) = 2^3 - 6 = 2$ non-surjections can be defined from $\{a, b, c\}$ to $\{x, y\}$. They are displayed in Figure 20.13.

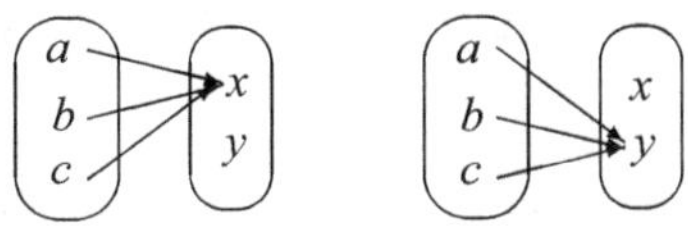

Figure 20.13 Non-surjections from $\{a, b, c\}$ to $\{x, y\}$

Ordered Trees and Stirling Numbers $S(n, r)$

The next example presents an interesting application of the Stirling numbers $S(n, r)$ to the theory of trees. It was proposed as a problem in 1997 by E. Deutsch of Polytechnic University, Brooklyn, New York. The proof given is based on the one given by R. J. Chapman of the University of Exeter, United Kingdom in 1999.

Before presenting the problem, we first give some additional graph-theoretic terminology. A graph is *acyclic* if it contains no cycles. A connected, acyclic graph is a *tree*. A *rooted tree* is a tree with a specifically designated vertex, called the *root*. (All trees can be drawn as rooted trees.) An *ordered tree* is a rooted tree in which the vertices at each level are ordered. The *height* of an ordered tree is the number of edges along a path of maximum length beginning at the root. Now we turn to the application.

Example 20.7 Let $a(n, k)$ denote the number of ordered trees with n edges and height k, where $n, k \geq 0$. Prove that $a(n, 2) = S(n, 2)$.

Proof: Notice that $a(n, 1) = 1 = S(n, 1)$. Let $b(n, k)$ denote the number of ordered trees with n edges and height $\leq k$. Then $b(0, k) = 1$ for $k \geq 0$.

We will let generating functions do the job for us. To this end, let $g_k(x) = \sum_{n\geq 0} b(n, k)x^n$. Since $g_0(x) = 1$, we let $k > 0$. An ordered tree of height $\leq k$ consists of the root v, r (= *degree* of v) edges emanating from it, and a sequence of r ordered trees of height $\leq k-1$ rooted at the children of v. So the generating function of ordered trees of height $\leq k$ with $\deg(v) = r$ is given by $g_{k-1}^r x^r$, where $g_i = g_i(x)$. Thus $g_k = \sum_{r\geq 0} g_{k-1}^r x^r = \frac{1}{1-g_{k-1}x}$. So $g_1 = \frac{1}{1-x}$ and hence:

$$g_2 = \frac{1}{1-g_1 x} = \frac{1-x}{1-2x} = 1 + \frac{x}{1-2x}$$

$$= 1 + x\sum_{n=0}^{\infty} 2^n x^n$$

so $b(n,2) = 2^{n-1}$, where $n \geq 1$. But $b(n,2) = a(n,1) + a(n,2) = 1 + a(n,2)$, so $a(n,2) = 2^{n-1} - 1 = S(n,2)$, as desired. ■

To pursue this example a bit further, it follows that $g_3(x) = \frac{1-2x}{1-3x+x^2}$ and

$$g_4 = \frac{1-3x+x^2}{1-4x+3x^2} = 1 + \frac{x}{2(1-x)} + \frac{x}{2(1-3x)}$$

$$= 1 + \frac{x}{2}\sum_{n=0}^{\infty} x^n + \frac{x}{2}\sum_{n=0}^{\infty} 3^n x^n$$

Consequently, $b(n,4) = \frac{1}{2} + \frac{3^{n-1}}{2} = \frac{3^{n-1}+1}{2}$. Hence:

$$a(n,3) + a(n,4) = b(n,4) - b(n,2)$$

$$= \frac{3^{n-1}+1}{2} - 2^{n-1} = \frac{3^{n-1} - 2^n + 1}{2}$$

Using partitions of the set $\{1, 2, 3, \cdots, n\}$ into at most three classes, Chapman showed that $b(n,4) = 1 + S(n,2) + S(n,3)$. This yields an explicit formula for $S(n,3)$:

$$S(n,3) = b(n,4) - S(n,2) - 1$$

$$= \frac{3^{n-1}+1}{2} - (2^{n-1} - 1) = \frac{3^{n-1} - 2^n + 1}{2}$$

We now add that the formula $a(n,3) + a(n,4) = S(n,3)$ was also discovered by Deutsch.

21

Bell's Triangles

This chapter presents another class of positive integers from combinatorics, called *Bell numbers*, named after the Scottish–American mathematician Eric Temple Bell (1883–1960). We will see how they are related to Stirling numbers of the second kind, pursued in the preceding chapter.

Bell Numbers

Recall that the Stirling number $S(n, r)$ denotes the different number of ways of partitioning a set of n elements into r nonempty subsets, where $0 \leq r \leq n$. The *Bell number* B_n, on the other hand, denotes the total number of ways of partitioning the set. Consequently,

$$B_n = S(n, 0) + S(n, 1) + \cdots + S(n, n) \tag{21.1}$$

n \ r	1	2	3	4	5	6	7	8	Row sums
1	1								1
2	1	1							2
3	1	3	1						5
4	1	7	6	1					15
5	1	15	25	10	1				52
6	1	31	90	65	15	1			203
7	1	63	301	350	140	21	1		877
8	1	127	960	1701	1050	266	28	1	4140

Figure 21.1 Stirling's Triangle for Partitions

For example, the set $\{a, b, c\}$ has five possible partitions:

$$\begin{array}{lll} abc & a \mid bc & b \mid ac \\ c \mid ab & a \mid b \mid c & \end{array}$$

So $B_3 = 5$.

By virtue of property (21.1), every Bell number B_n can be obtained by adding the various elements in row n in Stirling's triangle in Figure 21.1.

For example, $B_4 = 1 + 7 + 6 + 1 = 15$. You may confirm this using the set $\{a, b, c, d\}$.

Partitions and Equivalence Relations

Since there is a bijection between the set of partitions of a set S with n elements and the set of equivalence relations that can be defined on S,[†] it follows that the total number of equivalence relations that can be defined on S is given by B_n.

For example, let $S = \{a, b, c\}$. Exactly, $B_3 = 5$ partitions and hence five equivalence relations can be defined on S. The various partitions of S are displayed in Figure 21.2. The corresponding equivalence relations are

$$\{(a,a),(b,b),(c,c)\},\ \{(a,a),(a,b),(b,a),(b,b),(c,c)\},$$

$$\{(a,a),(a,c),(b,b),(c,a),(c,c)\},\ \{(a,a),(b,b),(b,c),(c,b),(c,c)\}$$

$$\{(a,a),(a,b),(a,c),(b,a),(b,b),(b,c),(c,a),(c,b),(c,c)\}$$

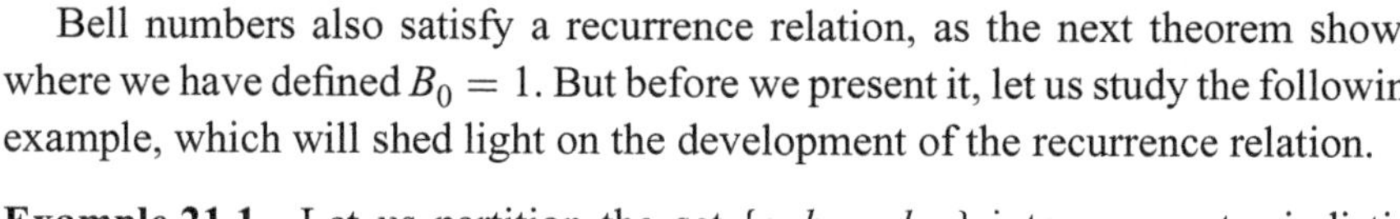

Figure 21.2

Bell numbers also satisfy a recurrence relation, as the next theorem shows, where we have defined $B_0 = 1$. But before we present it, let us study the following example, which will shed light on the development of the recurrence relation.

Example 21.1 Let us partition the set $\{a, b, c, d, e\}$ into nonempty, indistinguishable boxes; that is, into nonempty subsets. There must be a box containing at least the element e. That box must contain a subset X of $A = \{a, b, c, d\}$; X can be empty.

Case 1 Suppose X contains two elements.

[†]See the author's *Discrete Mathematics with Applications*.

Subcase 1 Let $X = \{a, b\}$. Then the corresponding partition of S has the form

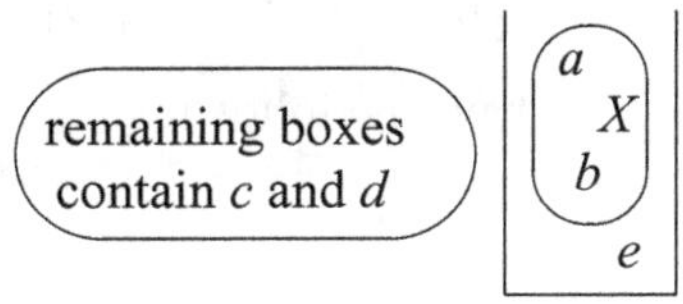

Then $A - X = \{c, d\}$ can be partitioned in two different ways: $c \mid d$ or cd. Thus, $X = \{a, b\}$ contributes two partitions of S:

$$c \mid d \mid abe$$
$$cd \mid abe$$

Subcase 2 Let $X = \{a, c\}$. Then $A - X = \{b, d\}$ can be partitioned in two different ways: $b \mid d$ or bd. Thus, $X = \{a, c\}$ contributes two partitions of S:

$$b \mid d \mid ace$$
$$bd \mid ace$$

Similarly, X can be $\{a, d\}$, $\{b, c\}$, $\{b, d\}$, $\{c, d\}$. Corresponding to each, we get two partitions of S, a total of 12 partitions. They are displayed in Figure 21.3

$a \mid b \mid cde$	$b \mid c \mid ade$
$ab \mid cde$	$bc \mid ade$
$a \mid c \mid bde$	$b \mid d \mid ace$
$ac \mid bde$	$bd \mid ace$
$a \mid d \mid bce$	$c \mid d \mid abe$
$ad \mid bce$	$cd \mid abe$

Figure 21.3

Before we look at the other cases, we make an important observation. In each case, X is a two-element subset of A, where $|A| = 4$. So X can be chosen in $\binom{4}{2} = 6$ different ways; each X contributes $B_2 = 2$ partitions; so, by the multiplication principle, the total number of partitions contributed by two-element subsets A equals $\binom{4}{2}B_2 = 6 \cdot 2 = 12$, as we just found.

There are four other cases remaining: X may contain 0, 1, 3, or 4 elements. As above, they contribute $\binom{4}{0}B_0$, $\binom{4}{1}B_1$, $\binom{4}{3}B_3$, and $\binom{4}{4}B_4$ partitions, respectively. Thus, by the addition principle, the total number of partitions of S is given by $B_5 = \binom{4}{0}B_0 + \binom{4}{1}B_1 + \binom{4}{2}B_2 + \binom{4}{3}B_3 + \binom{4}{4}B_4 = 52$; see Figure 21.1. ∎

More generally, we have the following result.

Theorem 21.1 Let $n \geq 1$. Then

$$B_n = \binom{n-1}{0}B_0 + \binom{n-1}{1}B_1 + \cdots + \binom{n-1}{n-1}B_{n-1}$$

Proof: We partition the set $S = \{1, 2, 3 \cdots n\}$ into nonempty indistinguishable boxes. Then there must be a box containing the element n. That box must contain a subset X of the set $\{1, 2, 3, \cdots, n-1\}$; notice that X can be empty.

Suppose X contains r elements, where $0 \leq r \leq n-1$. An r-element subset of the set $\{1, 2, 3, \cdots, n-1\}$ can be chosen in $\binom{n-1}{r}$ different ways. The remaining $n-1-r$ elements, by definition, can be partitioned into B_{n-1-r} nonempty indistinguishable boxes. Thus, by the multiplication principle, such subset X contribute $\binom{n-1}{r}B_{n-1-r}$ partitions.

Since $0 \leq r \leq n-1$, by the addition principle, the total number of partitions into nonempty indistinguishable boxes is given by

$$B_n = \sum_{r=0}^{n-1} \binom{n-1}{r} B_{n-1-r}$$

$$= \sum_{i=n-1}^{0} \binom{n-1}{n-1-i} B_i = \sum_{i=0}^{n-1} \binom{n-1}{i} B_i$$ ∎

For example, we have $B_0 = 1, B_1 = 1, B_2 = 2, B_3 = 5, B_4 = 15$, and $B_5 = 52$. So $B_6 = \binom{5}{0}B_0 + \binom{5}{1}B_1 + \binom{5}{2}B_2 + \binom{5}{3}B_3 + \binom{5}{4}B_4 + \binom{5}{5}B_5 = 203$.

Additional Bell's Triangles

The recurrence relation in Theorem 21.1 can be used to compute each Bell number B_n as a weighted sum of its predecessors B_i, where the weights are the binomial coefficients $\binom{n-1}{i}$ and $0 \leq i \leq n-1$. This is exactly how we just computed B_6. Bell numbers and their successive differences can be employed to construct the triangular array in Figure 21.4.

1	2	5	15	52	**203**	877	⋯	← Bell numbers
1	3	10						
2	7							
5								
⋱								

↖ Bell numbers

Figure 21.4

There is another triangular array B which can also be used to generate Bell numbers. We will keep it left-justified. The array B can be defined recursively, as follows:

$$B(1,1) = 1$$

$$B(n,j) = 0, \qquad n \geq 2$$

$$B(n,1) = B(n-1, n-1)$$

$$B(n,j) = B(n-1, j-1) + B(n, j-1), \qquad 2 \leq j \leq n$$

In words, the top row consists of 1. The last element of each row becomes the first element in the following row. Each succeeding element in a row is obtained by adding the neighboring element on the left and its northern neighbor. Figure 21.5 shows the resulting Bell's triangle.

1					
1	2				
2	3	5			
5	7	10	15		
15	20	27	37	52	
52	67	87	114	151	203

Figure 21.5

A Generating Function for Bell Numbers

Finally, Bell numbers can also be generated using the exponential generating function

$$e^{e^{x-1}} = 1 + x + 2\frac{x^2}{2!} + 5\frac{x^3}{3!} + \cdots$$

22

Euler's Triangles

In 1948, P. A. Piza of San Juan, Puerto Rico, studied a class of positive integers called *D-numbers*. This chapter investigates them and shows how closely linked they are to Stirling numbers $S(n, r)$ and Eulerian numbers.

D-numbers

D-numbers $D(n, r)$ are defined recursively as follows:

$$D(n, 1) = 1, \qquad n \geq 1$$

$$D(n, r) = 0, \qquad r < 1$$

$$D(n, r) = 0, \qquad r > n$$

$$D(n, r) = r[D(n-1, r) + D(n-1, r-1)], \qquad 1 \leq r \leq n \tag{22.1}$$

For example, $D(2, 2) = 2[D(1, 2) + D(1, 1)] = 2(0 + 1) = 2$. Similarly, $D(3, 2) = 6$.

The recursive definition can be employed to construct a triangular array of D-numbers, as in Figure 22.1. The recurrence relation provides us with a neat algorithm to compute each element $D(n, r)$: add the elements in positions r and $r - 1$ in the previous row and then multiply the sum by r. For example, $D(6, 4) = 4(240 + 150) = 1{,}560$.

An observation: $D(n, n) = n!$ for every positive integer n. For example, $D(8, 8) = 40320 = 8!$. The proof that $D(n, n) = n!$ for all n follows by induction from the fact that $D(n, n) = n[D(n-1, n) + D(n-1, n-1)] = n[0 + D(n-1, n-1)] = nD(n-1, n-1)$.

n \ r	1	2	3	4	5	6	7	8
1	1							
2	1	2						
3	1	6	6					
4	1	14	36	24				
5	1	30	150	240	120			
6	1	62	540	1560	1800	720		
7	1	126	1806	8400	16800	15120	5040	
8	1	254	5796	40824	126000	191520	141120	40320

Figure 22.1 A Triangular Array of D-numbers

By the recurrence relation, $D(1, 1) = D(0, 1) + D(0, 0)$; that is, $1 = D(0, 1) + D(0, 0)$. But $D(0, 1) = 0$. So $D(0, 0)$ can be defined as 1.

Another observation: $2! \mid D(3, 2)$, $3! \mid D(5, 3)$, and $4! \mid D(6, 4)$.

More generally, we have the following result, which we will establish using induction.

Theorem 22.1 $r! \mid D(n, r)$ for all integers $n \geq 1$, where $r \geq 1$.

Proof (by strong induction): Since $D(n, 1) = 1$, clearly, $1! \mid D(n, 1)$ for every $n \geq 1$.

Now, assume that $k! \mid D(n, k)$ for every positive integer $k < r$, where n is an arbitrary positive integer. By the recurrence relation (22.1), we have $D(n, r) = rD(n-1, r) + rD(n-1, r-1)$. By the inductive hypothesis, $(r-1)! \mid D(n-1, r-1)$ and $r! \mid D(n-1, r)$. So $r! \mid rD(n-1, r-1)$ and $r! \mid D(n-1, r)$. Thus $r! \mid [rD(n-1, r) + rD(n-1, r-1)]$; that is, $r! \mid D(n, r)$. ■

n \ r	1	2	3	4	5	6	7	8
1	1							
2	1	1						
3	1	3	1					
4	1	7	6	1				
5	1	15	25	10	1			
6	1	31	90	65	15	1		
7	1	63	301	350	140	21	1	
8	1	127	966	1701	1050	266	28	1

Figure 22.2

It follows by this theorem that $\frac{D(n,r)}{r!}$ is a positive integer. Dividing each $D(n, r)$ by $r!$ in Figure 22.1 yields the array in Figure 22.2. This is exactly the same as Stirling's triangle for partitions in Figure 20.10.

Consequently, we conjecture that $S(n, r) = \frac{D(n,r)}{r!}$. The following theorem confirms this result.

Theorem 22.2 Let $n, r \geq 0$. Then $\frac{D(n,r)}{r!} = S(n, r)$.

Proof: It suffices to show that $\frac{D(n,r)}{r!}$ satisfies the recursive definition of $S(n, r)$ in Chapter 20.

- Since $D(n, 1) = 1$, it follows that $S(n, 1) = 1$.
- $D(n, n) = n!$; so $\frac{D(n,n)}{n!} = 1 = S(n, n)$.
- Since $D(n, 0) = 0$, it follows that $\frac{D(n,0)}{0!} = 0$, where $n \geq 1$.
- When $r > n$, $D(n, r) = 0$. So $\frac{D(n,r)}{r!} = 0$, when $r > n$.
- Since $D(0, 0) = 1$, it follows that $\frac{D(n,r)}{r!} = 1$.
- It now remains to show that $\frac{D(n,r)}{r!}$ satisfies the same recurrence relation as $S(n, r)$. To this end, we have:

$$\frac{D(n-1, r-1)}{(r-1)!} + r \cdot \frac{D(n-1, r)}{r!} = \frac{1}{(r-1)!}[D(n-1, r-1) + D(n-1, r)]$$

$$= \frac{D(n, r)}{r \cdot (r-1)!} = \frac{D(n, r)}{r!}$$

So $\frac{D(n,r)}{r!}$ satisfies exactly the same recurrence relation as $S(n, r)$.

Thus $\frac{D(n,r)}{r!}$ satisfies the same recursive definition as $S(n, r)$. Consequently, $\frac{D(n,r)}{r!} = S(n, r)$, where $0 \leq r \leq n$. ■

Piza established the following interesting properties of D-numbers:

$$m^n - (m-1)^n = \sum_{r=1}^{n} D(n, r)\binom{m-1}{r-1} \tag{22.2}$$

$$m^n = \sum_{r=1}^{n} D(n, r)\binom{m}{r} \tag{22.3}$$

$$\sum_{i=1}^{m} i^n = \sum_{r=1}^{n} D(n, r)\binom{m+1}{r+1} \tag{22.4}$$

For example, let $m = 5$ and $n = 4$. Then:

$$\sum_{r=1}^{4} D(4, r)\binom{4}{r-1} = D(4, 1)\binom{4}{0} + D(4, 2)\binom{4}{1} + D(4, 3)\binom{4}{2} + D(4, 4)\binom{4}{3}$$

$$= 1 \cdot 1 + 14 \cdot 1 + 36 \cdot 6 + 24 \cdot 4 = 369 = 5^4 - 4^4$$

$$\sum_{r=1}^{4} D(4,r)\binom{5}{r} = D(4,1)\binom{5}{1} + D(4,2)\binom{5}{2} + D(4,3)\binom{5}{3} + D(4,4)\binom{5}{4}$$

$$= 1 \cdot 5 + 14 \cdot 10 + 36 \cdot 10 + 24 \cdot 5 = 625 = 5^4$$

Formula (22.4) gives a delightful mechanism for computing the sum of the nth powers of the first m positive integers.

For example,

$$\sum_{i=1}^{6} i^{10} = \sum_{r=1}^{10} D(10,r)\binom{7}{r+1}$$

$$= D(10,1)\binom{7}{2} + D(10,2)\binom{7}{3} + D(10,3)\binom{7}{4} + D(10,4)\binom{7}{5}$$

$$+ D(10,5)\binom{7}{6} + D(10,3)\binom{7}{7} + 0$$

$$= 71{,}340{,}451$$

Using this technique, Piza computed the sum of the 10th powers of the first 1000 positive integers fairly easily:

$$\sum_{i=1}^{1000} i^{10} = \sum_{r=1}^{1000} D(10,r)\binom{1001}{r+1}$$

$$= 91, 409, 924, 241, 424, 243, 424, 241, 924, 242, 500$$

You may confirm this using formula (22.4).

Formula (22.4), for example, can easily be established from formula (22.3), using the fact that $\sum_{i=r}^{m} \binom{i}{r} = \binom{m+1}{r+1}$, as follows. We have $i^n = \sum_{r=1}^{n} D(n,r)\binom{i}{r}$. Therefore:

$$\sum_{i=1}^{m} i^n = \sum_{i=1}^{m}\sum_{r=1}^{n} D(n,r)\binom{i}{r} = \sum_{r=1}^{n} D(n,r)\sum_{i=1}^{m}\binom{i}{r}$$

$$= \sum_{r=1}^{n} D(n,r)\sum_{i=r}^{m}\binom{i}{r} = \sum_{r=1}^{n} D(n,r)\binom{m+1}{r+1}$$

Summation Formulas

Formula (22.4) can be employed to develop the familiar summation formulas. For example:

$$\sum_{i=1}^{m} i = D(1,1)\binom{m+1}{2} = 1\binom{m+1}{2} = \frac{m(m+1)}{2};$$

$$\sum_{i=1}^{m} i^2 = \sum_{r=1}^{2} D(2,r)\binom{m+1}{r+1} = D(2,1)\binom{m+1}{2} + D(2,2)\binom{m+1}{3}$$

$$= \binom{m+1}{2} + 2\binom{m+1}{3} = \frac{m(m+1)}{2} + 2\cdot\frac{(m-1)m(m+1)}{6}$$

$$= \frac{m(m+1)(2m+1)}{6}$$

Similarly, $\sum_{i=1}^{m} i^3 = \left[\frac{m(m+1)}{2}\right]^2$ and $\sum_{i=1}^{m} i^4 = \frac{m(m+1)(6m^3+9m^2+m-1)}{30}$.

D-numbers and *S*(*n*, *r*)

Earlier we encountered a close relationship between D-numbers and Stirling numbers $S(n, r)$; namely,

$$D(n,r) = r!S(n,r) \tag{22.5}$$

For example, $3!S(5,3) = 6\cdot 25 = 150 = D(5,3)$.

By virtue of the relationship (22.5), the summation formulas (22.2) through (22.4) can be rewritten in terms of $S(n, r)$:

$$m^n - (m-1)^n = \sum_{r=1}^{n}\binom{m-1}{r-1} r!S(n,r) \tag{22.6}$$

$$m^n = \sum_{r=1}^{n}\binom{m}{r} r!S(n,r) \tag{22.7}$$

$$\sum_{i=1}^{m} i^n = \sum_{r=1}^{n}\binom{m+1}{r+1} r!S(n,r) \tag{22.8}$$

Eulerian Numbers Revisited

Interestingly, D-numbers can be used to generate Eulerian numbers. Before we formally introduce them, let us return to formula (22.3): $m^n = \sum_{r=1}^{n} D(n, r)\binom{m}{m-r}$. Then, by Pascal's identity, we have:

$$\begin{aligned}
5^5 &= \sum_{r=1}^{5} D(5, r)\binom{5}{5-r} \\
&= 1\binom{5}{4} + 30\binom{5}{3} + 150\binom{5}{2} + 240\binom{5}{1} + 120\binom{5}{0} \\
&= 1\binom{9}{5} + 26\binom{8}{5} + 66\binom{7}{5} + 26\binom{6}{5} + 1\binom{5}{5}
\end{aligned}$$

where we have omitted a number of steps for the sake of brevity. The coefficients in this sum are symmetrically located and form a partition of $5! = 120 : 5! = 1 + 26 + 66 + 26 + 1$.

The same technique can be employed to find the partitions of $n!$, where $n \geq 1$. For example:

$1^1 = 1\binom{1}{1}$ $\qquad$ $2^2 = 1\binom{3}{2} + 1\binom{2}{2}$

$3^3 = 1\binom{5}{3} + 4\binom{4}{3} + 1\binom{3}{3}$ $\qquad$ $4^4 = 1\binom{7}{4} + 11\binom{6}{4} + 11\binom{5}{4} + 1\binom{4}{4}$

$5^5 = 1\binom{9}{5} + 26\binom{8}{5} + 66\binom{7}{5} + 26\binom{6}{5} + 1\binom{5}{5}$ $\qquad$ $6^6 = 1\binom{11}{6} + 57\binom{10}{6} + 302\binom{9}{6} + 57\binom{7}{6} + 1\binom{6}{6}$

Collecting the various coefficients, we have:

$$\begin{aligned}
1! &= 1 & 4! &= 1 + 11 + 11 + 1 \\
2! &= 1 + 1 & 5! &= 1 + 26 + 66 + 26 + 1 \\
3! &= 1 + 4 + 1 & 6! &= 1 + 57 + 302 + 302 + 57 + 1
\end{aligned}$$

Recursive Definition of Eulerian Numbers

The numbers in these partitions are *Eulerian numbers* $E(n, r)$. They can be defined recursively as follows:

$$E(n, 1) = 1, \quad E(n, n) = 1$$

$$E(n, r) = 0, \quad r > n$$

$$E(n, r) = 0, \quad r < 1$$

$$E(n, r) = rE(n-1, r) + (n+1-r)E(n-1, r-1), \qquad 1 < r < n \tag{22.9}$$

This recursive definition can be used to construct a triangular array of Eulerian numbers, as in Figure 22.3.

n \ r	1	2	3	4	5	6	7	8	Row sums
1	1								1!
2	1	1							2!
3	1	4	1						3!
4	1	11	11	1					4!
5	1	26	66	26	1				5!
6	1	57	302	302	57	1			6!
7	1	120	1191	2416	1191	120	1		7!
8	1	247	4293	15619	15619	4293	247	1	8!

Figure 22.3 Euler's Triangle

As in Pascal's triangle, each row begins with and ends in a 1; and each row is symmetric about the vertical line through the middle. Also, each element $E(n, r)$ can be obtained by multiplying $E(n-r, r)$ by r, multiplying $E(n-1, r-1)$ by $(n+1-r)$, and then adding the two products.

For example, $E(6, 3) = 3E(5, 3) + (6+1-3)E(5, 2) = 3 \cdot 66 + 4 \cdot 26 = 302$.

It appears from the above discussion that the sum of the Eulerian numbers in row n is $n!$, where $n \geq 1$. The next theorem in fact confirms it.

Theorem 22.3 $\sum_{r=1}^{n} E(n, r) = n!$

Proof (by PMI): Since $E(1, 1) = 1!$ and $\sum_{r=1}^{2} E(2, r) = 2!$, the result is true when $n = 1$ and $n = 2$.

Now, assume that it is true for an arbitrary integer $n \geq 2$. Then, by the recurrence relation, we have:

$$\sum_{r=1}^{n} E(n,r) = \sum_{r=1}^{n} rE(n-1,r) + \sum_{r=1}^{n} (n+1-r)E(n-1,r-1)$$

$$= \sum_{r=1}^{n-1} rE(n-1,r) + (n+1)\sum_{r=1}^{n} E(n-1,r-1) - \sum_{r=1}^{n} rE(n-1,r-1)$$

$$= \sum_{r=1}^{n-1} rE(n-1,r) + (n+1)(n-1)! - \sum_{s=0}^{n-1} (s+1)E(n-1,s)$$

$$= \sum_{r=1}^{n-1} rE(n-1,r) + (n+1)(n-1)! - \sum_{s=1}^{n-1} sE(n-1,s) - \sum_{s=1}^{n-1} E(n-1,s)$$

$$= \sum_{r=1}^{n-1} rE(n-1,r) + (n+1)(n-1)! - \sum_{s=1}^{n-1} sE(n-1,s) - (n-1)!$$

$$= (n+1)!(n-1)! - (n-1)! = n!$$

Thus, by induction, the result is true for every $n \geq 1$. ∎

Since $E(n, r) = E(n, n - r + 1)$, the entries in row n in Euler's triangle are symmetric about the vertical line through the middle. For example, $E(8, 3) = 4293 = E(8, 6)$. This also implies that column r in the triangle consists of the same numbers as those in the rth diagonal, where $r = 1$ corresponds to the main southeast diagonal.

It follows from the recurrence relation (22.9) that:

$$\begin{aligned} E(n, 2) &= 2E(n-1, 2) + (n+1-2)E(n-1, 1) \\ &= 2E(n-1, 2) + (n-1) \cdot 1 \\ &= 2E(n-1, 2) + (n-1) \end{aligned} \tag{22.10}$$

This recurrence relation can be used to compute $E(n, 2)$ directly from its predecessor $E(n - 1, 2)$.

For example, $E(6, 2) = 2E(5, 2) + 5 = 2 \cdot 26 + 5 = 57$.

The recurrence relation (22.10) can be employed to establish an explicit formula for $E(n, 2)$, as the following theorem shows.

Theorem 22.4 $E(n, 2) = 2^n - n - 1$

Proof (by PMI): Since $2^2 - 2 - 1 = 1 = E(2, 2)$, the formula works for $n = 2$.

Now, assume that it is true for an arbitrary integer $n \geq 2$. Then, by the recurrence relation (22.10), $E(n + 1, 2) = 2E(n, 2) + n = 2(2^n - n - 1) + n = 2^{n+1} - (n + 1) - 1$. So the formula works for $n + 1$ also.

Thus, by induction, the theorem is true for every $n \geq 1$. ∎

For example, $E(8, 2) = 2^8 - 8 - 1 = 247$.

The summation formulas (22.2) through (22.4) have analogous results for Eulerian numbers:

$$m^n - (m-1)^n = \sum_{r=1}^{n} \binom{m+n-r-1}{n-1} E(n, r) \tag{22.11}$$

$$m^n = \sum_{r=1}^{n} \binom{m+n-r}{n} E(n, r) \tag{22.12}$$

$$\sum_{i=1}^{m} i^n = \sum_{r=1}^{n} \binom{m+n-r+1}{n+1} E(n, r) \tag{22.13}$$

For example, let $m = 5$ and $n = 4$. Then:

$$\sum_{r=1}^{4}\binom{5+4-r-1}{4-1}E(4,r) = 369 = 5^4 - 4^4$$

$$\sum_{r=1}^{4}\binom{5+4-r}{4}E(4,r) = 1\cdot 70 + 11\cdot 35 + 11\cdot 15 + 1\cdot 5 = 625 = 5^4$$

$$\sum_{r=1}^{4}\binom{5+4-r+1}{4+1}E(4,r) = 1\cdot 126 + 11\cdot 56 + 11\cdot 21 + 1\cdot 6 = 979 = \sum_{i=1}^{5} i^4$$

Formula (22.12) is known as *Worpitzky's identity*, after the German mathematician Julius Daniel Theodor Worpitsky, who developed it in 1883.

In the same year, he also established the following relationship between the Stirling numbers $S(n, r)$ and the Eulerian numbers $E(n, i)$: $S(n, r)r! = \sum_{i=0}^{n}\binom{i-1}{n-r}E(n, i)$.

For example, $\sum_{i=0}^{5}\binom{i-1}{5-3}E(5, i) = 1\cdot 66 + 3\cdot 26 + 6\cdot 1 = 150 = 25\cdot 3! = S(5, 3)3!$.

Eulerian Numbers and Summation Formulas

Just as D-numbers can be used to develop the familiar summation formulas, so can Eulerian numbers. For example, using formula (22.13),

$$\sum_{i=1}^{m} i = \binom{m+1-1+1}{1+1}E(1,1) = \binom{m+1}{2} = \frac{m(m+1)}{2}$$

and

$$\sum_{i=1}^{m} i^3 = \sum_{r=1}^{3}\binom{m+3-r+1}{4}E(3,r) = \sum_{r=1}^{3}\binom{m-r+4}{4}E(3,r)$$

$$= \binom{m+3}{4}E(3,1) + \binom{m+2}{4}E(3,2) + \binom{m+1}{4}E(3,3)$$

$$= \binom{m+3}{4} + 4\binom{m+2}{4} + \binom{m+1}{4}$$

$$= \frac{m(m+1)}{24}[(m+3)(m+2) + 4(m+2)(m-1) + (m-1)(m-2)]$$

$$= \frac{m(m+1)}{24} \cdot 6m(m+1) = \left[\frac{m(m+1)}{2}\right]^2$$

An Explicit Formula for $E(n, r)$

In 1994, D. Neal of Western Kentucky University developed an explicit formula for the Eulerian number $E(n, r)$. To establish it, we need the following theorem. The proof given here is based on the one by Neal.

Theorem 22.5 Let $|x| < 1$ and $n \geq 1$. Then

$$\sum_{m=1}^{\infty} m^n x^m = \sum_{r=1}^{n} \frac{E(n, r)x^r}{(1-x)^{n+1}}$$

Proof (by PMI): When $n = 1$:

$$\text{RHS} = \frac{E(1, 1)x}{(1-x)^2} = \frac{x}{(1-x)^2} = \sum_{m=1}^{\infty} mx^m = \text{LHS}$$

So the result is true when $n = 1$.

When $n = 2$:

$$\text{RHS} = \sum_{r=1}^{2} \frac{E(2, r)x^r}{(1-x)^3} = \frac{x + x^2}{(1-x)^3} = \sum_{m=1}^{\infty} m^2 x^m$$

So the result is also true when $n = 2$.

Now, assume that the theorem holds for $n - 1$:

$$\sum_{m=1}^{\infty} m^{n-1} x^m = \sum_{r=1}^{n-1} \frac{E(n-1, r)x^r}{(1-x)^n}$$

Using the quotient rule in calculus, this yields:

$$\sum_{m=1}^{\infty} m^n x^m = \frac{x}{(1-x)^{2n}}\left[\left(\sum_{r=1}^{n-1} rE(n-1, r)x^{r-1}\right)(1-x)^n - \left(\sum_{r=1}^{n-1} E(n-1, r)\right)(-n)(1-x)^{n-1}\right]$$

$$= \frac{1}{(1-x)^{n+1}}\left[\left(\sum_{r=1}^{n-1} rE(n-1, r)x^r\right)(1-x) + \sum_{r=1}^{n-1} nE(n-1, r)x^{r+1}\right]$$

$$= \frac{1}{(1-x)^{n+1}}\left[\sum_{r=1}^{n-1} rE(n-1, r)x^r + \sum_{r=1}^{n-1}(n-r)E(n-1, r)x^{r+1}\right]$$

$$= \frac{1}{(1-x)^{n+1}} \left\{ E(n-1,1)x + \sum_{r=2}^{n-1} [rE(n-1,r) + (n-r+1)E(n-1,r-1)]x^r \right.$$

$$\left. + E(n-1,n-1)x^n \right\}$$

$$= \frac{1}{(1-x)^{n+1}} \left(x + \sum_{r=2}^{n-1} E(n,r)x^r + x^n \right) = \sum_{r=1}^{n} \frac{E(n,r)x^r}{(1-x)^{n+1}}$$

Thus the theorem is true for every positive integer n. ■

This theorem gives us a way to develop a formula for the infinite series $\sum_{m=1}^{\infty} m^n x^m$ using row n of Euler's triangle in Figure 22.3, where $|x| < 1$. For example,

$$\sum_{m=1}^{\infty} m^5 x^m = \sum_{r=1}^{5} \frac{E(5,r)x^r}{(1-x)^6} = \frac{x + 26x^2 + 66x^3 + 26x^4 + x^5}{(1-x)^6}$$

where $|x| < 1$.

We are now ready to present the explicit formula for $E(n,r)$, developed by Neal.

Theorem 22.6

$$E(n,r) = \sum_{i=0}^{r-1} (-1)^i \binom{n+1}{i} (r-i)^n$$

Proof: By Theorem 22.5,

$$\sum_{r=1}^{n} E(n,r)x^r = \sum_{m=1}^{\infty} m^n x^m (1-x)^{n+1}$$

Now we would like to find the coefficient of x^r on the RHS. To this end, we apply the binomial theorem on the RHS. Then

$$\sum_{r=1}^{n} E(n,r)x^r = \sum_{m=1}^{\infty} \sum_{i=0}^{n+1} n^m (-1)^i \binom{n+1}{i} x^{m+i} \tag{22.14}$$

To find the coefficient of x^r on the RHS, we let $m = r - i$. So, when $m = 1$, $i = r - 1$; when $m = 2$, $i = r - 2$, and so on; and when $m = r$, $i = 0$. Consequently, the coefficient of x^r on the RHS is given by $\sum_{i=0}^{r-1} (-1)^i \binom{n+1}{i} (r-i)^n$. But the coefficient of x^r on the LHS is $E(n,r)$. Equating the two coefficients, we get the desired formula: $E(n,r) = \sum_{i=0}^{r-1} (-1)^i \binom{n+1}{i} (r-i)^n$. ■

For example, $\sum_{i=0}^{3} (-1)^i \binom{6}{i} (4-i)^5 = \binom{6}{0} 4^5 - \binom{6}{1} 3^5 + \binom{6}{2} 2^5 - \binom{6}{3} 1^5 = 26 = E(5,4)$, as expected.

The formula for $E(n, 2)$ in Theorem 22.4 can now be deduced from this theorem.

Corollary 22.1 $E(n, 2) = 2^n - n - 1$

Proof: By Theorem 22.6,

$$E(n, 2) = \sum_{i=0}^{1}(-1)^i \binom{n+1}{i}(r-i)^n = \binom{n+1}{0}2^n - \binom{n+1}{1} = 2^n - n - 1$$

as desired. ■

A Different Notation for *E*(*n*, *r*)

Eulerian numbers $E(n, r)$ are often denoted by $\left\langle {n \atop r} \right\rangle$, using angle brackets, as we saw in Chapter 7:

$$\left\langle {n \atop r} \right\rangle = E(n, r)$$

Using this notation, Euler's triangle can rewritten as in Figure 22.4. It should remind us of Pascal's triangle written using the bi-level notation.

$$\left\langle {1 \atop 1} \right\rangle$$

$$\left\langle {2 \atop 1} \right\rangle \quad \left\langle {2 \atop 2} \right\rangle$$

$$\left\langle {3 \atop 1} \right\rangle \quad \left\langle {3 \atop 2} \right\rangle \quad \left\langle {3 \atop 3} \right\rangle$$

$$\left\langle {4 \atop 1} \right\rangle \quad \left\langle {4 \atop 2} \right\rangle \quad \left\langle {4 \atop 3} \right\rangle \quad \left\langle {4 \atop 4} \right\rangle$$

Figure 22.4 Euler's Triangle

Eulerian Numbers and Permutations

Eulerian numbers $E(n, r)$ have a delightful combinatorial interpretation: $E(n, r)$ denotes the number of permutations $p_1 p_2 \cdots p_n$ of the set $\{1, 2, 3, \cdots, n\}$ with exactly r *ascents*; that is, the permutation $p_1 p_2 \cdots p_n$ contains r places where $p_i < p_{i+1}$. (Using this interpretation, the angle brackets, which remind us of the *less than* and *greater than* relations, are a good choice for the notation.)

For example, the permutation 13245 has three ascents and one descent, and the permutation 54321 has zero ascents and four descents. Three interesting observations:

- The minimum number of ascents in a permutations of the elements $1, 2, \cdots, n$ is 0 and the maximum number of ascents is $n - 1$.
- Let r denote the number of ascents and s the number of descents in a permutation of the elements $1, 2, 3, \cdots, n$. Then $r + s = n - 1$.
- Let $\alpha = \alpha_1\alpha_2 \cdots \alpha_n$ be a permutation of the elements $1, 2, 3, \cdots, n$ with r ascents (and hence $n - r - 1$ descents). Then its reflection α' about the vertical line at the end of α has $n - r - 1$ ascents (and hence r descents).

For example, $\alpha = 13245$ has 3 ascents and 1 descent. Its reflection $\alpha' = 54231$ has 1 ascent and 3 descents.

Consider the $3! = 6$ permutations of the set $\{1, 2, 3\}$: 132, 213, 231, 312, 321. Exactly $1 = E(3, 1)$ of them contains 0 ascents: 321; exactly $4 = E(3, 2)$ of them contain 1 ascent: 132, 213, 231, 312; and exactly $1 = E(3, 3)$ of them contain 2 ascents: 123. They account for all the 3! permutations: $1 + 4 + 1 = 3!$.

As another example, consider the $4! = 24$ permutations of the set $\{1, 2, 3, 4\}$:

1234	2134	3124	4123
1243	2143	3142	4132
1324	2314	3214	4213
1342	2341	3241	4231
1423	2413	3412	4312
1432	2431	3421	4321

Exactly $1 = E(4, 1)$ of them contains 0 ascents: 4321; exactly $11 = E(4, 2)$ of them contain 1 ascent: 1432, 2143, 2431, 3142, 3214, 3241, 3421, 4132, 4213, 4231, 4312; exactly $11 = E(4, 3)$ of them contain 2 ascents: 1243, 1324, 1342, 1423, 2134, 2314, 2341, 2413, 3124, 3412, 4123; and exactly $1 = E(4, 4)$ of them contains 3 ascents: 1234.

We are now ready to present the theorem.

Theorem 22.7 The number of permutations of the set $S = \{1, 2, 3, \cdots, n\}$ with r ascents is given by $E(n, r + 1)$, where $0 \le r \le n - 1$.

Proof: Let $P(n, r)$ denote the number of permutations of S with r ascents. Since there is exactly one permutation with no ascents, $P(n, 0) = 1$; also, there is exactly one permutation with $n - 1$ ascents, namely, $123 \cdots (n - 1)n$, so $P(n, n - 1) = 1$.

We now develop a recurrence formula for $P(n, r)$. First, notice that each permutation $\alpha = \alpha_1\alpha_2 \cdots \alpha_{n-1}$ of $\{1, 2, 3, \cdots, n - 1\}$ contributes n permutations of S, since n can be inserted in α in n different locations.

Secondly, suppose α has r ascents. Suppose we insert n in position i in α to yield a permutation β of S: $\beta = \alpha_1\alpha_2 \cdots \alpha_{i-1}n\alpha_i \cdots \alpha_{n-1}$. If $i = 1$, then β still has r ascents since n is larger than every α_i; if $1 < i < n$ and $\alpha_{i-1} < \alpha_i$, then $\alpha_{i-1} < n$, but $n > \alpha_i$; then also β has r ascents. By definition, there are $P(n-1, r)$ permutations of $\{1, 2, 3, \cdots, n-1\}$ that have r ascents; each contributes $r+1$ permutations of S with r ascents. So, the total number of such permutations equals $(r+1)P(n-1, r)$.

On the other hand, suppose $i = n$ and α has $r-1$ ascents. Since $\alpha_{n-1} < n$, β has one more ascent than α, namely, r. If $1 < i < n$ and $\alpha_{i-1} > \alpha_i$, then $\alpha_{i-1} < n$; this introduces a new ascent between α_{i-1} and α_i, for a total of r ascents. Since $2 \leq i \leq n-1$ and α has $r-1$ ascents, it has $(n-2)-(r-1)$ descents; the corresponding β has $(n-2)-(r-1)+1 = n-r$ descents and hence r ascents. Since n can be inserted at each descent and each such insertion contributes a permutation of S with r ascents, we can obtain a total of $(n-r)P(n-1, r-1)$ permutations of S from those of $\{1, 2, 3 \cdots n-1\}$ with $r-1$ ascents.

Thus, by the addition principle, $P(n, r) = (r+1)P(n-1, r) + (n-r)P(n-1, r-1)$. It now follows from the recursive definition of $E(n, r)$ that $P(n, r) = E(n, r+1)$. In other words, there are $E(n, r+1)$ permutations of the set $\{1, 2, 3 \cdots n\}$ with exactly r ascents, where $0 \leq r \leq n-1$. ■

Eulerian Numbers of the Second Kind

n \ j	0	1	2	3	4	5
1	1					
2	1	2				
3	1	8	6			
4	1	22	58	24		
5	1	52	328	444	120	
6	1	114	1452	4400	3708	720

Figure 22.5

We close this chapter with another triangular array of nonnegative integers, *Eulerian numbers of the second kind* $\left\langle\!\left\langle {n \atop j} \right\rangle\!\right\rangle$. They are defined recursively as follows, where $n \geq 1$:

$$\left\langle\!\left\langle {n \atop 0} \right\rangle\!\right\rangle = 1, \quad \left\langle\!\left\langle {n \atop n} \right\rangle\!\right\rangle = 0$$

$$\left\langle\!\left\langle {n \atop j} \right\rangle\!\right\rangle = (2n-j-1)\left\langle\!\left\langle {n-1 \atop j-1} \right\rangle\!\right\rangle + (j+1)\left\langle\!\left\langle {n-1 \atop j} \right\rangle\!\right\rangle$$

Notice that $\left\langle\!\left\langle {n \atop n-1} \right\rangle\!\right\rangle = \text{n!}$.

These numbers can be used to construct the triangular array in Figure 22.5. They have an interesting combinatorial interpretation, discovered in 1978 by I. M. Gessel of Brandeis University, Waltham, Massachusetts, and R. P. Stanley of MIT: $\left\langle\!\left\langle {n \atop j} \right\rangle\!\right\rangle$ denotes the number of permutations of the multiset $\{1^2, 2^2, \cdots, n^2\}$ with j ascents, where all numbers between two non-adjacent occurrences of i are greater than i for every $1 \leq i \leq n$. For example, there are exactly $\left\langle\!\left\langle {3 \atop 2} \right\rangle\!\right\rangle = 6$ such permutations with 2 ascents: 112233, 112332, 122133, 122331, 123321, and 133122.

23

Lah's Triangle

Lah numbers $L(n, r)$, studied in 1955 by the Slovenian mathematician Ivo Lah (1896–1979), are closely related to Stirling numbers of both kinds. They were so named by J. Riordan in his 1958 book *An Introduction to Combinatorial Analysis*. They occur in the study of Hanoi graphs[†] and are defined by

$$L(n, r) = (-1)^n \binom{n-1}{r-1} \frac{n!}{r!}$$

where $1 \leq r \leq n$. Clearly, $L(n, 0) = 0$ and $L(n, n) = (-1)^n$. $\mid L(n, r) \mid$ is an *unsigned Lah number.*

Notice that:

$$\begin{aligned}
-(n+r)L(n,r)-L(n,r-1) &= \frac{(n+r)(-1)^{n+1}(n-1)!n!}{(r-1)!(n-r)!r!} + \frac{(-1)^{n+1}(n-1)!n!}{(r-2)!(n-r+1)!(r-1)!} \\
&= \frac{(-1)^{n+1}n!(n-1)!}{(r-1)!(n-r+1)!r!}[(n-r+1)(n+r)+r(r-1)] \\
&= \frac{(-1)^{n+1}n!(n-1)!}{(r-1)!(n-r+1)!r!}[n(n+1)] \\
&= \frac{(-1)^{n+1}n!(n+1)!}{(r-1)!(n-r+1)!r!} = (-1)^{n+1}\binom{n}{r-1}\frac{(n+1)!}{r!} \\
&= L(n+1,r)
\end{aligned}$$

[†]See S. Klavžar et al., 2005.

Thus, we have the following recursive definition of Lah numbers:

$$L(n, 0) = 0$$

$$L(n+1, r) = -(n+r)L(n, r) - L(n, r-1), \qquad 1 \le r \le n$$

This recursive definition can be used to construct *Lah's triangle*, as in Figure 23.1.

n \ r	1	2	3	4	5	6	7
1	−1						
2	2	1					
3	−6	−6	−1				
4	24	36	12	1			
5	−120	−240	−120	−20	−1		
6	720	1800	1200	300	30	1	
7	−5040	−15120	−12600	−4200	−630	−42	−1

Falling diagonal 1

Figure 23.1 Lah's Triangle of Signed Lah Numbers

Three Interesting Properties

- Since $L(n, n) = (-1)^n$, the rightmost southeast diagonal consists of ± 1 alternately, beginning with -1.
- $L(n, 1) = (-1)^n n!$; so $|L(n, 1)| = n!$; thus the absolute values of the entries in column 1 give us the various values of $n!$, where $n \ge 1$.
- Since $L(n, n-1) = (-1)^n \binom{n-1}{n-2} \frac{n!}{(n-1)!} = (-1)^n n(n-1)$, $|L(n, n-1)| = n(n-1)$, the product of two consecutive integers. In other words, the absolute value of each element on falling diagonal 1 in Lah's triangle consists of twice a triangular number.

Lah numbers are the coefficients that occur when $(-1)^n \langle x \rangle_n$ is expressed as a linear combination of the descending factorials $[x]_r$, where $\langle x \rangle_r = x(x+1)\cdots(x+r-1)$, $[x]_r = x(x-1)\cdots(x-r+1)$ (see Chapter 20), and $0 \le r \le n$: $(-1)^n \langle x \rangle_n = \sum_{r=0}^{n} L(n, r)[x]_r$.

For example,

$$(-1)^2 \langle x \rangle_2 = \langle x \rangle_2 = x(x+1) = x^2 + x = 0 \cdot 1 + 2 \cdot x + 1 \cdot x(x-1)$$

$$= L(2, 0)[x]_0 + L(2, 1)[x]_1 + L(2, 2)[x]_2 = \sum_{r=0}^{2} L(2, r)[x]_r$$

Similarly,

$$(-1)^3 \langle x \rangle_3 = -x(x+1)(x+2) = -0 \cdot [x]_0 - 6 \cdot [x]_1 - 6 \cdot [x]_2 - 1 \cdot [x]_3$$

$$= \sum_{r=0}^{3} L(3, r)[x]_r$$

Stirling and Lah Numbers

There is a relationship linking Lah numbers to Stirling numbers of the first and second kinds:

$$L(n, r) = \sum_{r \le j \le n} (-1)^n s(n, j) S(j, r)$$

For example:

$$\sum_{2 \le j \le 3} (-1)^3 s(3, j) S(j, 2) = -s(3, 2)S(2, 2) - s(3, 3)S(3, 2)$$

$$= -3 \cdot 1 - 1 \cdot 3 = -6 = L(3, 2)$$

Kronecker delta can be used to state yet another property of Lah numbers:

$$\sum_{m \le r \le n} L(n, r)L(r, m) = \delta_{nm}$$

In other words, the dot product of the elements m through n in row n and the elements m through n in column m in Lah's triangle is always 0 or 1.

For example,

$$\sum_{1 \le r \le 4} L(4, r)L(r, 1) = L(4, 1)L(1, 1) + L(4, 2)L(2, 1) + L(4, 3)L(3, 1) + L(4, 4)L(4, 1)$$

$$= 24 \cdot (-1) + 36 \cdot 2 + 12 \cdot (-6) + 1 \cdot 24 = 0$$

and

$$\sum_{4 \le r \le 4} L(4, r)L(r, 4) = L(4, 4)L(4, 4) = 1 \cdot 1 = 1$$

A Generating Function for Lah Numbers

Finally, Lah numbers can be generated using the function

$$\frac{1}{r!}\left(\frac{-t}{1+t}\right)^r = \sum_{n=0}^{\infty} L(n, r)\frac{t^n}{n!}$$

This follows from the fact that

$$\sum_{r=0}^{x} \binom{x}{r} \left(\frac{-t}{1+t} \right)^{r} = \sum_{n=0}^{\infty} [-x]_n \frac{t^n}{n!}$$

24

Fibonacci Convolution Triangles

This chapter presents a brief introduction to the convolution of sequences, shows how it is related to the product of generating functions, and then investigates the convolution of Fibonacci numbers.

Convolution and Generating Functions

Consider the sequences $\{a_n\}$ and $\{b_n\}$. Let $a(x)$ and $b(x)$ be their generating functions:

$$a(x) = a_0 + a_1 x + a_2 x^2 + \cdots$$

$$b(x) = b_0 + b_1 x + b_2 x^2 + \cdots$$

Then

$$a(x)b(x) = a_0 b_0 + (a_0 b_1 + a_1 b_0)x + (a_0 b_2 + a_1 b_1 + a_2 b_0)x^2 + \cdots$$

Let c_n be the coefficient of x^n on the RHS. Then:

$$c_0 = a_0 b_0$$

$$c_1 = a_0 b_1 + a_1 b_0$$

$$c_2 = a_0 b_2 + a_1 b_1 + a_2 b_0$$

$$\vdots$$

$$c_n = a_0 b_n + a_1 b_{n-1} + a_2 b_{n-2} + \cdots + a_{n-1} b_1 + a_n b_0 \qquad (24.1)$$

The sequence $\{c_n\}$ is called the *convolution* of the sequences $\{a_n\}$ and $\{b_n\}$.

Notice that c_n is the sum of all products $a_i b_j$, where $i + j = n$ and i, $j \geq 0$; it can be considered the dot product $(a_0, a_1, \cdots, a_n) \cdot (b_0, b_1, \cdots, b_n)^R$, where $(x_0, x_1, \cdots, x_n)^R$ denotes the *reverse* $(x_n, x_{n-1}, \cdots, x_0)$ of the vector $(x_0, x_1, \cdots, x_n)$. The generating function of the sequence $\{c_n\}$ is the product of the generating functions of the sequences $\{a_n\}$ and $\{b_n\}$; namely, $a(x)b(x)$.

In particular, let $a_n = 1$ and $b_n = n$, where $n \geq 0$. Then:

$$\begin{aligned} c_n &= a_0 b_n + a_1 b_{n-1} + a_2 b_{n-2} + \cdots + a_{n-1} b_1 + a_n b_0 \\ &= 1 \cdot n + 1 \cdot (n-1) + 1 \cdot (n-2) + \cdots + 1 \cdot 1 + 1 \cdot 0 \\ &= \frac{n(n+1)}{2} = t_n \end{aligned}$$

where t_k is the kth triangular number. Thus, the convolution of the sequences $\{1\}$ and $\{n\}$ is the sequence $\{t_n\}$ of triangular numbers.

Since

$$\begin{aligned} a(x) &= a_0 + a_1 x + a_2 x^2 + \cdots \\ &= 1 + x + x^2 + \cdots = \frac{1}{1-x} \end{aligned}$$

and

$$\begin{aligned} b(x) &= b_0 + b_1 x + b_2 x^2 + \cdots \\ &= 0 + 1x + 2x^2 + \cdots = \frac{x}{(1-x)^2} \end{aligned}$$

the convolution $\{c_n\} = \{t_n\}$ is generated by

$$c(x) = a(x) \cdot b(x) = \frac{1}{1-x} \cdot \frac{x}{(1-x)^2} = \frac{x}{(1-x)^3}$$

Thus,

$$\begin{aligned} \frac{x}{(1-x)^3} &= t_0 + t_1 x + t_2 x^2 + t_3 x^3 + \cdots \\ \frac{1}{(1-x)^3} &= 1 + 3x + 6x^2 + 10x^3 + \cdots \end{aligned}$$

We now generalize this example. Let:

$$\begin{aligned} a(x) &= a_0 + a_1 x + a_2 x^2 + a_3 x^3 + \cdots \\ b(x) &= 1 + x + x^2 + x^3 + \cdots = \frac{1}{1-x} \end{aligned}$$

The convolution of the sequences $\{a_n\}$ and $\{b_n\}$ is generated by

$$\frac{a(x)}{1-x} = a_0 + (a_0 + a_1)x + (a_0 + a_1 + a_2)x^2 + \cdots$$

Notice that the coefficients $a_0 + a_1 + a_2 + \cdots + a_n$ on the RHS are the various partial sums of the terms in the sequence $\{a_n\}$; this is the effect of dividing $a(x)$ by $1 - x$, a very useful fact indeed.

In particular, let $a(x) = \frac{1}{1-x}$. Then

$$\frac{1}{1-x} \cdot \frac{1}{1-x} = 1 + (1+1)x + (1+1+1)x^2 + (1+1+1+1)x^3 + \cdots$$

That is,

$$\frac{1}{(1-x)^2} = 1 + 2x + 3x^2 + 4x^3 + \cdots$$

Fibonacci Convolutions

We now turn to the convolution of Fibonacci numbers F_n, which are generated by the generating function $\frac{x}{1-x-x^2}$:

$$\frac{x}{1-x-x^2} = F_0 + F_1x + F_2x^2 + F_3x^3 + \cdots$$

Since $F_0 = 0$, this yields

$$\frac{1}{1-x-x^2} = F_1 + F_2x + F_3x^2 + \cdots + F_{n+1}x^n + \cdots$$

In 1970, Hoggatt studied convolutions of Fibonacci numbers. Let $\{F_n^{(r)}\}$ denote their rth convolution, where $r \geq 0$. Notice that $\{F_n^{(0)}\} = F_n$. We have:

$$\begin{aligned}\frac{1}{1-x-x^2} \cdot \frac{1}{1-x-x^2} &= (F_1+F_2x+F_3x^2+F_4x^3+\cdots)\\ &\quad (F_1+F_2x+F_3x^2+F_4x^3+\cdots)\\ &= F_1F_1+(F_1F_2+F_2F_1)x+(F_1F_3+F_2F_2+F_3F_1)x^2+\cdots\\ \frac{1}{(1-x-x^2)^2} &= F_1^{(1)}+F_2^{(1)}x+F_3^{(1)}x^2+\cdots\end{aligned}$$

where

$$\begin{aligned}F_1^{(1)} &= F_1F_1 &= 1 \qquad F_4^{(1)} &= \sum_{i=1}^{4} F_iF_{5-i} = 10\\ F_2^{(1)} &= F_1F_2 + F_2F_1 &= 2 \qquad F_5^{(1)} &= \sum_{i=1}^{5} F_iF_{6-i} = 20\end{aligned}$$

$$F_3^{(1)} = F_1F_3 + F_2F_2 + F_3F_1 = 5 \qquad F_6^{(1)} = \sum_{i=1}^{6} F_iF_{7-i} = 38$$

$$F_7^{(1)} = \sum_{i=1}^{7} F_iF_{8-i} = 71$$

Thus, the *first Fibonacci convolution sequence* $\{F_n^{(1)}\}$ is $1, 2, 5, 10, 20, 38, 71, \cdots$.

The *second Fibonacci convolution sequence* $\{F_n^{(2)}\}$ is obtained by taking the convolution of the sequences $\{F_n\}$ and $\{F_n^{(1)}\}$:

$$\frac{1}{1-x-x^2} \cdot \frac{1}{(1-x-x^2)^2} = (F_1 + F_2x + F_3x^2 + \cdots)(F_1^{(1)} + F_2^{(1)}x + F_3^{(1)}x^2 + \cdots)$$

$$\frac{1}{(1-x-x^2)^3} = \sum_{n=0}^{\infty} F_{n+1}^{(2)} x^n$$

where

$$F_{n+1}^{(2)} = F_1F_{n+1}^{(1)} + F_2F_n^{(1)} + F_3F_{n-1}^{(1)} + \cdots + F_nF_2^{(1)} + F_{n+1}F_1^{(1)}$$

Consequently, we have:

$$F_1^{(2)} = \sum_{i=1}^{2} F_iF_{2-i}^{(1)} = 1 \qquad F_4^{(2)} = \sum_{i=1}^{4} F_iF_{5-i}^{(1)} = 22$$

$$F_2^{(2)} = \sum_{i=1}^{4} F_iF_{3-i}^{(1)} = 3 \qquad F_5^{(2)} = \sum_{i=1}^{5} F_iF_{6-i}^{(1)} = 51$$

$$F_3^{(2)} = \sum_{i=1}^{3} F_iF_{4-i}^{(1)} = 9 \qquad F_6^{(2)} = \sum_{i=1}^{6} F_iF_{7-i}^{(1)} = 111$$

It follows that the rth convolution sequence $\{F_n^{(r)}\}$ is generated by $\frac{1}{(1-x-x^2)^{r+1}}$:

$$\frac{1}{\left(1-x-x^2\right)^{r+1}} = \sum_{n=0}^{\infty} F_{n+1}^{(r)} x^n$$

Fibonacci Convolution Triangle

Next, we present a left-justified Pascal-like triangular array A that generates the rth convolution sequence. Hoggatt called it the *Fibonacci convolution triangle*.

It is defined recursively as follows:

$$A(n, 0) = F_n, \qquad n \geq 1$$

$$A(n, 1) = 0, \qquad 1 \leq n \leq 3$$

$$A(n, r) = A(n-1, r) + A(n-2, r) + A(n-3, r-1), \qquad n \geq 3$$

In other words, column 0 consists of Fibonacci numbers, and each remaining element in column j is obtained by adding the two elements $A(n-1, r)$ and $A(n-2, r)$ above it to the element $A(n-3, r-1)$ diagonally to the left of $A(n-2, r)$. As usual, it is assumed that $A(n, r) = 0$ if $r < 0$.

For example, $A(7, 1) = A(6, 1) + A(5, 1) + A(4, 1) = 5 + 2 + 3 = 10$.

Figure 24.1 shows the Fibonacci convolution triangle. Column 0 gives the 0th convolution numbers $F_n^{(0)} = F_n$, column 1 gives the first Fibonacci convolution numbers, and so on.

Column Generators of the Convolution Triangle

Before we find the column generators of the Fibonacci convolution triangle, we return to the left-justified Pascal triangle, where each column is moved down by two positions; see Figure 24.2. Column 0 consists of 1s. Each remaining element is obtained by adding the element above it to its northwestern neighbor. For example, $10 = 4 + 6$; see Figure 24.2. This is the same array we encountered in Figure 12.2.

Since the column generating functions for Pascal's triangle are

$$g_r(x) = \sum_{n=0}^{\infty} \binom{n}{r} x^n = \frac{x^r}{(1-x)^r} \tag{24.2}$$

$n \backslash r$	0	1	2	3
1	1			
2	1			
3	2			
4	3	1		
5	5	2		
6	8	5		
7	13	10	1	
8	21	20	3	

Figure 24.1 Fibonacci Convolution Triangle

$n \backslash r$	0	1	2	3	Row sums
0	1				1
1	1				1
2	1	1			2
3	1	2			3
4	1	3	1		5
5	1	4	3		8
6	1	5	6	1	13
7	1	6	10	4	21

Figure 24.2

it follows that those for the array in Figure 24.2 are $g_r(x) = \frac{x^{2r}}{(1-x)^{r+1}}$, where $r \geq 0$.
For example,

$$g_1(x)=\frac{x^2}{(1-x)^2}=x^2(1+2x+3x^2+4x^3+\cdots)=x^2+2x^3+3x^4+4x^5+\cdots$$

Notice that the coefficients yield the entries in column 1; and the coefficients in

$$\begin{aligned} g_2(x) &= \frac{x^4}{(1-x)^3} = x^3 \cdot \frac{x}{(1-x)^3} \\ &= x^3(1 + 3x + 6x^2 + 10x^3 + \cdots) = x^3 + 3x^4 + 6x^5 + 10x^6 + \cdots \end{aligned}$$

yield the entries in column 2.

Notice that:

$$\begin{aligned} \sum_{r=0}^{\infty} g_r(x) &= \sum_{r=0}^{\infty} \frac{x^{2r}}{(1-x)^{r+1}} = \frac{1}{1-x} \sum_{r=0}^{\infty} \left(\frac{x^2}{1-x}\right)^r \\ &= \frac{1}{1-x} \cdot \frac{1}{1-\frac{x^2}{1-x}} = \frac{1}{1-x-x^2} \\ &= \sum_{n=0}^{\infty} F_{n+1} x^n \end{aligned}$$

Consequently, the row sums in Figure 24.2 yield the various Fibonacci numbers.

Pascal Meets Fibonacci

We now return to the rth column generator (24.2) of Pascal's triangle:

$$g_r(x) = \frac{x^r}{(1-x)^{r+1}}$$

Replacing x with $\frac{x^2}{1-x}$, we have:

$$g_r\left(\frac{x^2}{1-x}\right) = \frac{\left(\frac{x^2}{1-x}\right)^r}{\left(1-\frac{x^2}{1-x}\right)^{r+1}} = \frac{(1-x)x^{2r}}{(1-x-x^2)^{r+1}}$$

$$\frac{1}{1-x} g_r\left(\frac{x^2}{1-x}\right) = \frac{x^{2r}}{(1-x-x^2)^{r+1}}$$

The RHS is the generating function for the rth convolution of Fibonacci numbers. Consequently, by multiplying the column generators of the array in Figure 24.2 with proper weights from Pascal's triangle, we can obtain the remaining columns of the Fibonacci convolution triangle.

For example, multiply the columns of the array in Figure 24.2 by 0, 1, 2, 3, 4, , ⋯ (column 1 in Pascal's triangle). The resulting row sums yield the first Fibonacci convolution sequence $\{F_n^{(1)}\}$. See Figure 24.3.

Weights					Weighted row sums
0	1	2	3	4	
1					
1					
1	1				1 = 1
1	2				2 = 2
1	3	1			3·1 + 1 · 2 = 5
1	4	3			4·1 + 3 · 2 = 10
1	5	6	1		5·1 + 6 · 2 + 1 · 3 = 20
1	6	10	4		6·1 + 10 · 2 + 4 · 3 = 38

Figure 24.3

Weights						Weighted row sums
0	0	1	3	6	10	
1						
1						
1	1					
1	2					
1	3	1				1
1	4	3				3
1	5	6	1			9
1	6	10	4			22

Figure 24.4

Using the elements of column 2 in Pascal's triangle as weights, the row sums yield the second Fibonacci convergence $\{F_n^{(2)}\}$. See Figure 24.4.

Similarly, using the elements of column 3 in Pascal's triangle as weights, the row sums yield the third Fibonacci convolution sequence $\{F_n^{(3)}\}$: 1, 4, 14, 40, 105, 256, ⋯ .

In general, multiplying the columns of the array in Figure 24.2 with the elements of column r in Pascal's triangle as weights, the row sums yield the rth Fibonacci convolution sequence $\{F_n^{(r)}\}$, where $r \geq 0$.

Next, we pursue an interesting side trip, as Hoggatt did. We will discuss how Pell numbers can be extracted from the Fibonacci convolution triangle.

Fibonacci Meets Pell

In the Fibonacci convolution triangle, move column j up by $2j$ positions. Figure 24.5 shows the resulting array. Its row sums yield the various Pell numbers P_n: 1, 2, 5, 12, ⋯ .

To confirm this observation, recall that the column generators of the Fibonacci convolution triangle are $\frac{x^{2r}}{(1-x-x^2)^{r+1}}$, where $r \geq 0$. Now reposition the columns $r(\geq 1)$ by choosing the column generator $g_r(x) = \frac{x^r}{(1-x-x^2)^{r+1}}$. Then

$$\begin{aligned}\sum_{r=0}^{\infty} g_r(x) &= \frac{1}{1-x-x^2}\sum_{r=0}^{\infty}\left(\frac{x}{1-x-x^2}\right)^r\\ &= \frac{1}{1-x-x^2}\cdot\frac{1}{1-\frac{x}{1-x-x^2}} = \frac{1}{1-2x-x^2}\end{aligned}$$

								Row sums
1								1
1	1							2
2	2	1						5
3	5	3	1					12
5	10	9	4	1				29
8	20	22	14	5	1			70
13	38	51	40	20	6	1		169
21	71	111	105	65	27	7	1	408

Figure 24.5

			Row sums
1			1
1			1
2			2
3	1		4
5	2		7
8	5		13
13	10	1	24
21	20	3	44

Figure 24.6 Fibonacci Convolution Triangle

This is the generating function for Pell numbers P_n.

Thus, by moving column n up by $2n$ positions and then computing the row sums of the resulting array (see Figure 24.5), we can retrieve the Pell numbers P_n, where $n \geq 1$.

Fibonacci Meets Tribonacci

Next, we present an unexpected occurrence of the *tribonacci numbers*† $1, 1, 2, 4, 7, 13, \cdots$: The row sums of the Fibonacci convolution triangle are the various tribonacci numbers T_n; see Figure 24.6.

To confirm this, notice that the column generators of the Fibonacci convolution triangle are

$$g_r(x) = \frac{x^{3r}}{(1 - x - x^2)^{r+1}} \tag{24.3}$$

So

$$\begin{aligned}\sum_{r=0}^{\infty} g_r(x) &= \frac{1}{1 - x - x^2} \sum_{r=0}^{\infty} \left(\frac{x^3}{1 - x - x^2} \right)^r \\ &= \frac{1}{1 - x - x^2} \cdot \frac{1}{1 - \dfrac{x^3}{1 - x - x^2}} = \frac{1}{1 - x - x^2 - x^3}\end{aligned}$$

This is the generating function for tribonacci numbers. Thus, the row sums in the Fibonacci convolution triangle yield the tribonacci numbers.

† See the author's *Fibonacci and Lucas Numbers with Applications.*

Using the binomial coefficients $\binom{n}{r}$ as weights, function (24.3) yields:

$$\sum_{r=0}^{\infty} g_r(x) = \frac{1}{1-x-x^2} \sum_{r=0}^{\infty} \binom{n}{r} \left(\frac{x^3}{1-x-x^2}\right)^r$$

$$= \frac{1}{1-x-x^2} \cdot \frac{\left(\frac{x^3}{1-x-x^2}\right)^r}{\left(1-\frac{x^3}{1-x-x^2}\right)^{r+1}} = \frac{x^{3r}}{(1-x-x^2-x^3)^{r+1}}$$

This is the rth tribonacci convolution sequence generator, and the various coefficients form column r of the Fibonacci convolution triangle. Thus, all columns of the tribonacci convolution triangle can be obtained from the Fibonacci convolution triangle in exactly the same way we obtained the Fibonacci convolution triangle from Pascal's triangle.

For example, using the entries in column 1 of Pascal's triangle as weights, the row sums yield the sequence 1, 2, 5, 12, 26, 56, 118, 244, 499, 1010, $\cdots$. See Figure 24.7. This is indeed the *first tribonacci convolution sequence* $\{T_n^{(1)}\}$.

Weights				Weighted
0	1	2	3	row sums
1				
1				
2				
3	1			1
5	2			2
8	5			5
13	10	1		12
21	20	3		26
34	38	9		56

Figure 24.7

1		
1		
2		
4	1	
7	2	
13	5	
24	12	1
44	26	3
81	56	9

Figure 24.8

Using column 2 of Pascal's triangle as weights, it can be seen that the second tribonacci convolution sequence $\{T_n^{(2)}\}$ is 1, 3, 9, 25, 63, 153, 359, $\cdots$.

Figure 24.8 shows the tribonacci convolution triangle T. Column 0 consists of the tribonacci numbers $T_n = T_n^{(0)}$. In addition, it exhibits an interesting pattern: Each remaining number $T(n, r)$ can be obtained by adding its three northern neighbors and the western neighbor of the top entry in that sum:

$$T(n,r) = T(n-1,r) + T(n-2,r) + T(n-3,r) + T(n-4,r-1)$$

where $n \geq 4$ and $r \geq 1$. For example, $63 = 25 + 3 + 9 + 26$; see Figure 24.8.

Tribonacci Meets Fibonacci

Now that we can construct the tribonacci convolution triangle from the Fibonacci convolution triangle, can we undo the process? In other words, can we retrieve the Fibonacci convolution triangle from the tribonacci convolution triangle? Fortunately, this can be achieved fairly quickly and easily.

To this end, recall that $\sum_{n=0}^{\infty}\binom{n}{r}x^n = \frac{x^r}{(1-x)^{r+1}}$. Replacing x with $-x$, this yields

$$\sum_{n=0}^{\infty}\binom{n}{r}(-1)^n x^n = \frac{(-1)^r x^r}{(1+x)^{r+1}}$$

That is,

$$\sum_{n=0}^{\infty}\binom{n}{r}(-1)^{n+r} x^n = \frac{x^r}{(1-x)^{r+1}} \tag{24.4}$$

Using $\binom{n}{r}(-1)^{n+r}$ as weights, we can return from the tribonacci convolution triangle to the Fibonacci convolution triangle.

To confirm this, recall that the column generators of the tribonacci convolution triangle are

$$g_n(x) = \frac{x^{3n}}{(1-x-x^2-x^3)^{n+1}}$$

Then, by using formula (24.4),

$$\begin{aligned}\sum_{n=0}^{\infty}\binom{n}{r}(-1)^{n+r} g_n(x) &= \frac{1}{1-x-x^2-x^3}\sum_{n=0}^{\infty}\binom{n}{r}(-1)^{n+r}\left(\frac{x^3}{1-x-x^2-x^3}\right)^n \\ &= \frac{1}{1-x-x^2-x^3}\cdot\frac{\left(\frac{x^3}{1-x-x^2-x^3}\right)^r}{\left(1-\frac{x^3}{1-x-x^2-x^3}\right)^{r+1}} \\ &= \frac{x^{3r}}{(1-x-x^2-x^3)^{r+1}}\end{aligned}$$

Interestingly, these are the column generators of the Fibonacci convolution triangle.

For example, when $r=0$, $\binom{n}{r}(-1)^{n+r} = (-1)^n$, where $n \ge 0$. Using the weights $(-1)^n$, the weighted row sums of the tribonacci convolution triangle do indeed yield the first convolution sequence $\{F_n^{(0)}\} = \{F_n\}$; see Figure 24.9.

When $r=1$, $\binom{n}{r}(-1)^{n+r} = n(-1)^{n+1}$, where $n \ge 0$. Using the weights $n(-1)^{n+1}$, the weighted row sums of the tribonacci convolution triangle yield the first convolution sequence $\{F_n^{(1)}\} = \{F_n\}$; see Figure 24.10.

Weights				Weighted
1	−1	1	−1	row sums
1				1
1				1
2				2
4	1			3
7	2			5
13	5			8
24	12	1		13
44	26	3		21
81	56	9		34

Figure 24.9

Weights					Weighted
0	1	−2	3	−4	row sums
1					
1					
2					
4	1				1
7	2				2
13	5				5
24	12	1			10
44	26	3			20
81	56	9			38

Figure 24.10

Similarly, the weights $\binom{n}{2}(-1)^{n+2} = \binom{n}{2}(-1)^n$ yield the second Fibonacci convolution sequence $\{F_n^{(2)}\}$; see Figure 24.11.

Weights						Weighted
0	0	1	−3	6	−10	row sums
1						
1						
2						
4	1					
7	2					
13	5					
24	12	1				1
44	26	3				3
81	56	9				9
149	118	25	1			22
274	244	63	4			51

Figure 24.11

25

Tartaglia's Arrays

More than a century before Pascal's article, *Treatise on the Arithmetic Triangle*, was published posthumously in 1665, a strikingly similar rectangular array appeared in the *General Treatise on Number and Measure* by the Italian mathematician and scientist Niccolo Tartaglia (1499?–1557). Tartaglia used his rectangular array in computing the coefficients of the 12th power of a binomial. Like Pascal's work, his too was published posthumously, during 1550–1557.

Tartaglia's Array

Tartaglia's rectangular array T is given in Figure 25.1. The same array can be seen in the manuscript *De Numeris Triangularibus et inde De Progressionibus Arithmeticis*, by the English mathematician and astronomer Thomas Harriot (1560–1621), written around 1611 and still unpublished. Tartaglia arrived at this array in his study of figurate numbers,[†] where row m gives the figurate numbers in dimension m and $m \geq 0$.

In 1977, W. Fred Lunnon of the Mathematics Institute, Cardiff, Wales, called it the Pascal matrix. But in fairness to history, we will continue to call it Tartaglia's array, although Harriot discovered the same array independently.

It is worth noting that Pascal's triangle and hence Tartaglia's rectangular array were also investigated by the Russian mathematical logician Vladimir Andreevich Uspenski (1930–) of Moscow University.

Array T possesses several fascinating properties. It is clearly symmetric about the main diagonal; that is, $T(m, n) = T(n, m)$, where $m, n \geq 0$. Each northeast diagonal yields the corresponding row in the familiar Pascal's triangle;

[†] See A. H. Beiler, *Recreations in the Theory of Numbers*, 2nd edition, Dover, New York, 1966.

m \ n	0	1	2	3	4	5	6
0	1	1	1	1	1	1	1
1	1	2	3	4	5	6	7
2	1	3	6	10	15	21	28
3	1	4	10	20	35	56	84
4	1	5	15	35	70	126	210
5	1	6	21	56	126	252	462

Figure 25.1 Tartaglia's Rectangular Array

see Figure 25.2. The top row and the leftmost column consist of 1s. Each remaining element $T(m, n)$ can be obtained by adding its western and northern neighbors. For example, $21 = 15 + 6$; see Figure 25.2. Accordingly, the array can be defined recursively as follows:

$$T(0, 0) = 1$$
$$T(m, n) = T(m, n - 1) + T(m - 1, n)$$

where $m, n \geq 1$. Once again, blank spaces are considered zeros when needed.

1	1	1	1	1	1	1	
1	2	3	4	5	6	7	
1	3	6	10	15	21	28	← triangular numbers
1	4	10	20	35	56	84	← tetrahedral numbers
1	5	15	35	70	126	210	
1	6	21	56	126	252	462	
1	7	28	84	210	462	924	
							CBCs

Figure 25.2 CBCs

Each entry $T(m, n)$ gives the number of different routes a rook can take from the upper left-hand corner to the square (m, n) on a chessboard. This result is stated in Lucas' *Théorie des Nombres* (1891).

Using a repeated application of the recurrence relation, we have:

$$\begin{aligned}
T(m,n) &= T(m,n-1)+T(m-1,n)\\
&= T(m,n-1)+[T(m-1,n-1)+T(m-2,n)]\\
&= T(m,n-1)+T(m-1,n-1)+[T(m-2,n-1)+T(m-3,n)]\\
&\ \vdots\\
&= T(m,n-1)+T(m-1,n-1)+T(m-2,n-1)+\cdots+T(0,n-1)
\end{aligned} \tag{25.1}$$

This can be confirmed using PMI. Thus, each element $T(m, n)$ is the sum of the top $m+1$ elements in the preceding column. For example, $35 = 15+10+6+3+1$; see Figure 25.3.

m \ n	0	1	2	3	4	5	6
0	1	1	1	1	1	1	1
1	1	2	3	4	5	6	7
2	1	3	6	10	15	21	28
3	1	4	10	20	35	56	84
4	1	5	15	35	70	126	210
5	1	6	21	56	126	252	462
6	1	7	28	84	210	462	924

Figure 25.3

By symmetry, it follows that $T(m, n)$ is also given by

$$T(m,n)=T(m-1,n)+T(m-1,n-1)+T(m-1,n-2)+\cdots+T(m-1,0) \tag{25.2}$$

That is, $T(m, n)$ is the sum of the first $m+1$ elements in the preceding row. For example, $20 = 10+6+3+1$; see Figure 25.3.

It follows by property (25.2) that:

$$\begin{aligned} T(m,n) &= \sum_{j=0}^{n} T(m-1,j) = \sum_{j=0}^{n-1} T(m-1,j) + T(m-1,n) \\ &= \sum_{j=0}^{n-1} T(m-1,j) + \sum_{j=0}^{n-1} T(m-1,j) + T(m-2,n) \\ &\ \ \vdots \\ &= \sum_{j=0}^{n-1} T(m-1,j) + \sum_{j=0}^{n-1} T(m-1,j) + \cdots + \sum_{j=0}^{n-1} T(0,j) + T(0,n) \\ &= \sum_{i=1}^{m}\sum_{j=0}^{n-1} T(m-i,j) + 1 \end{aligned} \tag{25.3}$$

Thus each element $T(m, n)$ is one more than the sum of the elements in row 0 through $m-1$ and in columns 0 through $n-1$; that is, one more than the sum of the elements in the upper left-hand rectangle bounded by row $m-1$ and column $n-1$.

For example, $35 = 1 + (1+1+1+1) + (1+2+3+4) + (1+3+6+10)$; see Figure 25.4.

1	1	1	1	1	1	1
1	2	3	4	5	6	7
1	3	6	10	15	21	28
1	4	10	20	35	56	84
1	5	15	35	70	126	210
1	6	21	56	126	252	462
1	7	28	84	210	462	924

Figure 25.4

Since $\sum_{i=k}^{n} \binom{i}{k} = \binom{n+1}{k+1}$, it follows that

$$\begin{aligned}\binom{m+n-4}{m-1} + \binom{m+n-3}{m-1} + \binom{m+n-2}{m-1} + \binom{m+n-1}{m-1} &= \binom{(m-1)+(n-3)}{m-1} \\ &+ \binom{(m-1)+(n-2)}{m-1} + \binom{(m-1)+(n-1)}{m-1} + \binom{(m-1)+n}{m-1} \\ &= \binom{m+n}{m}\end{aligned}$$

Consequently, $T(m, n) = \binom{m+n}{m}$; that is, the figurate numbers in row m are the binomial coefficients $\binom{m+n}{m}$, where $m, n \geq 0$.

In particular, $T(m, m) = \binom{2m}{m}$, the CBC, as expected; see Figure 25.2.

Next, we will pursue two other intriguing properties of Tartaglia's matrix T_n, where T_n consists of the top n rows and the leftmost n columns of the array. First, we investigate the determinant $|T_n|$, a problem proposed in 1930 by C. A. Rupp of Pennsylvania State College.

Determinant of Tartaglia's Array

Notice that

$$|\,1\,| = \begin{vmatrix} 1 & 1 \\ 1 & 2 \end{vmatrix} = \begin{vmatrix} 1 & 1 & 1 \\ 1 & 2 & 3 \\ 1 & 3 & 6 \end{vmatrix} = \begin{vmatrix} 1 & 1 & 1 & 1 \\ 1 & 2 & 3 & 4 \\ 1 & 3 & 6 & 10 \\ 1 & 4 & 10 & 20 \end{vmatrix} = 1$$

More generally, consider T_n:

$$|T_n| = \begin{vmatrix} \binom{0}{0} & \binom{1}{0} & \binom{2}{0} & & \binom{n-1}{0} \\ \binom{1}{1} & \binom{2}{1} & \binom{3}{1} & & \binom{n}{1} \\ \binom{2}{2} & \binom{3}{2} & \binom{4}{2} & \cdots & \binom{n+1}{2} \\ & & \vdots & & \\ \binom{n-1}{n-1} & \binom{n}{n-1} & \binom{n+1}{n-1} & & \binom{2n-1}{n-1} \end{vmatrix}$$

Using Pascal's identity, we subtract each row from the preceding row. This yields

$$|T_n| = \begin{vmatrix} \binom{0}{0} & \binom{1}{0} & \binom{2}{0} & & \binom{n-1}{0} \\ 0 & \binom{1}{1} & \binom{2}{1} & & \binom{n-1}{1} \\ 0 & \binom{2}{2} & \binom{3}{2} & \cdots & \binom{n}{2} \\ & & \vdots & & \\ 0 & \binom{n}{n} & \binom{n+1}{n} & & \binom{2n-2}{n} \end{vmatrix} = \begin{vmatrix} \binom{1}{1} & \binom{2}{1} & & \binom{n-1}{1} \\ \binom{2}{2} & \binom{3}{2} & \cdots & \binom{n}{2} \\ & & \vdots & \\ \binom{n}{n} & \binom{n+1}{n} & & \binom{2n-2}{n} \end{vmatrix}$$

$$= |T_{n-1}|$$

Since $|T_1| = 1$, this implies that $|T_n| = 1$; that is, the determinant of the $n \times n$ Tartaglia's matrix is unity, as shown by H. T. R Aude of Colgate University in 1931 and by Lunnon in 1977.

Nearly the same problem was proposed in 1963 by J. E. Schneider of Franklin and Marshall College. It was solved in the following year by D. I. A. Cohen of Princeton University, among many others.

Inverse of Tartaglia's Matrix

Since $|T_n| \neq 0$, it follows that Tartaglia's matrix $T_n = (t_{ij})_{n\times n}$ is invertible, where $t_{ij} = T(i,j) = \binom{i+j}{i}$. Let C_{ij} denote the cofactor of t_{ij} for every i and j. Then the inverse of T_n is given by

$$T_n^{-1} = \frac{1}{|T_n|} adj(T_n) = adj(T_n)$$

where $adj(T_n) = (C_{ij})_{n\times n}$ denotes the adjoint of T_n. Since each C_{ij} is an integer, it follows that the elements in T_n^{-1} are all integers.

The next theorem gives an explicit formula for T_n^{-1}. It uses the following combinatorial formula:

$$\sum_{k=0}^{n-1}(-1)^k \binom{s+k}{u}\binom{t}{k} = (-1)^t \binom{s}{u-t} \tag{25.4}$$

Theorem 25.1 (Lunnon, 1977) The ij-th element in T_n^{-1} is given by

$$\left(T_n^{-1}\right)_{ij} = (-1)^{i+j}\sum_{k=0}^{n-1}\binom{k}{i}\binom{k}{j}, \quad 0 \le i,j \le n-1$$

Proof: Let $S = (s_{ij})_{n\times n}$, where $s_{ij} = (-1)^{i+j}\sum_{k=0}^{n-1}\binom{k}{i}\binom{k}{j}$. Then:

$$\begin{aligned}
(T_nS)_{ij} &= \sum_{r=0}^{n-1}\binom{i+r}{i}\left[(-1)^{r+j}\sum_{k=0}^{n-1}\binom{k}{r}\binom{k}{j}\right]\\
&= \sum_{r=0}^{n-1}(-1)^j\binom{k}{j}\left[\sum_{r=0}^{n-1}(-1)^r\binom{i+r}{i}\binom{k}{r}\right]\\
&= (-1)^{i+j}\sum_{r=0}^{n-1}\binom{k}{j}\binom{i}{k}, \quad \text{by equation (25.4)}\\
&= (-1)^{i+j}\binom{0}{j-i}, \quad \text{by equation (25.4)}\\
&= \delta_{ij}
\end{aligned}$$

where δ_{ij} denotes the Kronecker delta. Thus $T_nS = I_n$, and hence $S = T_n^{-1}$, as desired. ■

For example,

$$T_4 = \begin{bmatrix} 1 & 1 & 1 & 1\\ 1 & 2 & 3 & 4\\ 1 & 3 & 6 & 10\\ 1 & 4 & 10 & 20 \end{bmatrix}$$

Then:

$$T_4^{-1} = \begin{bmatrix}
\sum_{k=0}^{3}\binom{k}{0}\binom{k}{0} & -\sum_{k=0}^{3}\binom{k}{0}\binom{k}{1} & \sum_{k=0}^{3}\binom{k}{0}\binom{k}{2} & -\sum_{k=0}^{3}\binom{k}{0}\binom{k}{3}\\
-\sum_{k=0}^{3}\binom{k}{1}\binom{k}{0} & \sum_{k=0}^{3}\binom{k}{1}\binom{k}{1} & -\sum_{k=0}^{3}\binom{k}{1}\binom{k}{2} & \sum_{k=0}^{3}\binom{k}{1}\binom{k}{3}\\
\sum_{k=0}^{3}\binom{k}{2}\binom{k}{0} & -\sum_{k=0}^{3}\binom{k}{2}\binom{k}{1} & \sum_{k=0}^{3}\binom{k}{2}\binom{k}{2} & -\sum_{k=0}^{3}\binom{k}{2}\binom{k}{3}\\
-\sum_{k=0}^{3}\binom{k}{3}\binom{k}{0} & \sum_{k=0}^{3}\binom{k}{3}\binom{k}{1} & -\sum_{k=0}^{3}\binom{k}{3}\binom{k}{2} & \sum_{k=0}^{3}\binom{k}{3}\binom{k}{3}
\end{bmatrix}$$

$$= \begin{bmatrix} 4 & -6 & 4 & -1\\ -6 & 14 & -11 & 3\\ 4 & -11 & 10 & -3\\ -1 & 3 & -3 & 1 \end{bmatrix}$$

Since $\left(T_n^{-1}\right)_{ij} = \left(T_n^{-1}\right)_{ji}$, it follows that the inverse T_n^{-1} of Tartaglia's matrix is also symmetric; for instance, see T_4^{-1}.

Harriot's Triangle

Harrriot also studied a rearrangement of Tartaglia's array in Figure 25.1, shown in Figure 25.5. To give at least some credit where it is due, we call it *Harriot's triangle*, although it is the left-justified Pascal's triangle.

$$
\begin{array}{llllllll}
1 \\
1 & 1 \\
1 & 2 & 1 \\
1 & 3 & 3 & 1 \\
1 & 4 & 6 \longrightarrow & 4 & 1 \\
1 & 5 & 10 & 10 & 5 & 1 \\
1 & 6 & 15 & 20 & 15 & 6 & 1 \\
1 & 7 & 21 & 35 & 35 & 21 & 7 & 1
\end{array}
$$

Figure 25.5 Harriot's Triangle

Montmort's Array

To Harriot's triangular array, add 0s in the blank spaces on the right to form the square array in Figure 25.6, as the French mathematician Pierre Raymond De Montmort (1678–1719) did in 1708. In 1982, A. W. F. Edwards of Gonville and Caius College, Cambridge, England, called it Pascal matrix, although matrices were not developed until 1857 by Cayley and the English mathematician James Joseph Sylvester (1814–1897).

$$
\begin{array}{cccccccc}
1 & 0 & 0 & 0 & 0 & 0 & 0 & 0 \\
1 & 1 & 0 & 0 & 0 & 0 & 0 & 0 \\
1 & 2 & 1 & 0 & 0 & 0 & 0 & 0 \\
1 & 3 & 3 & 1 & 0 & 0 & 0 & 0 \\
1 & 4 & 6 & 4 & 1 & 0 & 0 & 0 \\
1 & 5 & 10 & 10 & 5 & 1 & 0 & 0 \\
1 & 6 & 15 & 20 & 15 & 6 & 1 & 0 \\
1 & 7 & 21 & 35 & 35 & 21 & 7 & 1
\end{array}
$$

Figure 25.6 Montmort's Array

Interestingly, the same array can be found in Bernoulli's *Ars Conjectandi*. Recently, G. S. Call and D. J. Velleman of Amherst College, Amherst, Massachusetts, encountered such matrices in their pursuit of probability; and R. Aggarwala and M. P. Lamoureux of the University of Calgary, Alberta, Canada encountered them in order statistics.

Like Pascal's triangle, this array also possesses several intriguing properties.

For example, let $M = (m_{ij})_{n\times n}$ denote the $n \times n$ Montmort's array. Then:

$$\begin{vmatrix} 1 \end{vmatrix} = \begin{vmatrix} 1 & 0 \\ 1 & 1 \end{vmatrix} = \begin{vmatrix} 1 & 0 & 0 \\ 1 & 1 & 0 \\ 1 & 2 & 1 \end{vmatrix} = \begin{vmatrix} 1 & 0 & 0 & 0 \\ 1 & 1 & 0 & 0 \\ 1 & 2 & 1 & 0 \\ 1 & 3 & 3 & 1 \end{vmatrix} = 1$$

More generally, since M is a lower triangular matrix, $|M|$ equals the product of the main diagonal elements, namely, 1.

Inverse of Montmort's Matrix

Since $|M| \neq 0$, it follows that the matrix M is invertible. Its inverse is given by

$$M^{-1} = \begin{bmatrix} 1 & 0 & 0 & 0 & 0 & \\ -1 & 1 & 0 & 0 & 0 & \\ 1 & -2 & 1 & 0 & 0 & \cdots \\ -1 & 3 & -3 & 1 & 0 & \\ 1 & -4 & 6 & -4 & 1 & \\ & & \vdots & & & \end{bmatrix}$$

It is obtained by taking the negative of every other non-zero entry in each column.

For example,

$$\begin{bmatrix} 1 & 0 & 0 & 0 \\ 1 & 1 & 0 & 0 \\ 1 & 2 & 1 & 0 \\ 1 & 3 & 3 & 1 \end{bmatrix} \begin{bmatrix} 1 & 0 & 0 & 0 \\ -1 & 1 & 0 & 0 \\ 1 & -2 & 1 & 0 \\ -1 & 3 & -3 & 1 \end{bmatrix} = I_4$$

Notice that M^{-1} also possess interesting properties:

- M^{-1} is also lower triangular.
- Its main diagonal consists of 1s.

- Every other falling diagonal below the main diagonal is obtained by taking the negatives of the elements on the corresponding diagonal in M; that is,

$$\left(M^{-1}\right)_{ij} = \begin{cases} m_{ij} & \text{if } i \geq j \\ (-1)^{i+j} m_{ij} & \text{otherwise} \end{cases}$$
$$= \begin{cases} \binom{i}{j} & \text{if } i \geq j \\ (-1)^{i+j}\binom{i}{j} & \text{otherwise} \end{cases}$$

To confirm this property, let $M^{-1} = A = (a_{ij})_{n \times n}$. Then, using property (25.4),

$$(MA)_{ij} = \sum_{r=0}^{n-1} m_{ir} a_{rj}$$
$$= \begin{cases} \sum_{r=0}^{n-1} \binom{i}{r}\binom{r}{j} & \text{if } i \geq j \\ \sum_{r=0}^{n-1} (-1)^{r+j}\binom{i}{r}\binom{r}{j} & \text{otherwise} \end{cases}$$
$$= \begin{cases} \binom{0}{j-i} & \text{if } i \geq j \\ (-1)^{r+j}\binom{0}{j-i} & \text{otherwise} \end{cases}$$
$$= \begin{cases} \delta_{ij} & \text{if } i \geq j \\ 0 & \text{otherwise} \end{cases}$$

Thus $MA = I_n$, so $M^{-1} = A$. Consequently, M^{-1} can be read directly from M.

Next, we compute $M \cdot M^T$. For example:

$$\begin{bmatrix} 1 & 0 \\ 1 & 1 \end{bmatrix}\begin{bmatrix} 1 & 0 \\ 1 & 1 \end{bmatrix} = \begin{bmatrix} 1 & 1 \\ 1 & 2 \end{bmatrix}$$

$$\begin{bmatrix} 1 & 0 & 0 \\ 1 & 1 & 0 \\ 1 & 2 & 1 \end{bmatrix}\begin{bmatrix} 1 & 1 & 1 \\ 0 & 1 & 2 \\ 0 & 0 & 1 \end{bmatrix} = \begin{bmatrix} 1 & 1 & 1 \\ 1 & 2 & 3 \\ 1 & 3 & 6 \end{bmatrix}$$

$$\begin{bmatrix} 1 & 0 & 0 & 0 \\ 1 & 1 & 0 & 0 \\ 1 & 2 & 1 & 0 \\ 1 & 3 & 3 & 1 \end{bmatrix}\begin{bmatrix} 1 & 1 & 1 & 1 \\ 0 & 1 & 2 & 3 \\ 0 & 0 & 1 & 3 \\ 0 & 0 & 0 & 1 \end{bmatrix} = \begin{bmatrix} 1 & 1 & 1 & 1 \\ 1 & 2 & 3 & 4 \\ 1 & 3 & 6 & 10 \\ 1 & 4 & 10 & 20 \end{bmatrix}$$

Slowly, but surely, a clear pattern emerges: The product MM^T is the same as Tartaglia's rectangular array in Figure 25.1.

To confirm this observation, let $0 \leq i, j \leq n-1$. Then the ij-th element in MM^T is given by $\sum_{r=0}^{n-1} \binom{i}{r}\binom{j}{r}$. This equals

$$\sum_{r=0}^{n} \binom{i}{r}\binom{j}{n-r} = \binom{i+j}{j} = \binom{i+j}{i} = T(i,j)$$

Thus, MM^T is exactly the same as Tartaglia's rectangular array T_n.

Since $|M| = 1$ and $T_n = MM^T$, it follows that $|T_n| = |MM^T| = |M| \cdot |M^T| = 1 \cdot 1 = 1$, as we found earlier.

What can we say about $M^T M$? Notice that:

$$\begin{bmatrix} 1 & 1 \\ 0 & 1 \end{bmatrix} \begin{bmatrix} 1 & 0 \\ 1 & 1 \end{bmatrix} = \begin{bmatrix} 2 & 1 \\ 1 & 1 \end{bmatrix}$$

$$\begin{bmatrix} 1 & 1 & 1 \\ 0 & 1 & 2 \\ 0 & 0 & 1 \end{bmatrix} \begin{bmatrix} 1 & 0 & 0 \\ 1 & 1 & 0 \\ 1 & 2 & 1 \end{bmatrix} = \begin{bmatrix} 3 & 3 & 1 \\ 3 & 5 & 2 \\ 1 & 2 & 1 \end{bmatrix}$$

$$\begin{bmatrix} 1 & 1 & 1 & 1 \\ 0 & 1 & 2 & 3 \\ 0 & 0 & 1 & 3 \\ 0 & 0 & 0 & 1 \end{bmatrix} \begin{bmatrix} 1 & 0 & 0 & 0 \\ 1 & 1 & 0 & 0 \\ 1 & 2 & 1 & 0 \\ 1 & 3 & 3 & 1 \end{bmatrix} = \begin{bmatrix} 4 & 6 & 4 & 1 \\ 6 & 14 & 11 & 3 \\ 4 & 11 & 10 & 3 \\ 1 & 3 & 3 & 1 \end{bmatrix}$$

It now follows that, in general, $M^T M$ is *not* Tartaglia's array. But interestingly, the product is symmetric. This should not come as a surprise, since $(M^T M)^T = M^T (M^T)^T = M^T M$.

To pursue yet another property of Montmort's array, let $N = (n_{ij})_{n \times n}$ be the array obtained from M by changing the signs of the entries in even-numbered rows:

$$n_{ij} = \begin{cases} m_{ij} & \text{if } i \text{ is odd} \\ -m_{ij} & \text{otherwise} \end{cases}$$

For example, the 4×4 array N is

$$N = \begin{bmatrix} -1 & 0 & 0 & 0 \\ 1 & 1 & 0 & 0 \\ -1 & -2 & -1 & 0 \\ 1 & 3 & 3 & 1 \end{bmatrix}$$

Notice that $N^2 = I_4$, where I_n denotes the $n \times n$ identity matrix. More generally, let N be $n \times n$. We then claim that $N^2 = I_n$.

To establish this intriguing result, we shall use identity (25.4). First, notice that

$$n_{ij} = \begin{cases} \binom{i}{j} & \text{if } i \text{ is odd} \\ -\binom{i}{j} & \text{otherwise} \end{cases}$$

where $0 \leq j \leq i$. Then:

$$\begin{aligned} (N^2)_{ij} &= \sum_{r=0}^{n-1} (-1)^{i+1} \binom{i}{r} \cdot (-1)^{r+1} \binom{r}{j} \\ &= \sum_{r=0}^{n-1} (-1)^{i+r} \binom{i}{r} \binom{r}{j} \end{aligned}$$

$$= (-1)^i \sum_{r=0}^{n-1} (-1)^r \binom{r}{j} \binom{i}{r}$$

$$= (-1)^i \binom{0}{j-i} (-1)^i, \quad \text{by equation (25.4)}$$

$$= \delta_{ij}$$

Thus $N^2 = I_n$, as desired.

Next, we pursue an interesting problem that combines Monmort's matrix M_n and Tartaglia's matrix T_n. It was proposed by Hoggatt in 1973 as a problem to show that $M_n \cdot T_n = H_n$, where

$$H_n = \begin{bmatrix} 1 & 1 & 1 & 1 & 1 & \\ 1 & 2 & 3 & 4 & 5 & \\ 2 & 5 & 9 & 14 & 20 & \cdots \\ 2 & 7 & 16 & 30 & 50 & \\ 2 & 9 & 25 & 55 & 105 & \\ & & \vdots & & & \end{bmatrix}$$

(We encountered this array in Figure 16.5.) The solution presented here is based on the one given in 1975 by P. S. Bruckman of the University of Illinois at Chicago.

Let $M = (m_{ij})_{n \times n}$ and $T = (t_{ij})_{n \times n}$. Then

$$m_{ij} = \begin{cases} \binom{j-1}{i-j} & \text{if } j \le i \le 2j-1 \\ 0 & \text{otherwise} \end{cases}$$

and $t_{ij} = \binom{i+j-2}{j-1}$. Let $D = (d_{ij}) = M_n T_n$. Then

$$d_{ij} = \sum_{k=\lfloor i/2 \rfloor + 1}^{i} \binom{k-1}{i-k} \binom{k+j-2}{j-1}$$

For convenience, let $r = i - 1$, $s = j - 1$, and $m = i - k$. Then

$$d_{ij} = \sum_{m=0}^{\lfloor r/2 \rfloor} \binom{r-m}{m} \binom{r-m+s}{s}$$

Let $f_j(x) = \sum_{i=1}^{\infty} d_{ij}x^{i-1}$ be the generating function of column j of D. Then:

$$
\begin{aligned}
f_j(x) &= \sum_{r=0}^{\infty} x^r \left[\sum_{m=0}^{\lfloor r/2 \rfloor} \binom{r-m}{m}\binom{r-m+s}{s} \right] \\
&= \sum_{m=0}^{\infty} x^{2m} \sum_{r=0}^{\infty} \binom{r+m}{m}\binom{r+s+m}{r+m} x^r \\
&= \sum_{m=0}^{\infty} x^{2m} \sum_{r=0}^{\infty} \binom{s+m}{m}\binom{r+s+m}{r} x^r \\
&= \sum_{m=0}^{\infty} \binom{-s-1}{m}(-x^2)^m \sum_{r=0}^{\infty} \binom{-s-m-1}{r}(-x)^r \\
&= \sum_{m=0}^{\infty} \binom{-s-1}{m}(-x^2)^m (1-x)^{-s-m-1} \\
&= (1-x)^{-s-1} \sum_{m=0}^{\infty} \binom{-s-1}{m}\left(\frac{-x^2}{1-x}\right)^m \\
&= (1-x)^{-s-1}\left(1 - \frac{x^2}{1-x}\right)^{-s-1} \\
&= (1-x-x^2)^{-s-1} = (1-x-x^2)^{-j}
\end{aligned}
$$

But $f_1(x) = \frac{1}{1-x-x^2}$ is the generating function for Fibonacci numbers, so $f_j(x)$ is the column generator for the Fibonacci convolution matrix, namely H_n. Thus, $D_n = M_n \cdot T_n = H_n$, as desired.

Brennan's Matrices

Closely related to Montmort's array is the $(n+1) \times (n+1)$ *Brennan matrix* P_n:

$$
P_n = \begin{bmatrix}
 & 0 & 0 & 0 & 0 & 1 \\
 & 0 & 0 & 0 & 1 & 1 \\
\cdots & 0 & 0 & 1 & 2 & 1 \\
 & 0 & 1 & 3 & 3 & 1 \\
 & 1 & 4 & 6 & 4 & 1 \\
 & & & \vdots & &
\end{bmatrix}
$$

It was studied in 1962 by T. A. Brennan of Lockheed Missiles and Space Co., Sunnyvale, California.

Notice that $|P_0| = 1$, $|P_1| = -1 = P_2$, and $|P_3| = 1 = |P_4|$. More generally, we conjecture that $|P_n| = (-1)^{n(n+1)/2}$.

To confirm this, first notice that the formula works when $n = 0$. Now, assume that it works for an arbitrary nonnegative integer $< n$. Expanding $|P_n|$ with respect to column 0,

$$\begin{aligned}|P_n| &= (-1)^n |P_{n-1}| \\ &= (-1)^n \cdot (-1)^{(n-1)n/2} = (-1)^{n(n+1)/2}\end{aligned}$$

Thus, by PMI, $|P_n| = (-1)^{n(n+1)/2}$ for every $n \geq 0$.

In particular,

$$P_2 = \begin{bmatrix} 0 & 0 & 1 \\ 0 & 1 & 1 \\ 1 & 2 & 1 \end{bmatrix},$$

studied by Brennan and Bicknell in 1964.

Notice that

$$P_2 = \begin{bmatrix} F_0^2 & 2F_0F_1 & F_1^2 \\ F_0F_1 & F_2^2 + F_0F_1 & F_1F_2 \\ F_1^2 & 2F_1F_2 & F_2^2 \end{bmatrix}$$

It follows by PMI that

$$P_2^n = \begin{bmatrix} F_{n-1}^2 & 2F_{n-1}F_n & F_n^2 \\ F_{n-1}F_n & F_{n+1}^2 + F_{n-1}F_n & F_nF_{n+1} \\ F_n^2 & 2F_nF_{n+1} & F_{n+1}^2 \end{bmatrix}$$

where $n \geq 1$.

Since $|P_2| = -1$, it follows that

$$\begin{vmatrix} F_{n-1}^2 & 2F_{n-1}F_n & F_n^2 \\ F_{n-1}F_n & F_{n+1}^2 + F_{n-1}F_n & F_nF_{n+1} \\ F_n^2 & 2F_nF_{n+1} & F_{n+1}^2 \end{vmatrix} = (-1)^n$$

For example,

$$P_2^3 = \begin{bmatrix} 1 & 2 & 4 \\ 4 & 7 & 12 \\ 4 & 6 & 9 \end{bmatrix}$$

So

$$|P_2^3| = \begin{vmatrix} 1 & 2 & 4 \\ 4 & 7 & 12 \\ 4 & 6 & 9 \end{vmatrix} = -1 = (-1)^3$$

as expected.

Closely related to P_2 is the matrix

$$R_2 = \begin{bmatrix} 0 & 0 & 1 \\ 0 & 1 & 2 \\ 1 & 1 & 1 \end{bmatrix},$$

again studied by Brennan and Bicknell in 1963. It follows by PMI that

$$R_2^n = \begin{bmatrix} F_{n-1}^2 & F_{n-1}F_n & F_n^2 \\ 2F_{n-1}F_n & F_{n+1}^2 + F_{n-1}F_n & 2F_nF_{n+1} \\ F_n^2 & F_nF_{n+1} & F_{n+1}^2 \end{bmatrix}$$

where $n \geq 1$.

Since $|R_2| = -1$, it follows that

$$\begin{vmatrix} F_{n-1}^2 & F_{n-1}F_n & F_n^2 \\ 2F_{n-1}F_n & F_{n+1}^2 + F_{n-1}F_n & 2F_nF_{n+1} \\ F_n^2 & F_nF_{n+1} & F_{n+1}^2 \end{vmatrix} = (-1)^n$$

In his study, Brennan also introduced the $(n + 1) \times (n + 1)$ matrix Q_n, given by

$$Q_n = \begin{bmatrix} & 1 & 1 & 1 & 1 & 1 \\ & 4 & 3 & 2 & 1 & 0 \\ \cdots & 6 & 3 & 1 & 0 & 0 \\ & 4 & 1 & 0 & 0 & 0 \\ & 1 & 0 & 0 & 0 & 0 \\ & & & \vdots & & \end{bmatrix}$$

As in the case of P_n, it can be shown that $|Q_n| = (-1)^{n(n+1)/2}$ for every $n \geq 0$.

A Modified Montmort's Triangle

Over the centuries, many attempts have been made to develop an explicit formula for $S_n(i) = \sum_{j=1}^{n} j^i$. For instance, in 1636, Fermat, using figurate numbers, discovered a recurrence relation for $\sum_{j=1}^{n} j^i$ in terms of $S_n(i-1), S_n(i-2), \cdots$; but his method turned out to be *not* practical. In 1654, Pascal developed a more practical method using the binomial theorem. Both Fermat and Pascal overlooked the work of the German mathematician Johann Faulhaber (1580–1635), who by 1631 had developed formulas for up to $\sum_{j=1}^{n} j^{17}$. Based on them, Faulhaber in 1631 discovered the following interesting properties, where $1 \leq j \leq n$:

$$\sum_{i \text{ odd}} j^i = \text{polynomial in } n(n+1)$$

$$\sum_{i \text{ even}} j^i = (2n+1) \cdot \text{ polynomial in } n(n+1)$$

In each case, the sum is a polynomial in n of degree $i+1$.

In his *Ars Conjectandi*, the Swiss mathematician James Bernoulli (1654–1705) gave formulas for $S_n(i)$ for up to $i = 10$. His formula for $\sum_{j=1}^{n} j^9$ incorrectly gave the coefficient of n^2 as $-\frac{1}{12}$; see the corrected list in Table 25.1.

Table 25.1 Bernoulli's Corrected Summation Formulas

$$\begin{aligned}
\sum_{i=1}^{n} i &= \tfrac{1}{2}n^2 + \tfrac{1}{2}n \\
\sum_{i=1}^{n} i^2 &= \tfrac{1}{3}n^3 + \tfrac{1}{2}n^2 + \tfrac{1}{6}n \\
\sum_{i=1}^{n} i^3 &= \tfrac{1}{4}n^4 + \tfrac{1}{2}n^3 + \tfrac{1}{4}n^2 \\
\sum_{i=1}^{n} i^4 &= \tfrac{1}{5}n^5 + \tfrac{1}{2}n^4 + \tfrac{1}{3}n^3 - \tfrac{1}{30}n \\
\sum_{i=1}^{n} i^5 &= \tfrac{1}{6}n^6 + \tfrac{1}{2}n^5 + \tfrac{5}{12}n^4 - \tfrac{1}{12}n^2 \\
\sum_{i=1}^{n} i^6 &= \tfrac{1}{7}n^7 + \tfrac{1}{2}n^6 + \tfrac{1}{2}n^5 - \tfrac{1}{6}n^3 + \tfrac{1}{42}n \\
\sum_{i=1}^{n} i^7 &= \tfrac{1}{8}n^8 + \tfrac{1}{2}n^7 + \tfrac{7}{12}n^6 - \tfrac{7}{24}n^4 + \tfrac{1}{12}n^2 \\
\sum_{i=1}^{n} i^8 &= \tfrac{1}{9}n^9 + \tfrac{1}{2}n^8 + \tfrac{2}{3}n^7 - \tfrac{7}{15}n^5 + \tfrac{2}{9}n^3 - \tfrac{1}{30}n \\
\sum_{i=1}^{n} i^9 &= \tfrac{1}{10}n^9 + \tfrac{1}{2}n^9 + \tfrac{3}{4}n^8 - \tfrac{7}{10}n^6 + \tfrac{1}{2}n^4 - \tfrac{3}{20}n^2 \\
\sum_{i=1}^{n} i^{10} &= \tfrac{1}{11}n^{11} + \tfrac{1}{2}n^{10} + \tfrac{5}{6}n^9 - n^7 + n^5 - \tfrac{1}{2}n^3 - \tfrac{5}{66}n
\end{aligned}$$

Familiar with this historical background, we now return to Pascal's work of 1654. He established that

$$(n+1)^i = (n+1) + \binom{i}{1}\sum_{j=1}^{n} j + \binom{i}{2}\sum_{j=1}^{n} j^2 + \binom{i}{3}\sum_{j=1}^{n} j^3 + \cdots + \binom{i}{i-1}\sum_{j=1}^{n} j^i \tag{25.5}$$

For example:

$$n+1 = n+1$$

$$(n+1)^2 = (n+1) + (n^2+n)$$

$$= (n+1) + \binom{2}{1}\frac{n(n+1)}{2} = (n+1) + \binom{2}{1}\sum_{j=1}^{n} j$$

and similarly,

$$(n+1)^3 = (n+1) + \binom{3}{1}\sum_{j=1}^{n} j + \binom{3}{2}\sum_{j=1}^{n} j^2$$

The various formulas resulting from (25.5) corresponding to different values of i yield the following matrix equation:

$$\begin{bmatrix} n+1 \\ (n+1)^2 \\ (n+1)^3 \\ (n+1)^4 \\ (n+1)^5 \\ \vdots \end{bmatrix} = \begin{bmatrix} 1 & 0 & 0 & 0 & 0 & \\ 1 & 2 & 0 & 0 & 0 & \\ 1 & 3 & 3 & 0 & 0 & \cdots \\ 1 & 4 & 6 & 4 & 0 & \\ 1 & 5 & 10 & 10 & 5 & \\ & & \vdots & & & \end{bmatrix} \begin{bmatrix} n+1 \\ \sum i \\ \sum i^2 \\ \sum i^3 \\ \sum i^4 \\ \vdots \end{bmatrix} \tag{25.6}$$

where the limits of the sums are 1 and n. The coefficient matrix Q on the RHS is precisely the same as Pascal's matrix with the main diagonal deleted.

Let Q_n denote the $n \times n$ submatrix in the upper left-hand corner of Q. Since $|Q_n| = n!$, Q_n is invertible. In 1980, S. H. Scott computed Q_n^{-1}:

$$Q_n^{-1} = \begin{bmatrix} 1 & 0 & 0 & 0 & 0 & \\ -\frac{1}{2} & \frac{1}{2} & 0 & 0 & 0 & \\ \frac{1}{6} & -\frac{1}{2} & \frac{1}{3} & 0 & 0 & \cdots \\ 0 & \frac{1}{4} & -\frac{1}{2} & \frac{1}{4} & 0 & \\ -\frac{1}{30} & 0 & \frac{1}{3} & -\frac{1}{2} & \frac{1}{5} & \\ & & \vdots & & & \end{bmatrix}$$

↑
Bernoulli numbers

Column 1 of Q_n^{-1} consists of the *Bernoulli numbers* $B_0, B_1, B_2, \cdots, B_{n-1}$, defined recursively as follows:

$$B_0 = 1$$

$$B_n = \sum_{r=0}^{n} \binom{n}{r} B_r, \quad n \geq 2$$

Using Q^{-1} and changing n to $n-1$, the matrix equation (25.6) can be rewritten as follows:

$$\begin{bmatrix} n \\ \sum(i-1) \\ \sum(i-1)^2 \\ \sum(i-1)^3 \\ \sum(i-1)^4 \\ \vdots \end{bmatrix} = \begin{bmatrix} 1 & 0 & 0 & 0 & 0 & \\ -\frac{1}{2} & \frac{1}{2} & 0 & 0 & 0 & \\ \frac{1}{6} & -\frac{1}{2} & \frac{1}{3} & 0 & 0 & \cdots \\ 0 & \frac{1}{4} & -\frac{1}{2} & \frac{1}{4} & 0 & \\ -\frac{1}{30} & 0 & \frac{1}{3} & -\frac{1}{2} & \frac{1}{5} & \\ & & \vdots & & & \end{bmatrix} \begin{bmatrix} n \\ n^2 \\ n^3 \\ n^4 \\ n^5 \\ \vdots \end{bmatrix}$$

where the sum limits are 1 and n.

Returning to Tartaglia's array (or Pascal's triangle), we replace each element by the partial sum of the elements in its northeast diagonal (or row). Figure 25.7 shows the resulting array B. It can be found in Bernoulli's *Ars Conjectandi*. Bernoulli arrived at array† B in his investigation of the *problem of points*. This famous gambling problem of the day led Bernoulli to develop the theory of probability. In 1987, A. W. F. Edwards christened it *Bernoulli's triangle*.

n \ j	0	1	2	3	4	5	6
0	1	2	4	8	16	32	64
1	1	3	7	15	31	63	
2	1	4	11	26	57		
3	1	5	16	42			
4	1	6	22				
5	1	7					
6	1						

Figure 25.7 Bernoulli's Triangle

Interestingly, Bernoulli's triangle B preserves many properties of Tartaglia's array. For example, each entry $B(n,j)$ can be obtained by adding its western neighbor to the northern neighbor: $B(n,j) = B(n,j-1) + B(n-1,j)$; see Figure 25.7. Consequently, $B(n,j)$ equals the sum of its northern neighbor and the elements to its left in the preceding row: $B(n,j) = \sum_{r=0}^{j} B(n-1,r)$.

†See A. W. F. Edwards, "Pascal's Triangle – and Bernoulli's and Vieta's," *Mathematical Spectrum* 20 (1987–1988), 33–37.

For example,

$$\begin{aligned}42 &= 16 + 26 = (5 + 11) + 26 \\ &= (1 + 4) + 11 + 26 = (0 + 1) + 4 + 11 + 26 \\ &= 1 + 4 + 11 + 26\end{aligned}$$

See Figure 25.7.

Since $T(m, n) = \binom{m+n}{m}$, it follows that

$$\begin{aligned}B(n, j) &= T(n + j, 0) + T(n + j - 1, 1) + T(n + j - 2, 2) + \cdots + T(n, j) \\ &= \sum_{r=0}^{j} T(n + j - r, r) = \sum_{r=0}^{j} \binom{n + j}{r} \qquad (25.7)\end{aligned}$$

For example, $B(5, 4) = \sum_{r=0}^{4} \binom{9}{r} = \binom{9}{0} + \binom{9}{1} + \binom{9}{2} + \binom{9}{3} + \binom{9}{4} = 256.$

Since the sum of the entries in row n in Pascal's triangle is 2^n, it follows that row 0 in Bernoulli's triangle consists of powers of 2; see Figure 25.8. As a bonus, row 1 gives the various Mersenne numbers. This is true since $\sum_{r=0}^{m-1} \binom{m}{r} = \sum_{r=0}^{m} \binom{m}{r} - \binom{m}{m} = 2^m - 1.$

1	2	4	8	16	32	64	← powers of 2
1	3	7	15	31	63		← Mersenne numbers
1	4	11	26	57			
1	5	16	42				
1	6	22					
1	7						
1							

powers of 4

Figure 25.8 Bernoulli's Triangle

Yet another bonus: The elements on the descending diagonal beginning with $1 = B(1, 0)$ are powers of 4; see Figure 25.8. In other words, $B(n, n - 1) = 4^{n-1}$, where $n \geq 1$.

This can be established fairly easily. By formula (25.7) above, $B(n, n - 1) = \sum_{r=0}^{n-1} \binom{2n-1}{r}$. But $\sum_{r=0}^{2n-1} \binom{2n-1}{r} = 2^{2n-1}$. That is:

$$\sum_{r=0}^{n-1} \binom{2n - 1}{r} + \sum_{r=n}^{2n-1} \binom{2n - 1}{r} = 2^{2n-1}$$

$$\sum_{r=0}^{n-1}\binom{2n-1}{r}+\sum_{r=0}^{n-1}\binom{2n-1}{r}=2^{2n-1}$$

$$2\sum_{r=0}^{n-1}\binom{2n-1}{r}=2^{2n-1}$$

$$\sum_{r=0}^{n-1}\binom{2n-1}{r}=4^{n-1}$$

Thus $B(n, n-1) = 4^{n-1}$, as desired.

26 Miscellaneous Arrays

This chapter presents an array of unrelated, but interesting, triangular arrays.

We begin with the triangular array in Figure 26.1. It consists of positive integers, where each row contains one element more than the previous row and the top row consists of the element 1.

$$
\begin{array}{ccccccccccc}
 & & & & & 1 & & & & & \leftarrow \text{ row } 1 \\
 & & & & 2 & & 3 & & & & \\
 & & & 4 & & 5 & & 6 & & & \\
 & & 7 & & 8 & & 9 & & 10 & & \\
 & 11 & & 12 & & 13 & & 14 & & 15 & \\
16 & & 17 & & 18 & & 19 & & 20 & & 21
\end{array}
$$

Figure 26.1 A Triangular Array of Positive Integers

Since row i in Figure 26.1 consists of i elements, it follows that the top n rows contain t_n elements. In other words, the rightmost element in row n is triangular number t_n. Consequently, the rightmost southeast diagonal of the array consists of triangular numbers.

Let S_n denote the sum of the elements in row n. We will find a formula for S_n.

Let A_n denote the sum of the elements in the top n rows. Then:

$$
\begin{aligned}
A_n &= \sum_{i=1}^{t_n} i = \sum_{i=1}^{n(n+1)/2} i \\
&= \frac{1}{2} \cdot \frac{n(n+1)}{2} \left[\frac{n(n+1)}{2} + 1 \right] = \frac{(n^2+n)(n^2+n+2)}{8}
\end{aligned}
$$

Since $n(n-1) = n^2 - n$, A_{n-1} can be obtained by changing n to $-n$ in A_n. So

$$A_{n-1} = \sum_{i=1}^{t_{n-1}} i = \frac{t_{n-1}\left(t_{n-1}+1\right)}{2}$$
$$= \frac{(n^2-n)(n^2-n+2)}{8}$$

Thus:

$$S_n = A_n - A_{n-1}$$
$$= \frac{(n^2+n)(n^2+n+2)}{8} - \frac{(n^2-n)(n^2-n+2)}{8}$$
$$= \frac{n}{8}\left[(n+1)(n^2+n+2)-(n-1)(n^2-n+2)\right]$$
$$= \frac{n}{8}(4n^2+2) = \frac{n(n^2+1)}{2}$$

An Alternate Method

There is a shorter and more elegant way of deriving this formula. Since row n contains n elements and the nth element is t_n, the various elements in row n are $t_n-(n-1), t_n-(n-2), \cdots, t_n-2, t_n-1$, and t_n-0. So:

$$S_n = [t_n-(n-1)] + [t_n-(n-2)] + \cdots + (t_n-2) + (t_n-1) + t_n$$
$$= nt_n - \sum_{i=1}^{n-1} i = nt_n - \frac{n(n-1)}{2}$$
$$= n\cdot\frac{n(n+1)}{2} - \frac{n(n-1)}{2} = \frac{n(n^2+1)}{2}, \quad \text{as desired}$$

For example, the sum of the elements in row 5 equals $\frac{5(5^2+1)}{2} = 65$.

The sum $S_n = \frac{n(n^2+1)}{2}$ is the *magic constant* of an $n \times n$ magic square,† where $n \geq 3$.

Generating Functions for Rising Diagonals

Next, we would like to find the generating function $g_k(x)$ for each northeast diagonal, as Hoggatt and Bicknell did in 1974, where $k \geq 0$. To this end, notice that the entries on diagonal 0 are one more than the triangular numbers t_n, where

†See the author's *Elementary Number Theory with Applications.*

$n \geq 0$. Since the generating function of triangular numbers is $\sum_{n=0}^{\infty} t_n x^n = \frac{x}{(1-x)^3}$ and that for diagonal 0 is given by

$$g_0(x) = \sum_{n=0}^{\infty} t_n x^n + \sum_{n=0}^{\infty} x^n$$
$$= \frac{x}{(1-x)^3} + \frac{x}{1-x} = \frac{1-x+x^2}{(1-x)^3}$$

it follows by PMI that $g_k(x) = \frac{t_{k+1} - (k+1)^2 x + (t_k+1)x^2}{(1-x)^3}$.

Notice that $g_k(x)$ is the generating function for the columns of the left-justified array in Figure 26.2.

1				
2	3			
4	5	6		
7	8	9	10	
11	12	13	14	15

Figure 26.2

Suppose we would like to find the 1000th element, or more generally, the mth element in the sequence $1, 2, 2, 3, 3, 3, 4, 4, 4, 4, 5, \cdots$. Although this problem appears to be a bit confusing, it can be solved nicely and easily by arranging the elements in a triangular array, as in Figure 26.3.

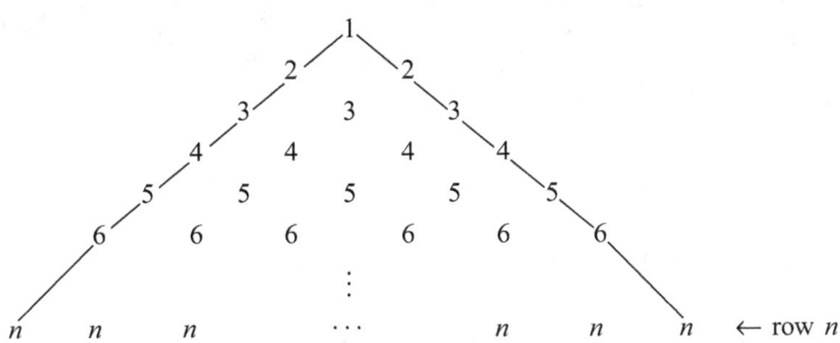

Figure 26.3

Clearly, row n consists of n ns, where $n \geq 1$. So, the top k rows contain a total of t_k elements; consequently, the t_kth element in the sequence is k.

Suppose the mth element of the sequence is n, where $n \geq 1$. (For example, the 7th element of the sequence is 4 and it lies in row 4.) It lies in row n, so

$t_{n-1} < m \le t_n$. That is,

$$\frac{n(n-1)}{2} < m \le \frac{n(n+1)}{2}$$

Since $\frac{n(n-1)}{2} < m$, $n^2 - n - 2m < 0$. Solving the quadratic equation $n^2 - n - 2m = 0$ yields $n = \frac{1 \pm \sqrt{8m+1}}{2}$. Since $n^2 - n - 2m < 0$,

$$\frac{1-\sqrt{8m+1}}{2} < n < \frac{1+\sqrt{8m+1}}{2}$$

See Figure 26.4.

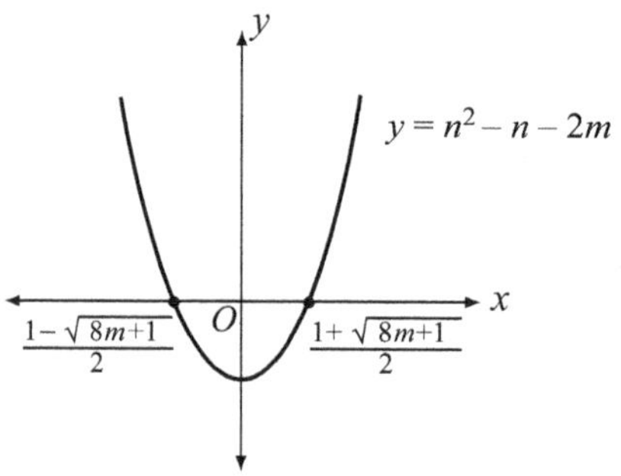

Figure 26.4

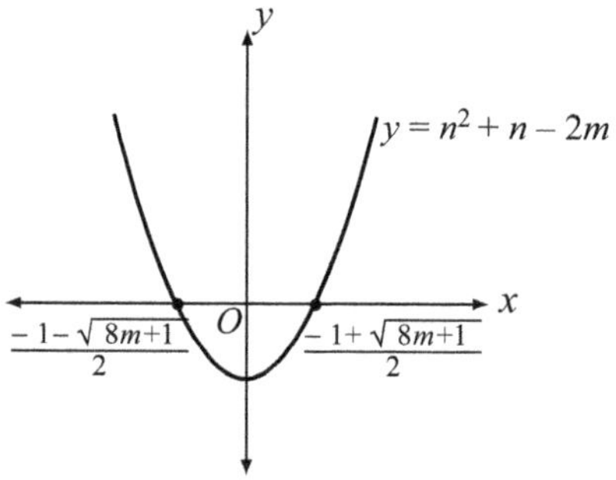

Figure 26.5

But $n \ge 1$. So $1 \le n < \frac{1+\sqrt{8m+1}}{2}$.

Since $m \le \frac{n(n+1)}{2}$, $n^2 + n - 2m \ge 0$. Solving the quadratic equation $n^2 + n - 2m = 0$ yields $n = \frac{-1 \pm \sqrt{8m+1}}{2}$. Since $n^2 - n - 2m \ge 0$, this implies that either $n \le \frac{-1-\sqrt{8m+1}}{2}$ or $n \ge \frac{-1+\sqrt{8m+1}}{2}$; see Figure 26.5. But $n \ge 1$; so $n \ge \frac{\sqrt{8m+1}-1}{2}$.

Since $m \ge 1$, $\frac{\sqrt{8m+1}-1}{2} \ge 1$. Thus

$$\frac{\sqrt{8m+1}-1}{2} \le n < \frac{\sqrt{8m+1}+1}{2}$$

Since n is an integer, this implies that

$$n = \left\lceil \frac{\sqrt{8m+1}-1}{2} \right\rceil = \left\lfloor \frac{\sqrt{8m+1}+1}{2} \right\rfloor$$

In particular, the 1000th element in the sequence lies in row

$$\left\lceil \frac{\sqrt{8 \cdot 1000 + 1} - 1}{2} \right\rceil = \lceil 44.224154\cdots \rceil = 45 = \left\lfloor \frac{\sqrt{8 \cdot 1000 + 1} + 1}{2} \right\rfloor$$

So the 1000th element in the sequence is 45.

Since the sum of the elements in row k is $k \cdot k = k^2$, it follows that the sum of the elements in the top n rows equals

$$\sum_{k=1}^{n} k^2 = \frac{n(n+1)(2n+1)}{6}$$

A Triangular Array of Odd Integers

Consecutive odd positive integers can be used to construct triangular arrays with interesting properties. They too provide excellent opportunities for exploration, experimentation, pattern recognition, and conjecturing. Figure 26.6, shows one such array, developed in 1995. Column 0 of the array consists of odd positive integers, and each successive column consists of the same numbers with one level pushed down from the preceding column.

$i \backslash j$	0	1	2	3	4	5
0	1					
1	3	1				
2	5	3	1			
3	7	5	3	1		
4	9	7	5	3	1	
5	11	9	7	5	3	1

Figure 26.6

$i \backslash j$	0	1	2	3	4	5	Row sums
0	1						1^2
1	3	1					2^2
2	5	3	1				3^2
3	7	5	3	1			4^2
4	9	7	5	3	1		5^2
5	11	9	7	5	3	1	6^2

Figure 26.7

Since $\sum_{i=1}^{n}(2i+1) = (n+1)^2$, it follows that the row sum in row n is $(n+1)^2$, where $n \geq 0$; see Figure 26.7.

Row Sums and Generating Functions

The fact that each row sum is a square can be confirmed using generating functions. Unfortunately, this approach is a bit long.

Let

$$g(x) = \sum_{n=0}^{\infty}(2n+1)x^n = 2\sum_{n=0}^{\infty} nx^n + \sum_{n=0}^{\infty} x^n$$

$$= 2x\sum_{n=1}^{\infty} nx^{n-1} + \sum_{n=0}^{\infty} x^n = 2x\sum_{n=0}^{\infty}(n+1)x^n + \sum_{n=0}^{\infty} x^n$$

$$= \frac{2x}{(1-x)^2} + \frac{1}{1-x} = \frac{1+x}{(1-x)^2}$$

So the function $g(x) = \frac{1+x}{(1-x)^2}$ generates all odd positive integers.

Then the function

$$g_r(x) = \frac{(1+x)x^r}{(1-x)^2}$$

generates column r, where $r \geq 0$ and $g_0(x) = g(x)$. The effect of x^r in the numerator of $g_r(x)$ is to push column 0 down by r spaces. For instance, $g_1(x) = \frac{(1+x)x}{(1-x)^2}$ generates column 1 and $g_2(x) = \frac{(1+x)x^2}{(1-x)^2}$ generates column 2.

So the various row sums are generated by the function:

$$\begin{aligned} G(x) &= \sum_{r=0}^{\infty} g_r(x) \\ &= \sum_{r=0}^{\infty} \frac{(1+x)x^r}{(1-x)^2} = \frac{1+x}{(1-x)^2} \sum_{r=0}^{\infty} x^r \\ &= \frac{1+x}{(1-x)^2} \cdot \frac{1}{1-x} = \frac{1+x}{(1-x)^3} \end{aligned}$$

The row sum in row n is given by the coefficient of x^n in the expansion of $\frac{1+x}{(1-x)^3}$. But

$$\frac{1+x}{(1-x)^3} = (1+x) \sum_{r=0}^{\infty} \binom{n+2}{2} x^n$$

Therefore, the coefficient of x^n on the RHS is given by

$$\binom{n+2}{2} + \binom{n+1}{2} = \frac{(n+2)(n+1)}{2} + \frac{(n+1)n}{2} = (n+1)^2$$

Thus, the row sum in row n is $(n+1)^2$, where $n \geq 0$.

Rising Diagonal Sums

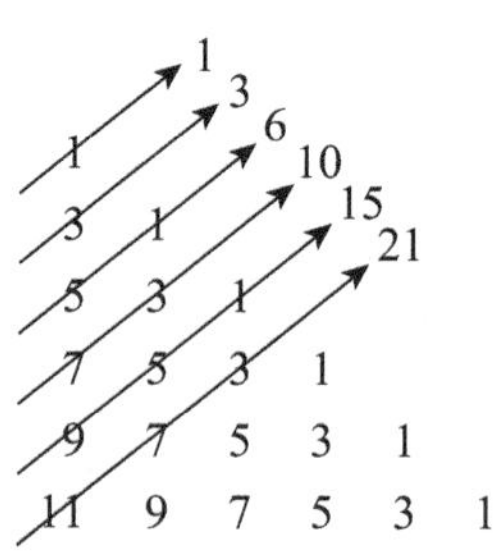

Figure 26.8

The rising diagonal sums yield an interesting dividend. From Figure 26.8, it appears that each rising diagonal sum is a triangular number. In fact, we conjecture that the nth rising diagonal sum D_n is t_{n+1}, where $n \geq 0$. This can be confirmed as follows.

We have $D_n = \sum_{j=0}^{\lfloor \frac{2n+1}{4} \rfloor} (2n + 1 - 4j)$. Noticing that diagonal n ends in a 1 if n is even, and in a 3 if n is odd, this sum can be rewritten as

$$D_n = \begin{cases} (2n + 1) + (2n - 3) + \cdots + 5 + 1 & \text{if } n \text{ is even} \\ (2n + 1) + (2n - 3) + \cdots + 7 + 3 & \text{otherwise} \end{cases}$$

Suppose n is even, say, $n = 2k$. Then $D_n = (4k + 1) + (4k - 3) + \cdots + 5 + 1$. The RHS is an arithmetic sum with the first term 1 and the last term $4k + 1$, and common difference 4, and it contains $k + 1$ terms. So:

$$\begin{aligned} D_n &= \frac{(k+1)}{2}[1 + (4k + 1)] = (k + 1)(2k + 1) \\ &= \frac{(n + 1)(n + 2)}{2} = t_{n+1} \end{aligned}$$

Similarly, when n is odd, $D_n = t_{n+1}$.

Thus, in both cases, $D_n = t_{n+1}$, as desired.

Rising Diagonal Sums and Generating Functions

We can apply generating functions to find the rising diagonal sums, although the process is a bit complicated.

n \ j	0	1	2	3	Row sums
0	1				1
1	3				3
2	5	1			6
3	7	3			10
4	9	5	1		15
5	11	7	3		21
6	13	9	5	1	28

Figure 26.9

The rising diagonals can be made horizontal by shifting column j down by j spaces. See Figure 26.9. So the nth diagonal sum in Figure 26.8 is given by the nth row sum in Figure 26.9, where $n \geq 0$.

Since column 0 is generated by $g_0^*(x) = g_0(x) = \frac{1+x}{(1-x)^2}$, column r is generated by $g_r^*(x) = \frac{(1+x)x^{2r}}{(1-x)^2}$. So the row sums of the array in Figure 26.9 are generated by:

$$
\begin{aligned}
G^*(x) &= \sum_{r=0}^{\infty} g_r^*(x) = \frac{1+x}{(1-x)^2} \sum_{r=0}^{\infty} x^{2r} \\
&= \frac{1+x}{(1-x)^2} \cdot \frac{1}{1-x^2} = \frac{1}{(1-x)^3} \\
&= \sum_{r=0}^{\infty} \binom{n+2}{2} x^n = \sum_{r=0}^{\infty} t_{n+1} x^n
\end{aligned}
$$

Thus, the nth row sum of the array in Figure 26.9 (or the nth rising diagonal sum in Figure 26.8) is t_{n+1}, where $n \geq 0$.

Fibonacci's Triangle

Consecutive odd positive integers can also be used to construct another triangular array, as in Figure 26.10. This array is attributed to Fibonacci, so we call it *Fibonacci's triangle*. As before, row n contains n elements, where $n \geq 1$.

	Row sums
$\boxed{1}$	1^3
3 5	2^3
7 $\boxed{9}$ 11	3^3
13 15 17 19	4^3
21 23 $\boxed{25}$ 27 29	5^3
31 33 35 37 39 41	6^3

Figure 26.10 Fibonacci's Triangle

A quick inspection of the elements in the leftmost rising diagonal reveals an interesting pattern:

$$
\begin{aligned}
1 &= 2 \cdot 0 + 1 \\
3 &= 2 \cdot 1 + 1 \\
7 &= 2 \cdot 3 + 1 \\
13 &= 2 \cdot 6 + 1 \\
21 &= 2 \cdot 10 + 1 \\
31 &= 2 \cdot 15 + 1
\end{aligned}
$$

$\uparrow$ triangular numbers

More generally, the leftmost element in row n is given by $2t_{n-1}+1=n^2-n+1$, where $n \geq 1$.

Since row $n+1$ begins with $2t_n+1$, it follows that row n ends in

$$(2t_n+1)-2=2\cdot\frac{n(n+1)}{2}+1-2=n^2+n-1$$

The array manifests another intriguing property: The central elements are squares (see Figure 26.10); they occur in rows n, where n is odd. The central element in row n is n^2 and it occurs in position $\frac{n+1}{2}$, where n is odd.

This can be confirmed fairly easily. Since row n begins with $2t_{n-1}+1$ and the difference between any two adjacent elements in the row is 2, the central element in row n is given by

$$2t_{n-1}+1+\frac{2(n-1)}{2}=2\cdot\frac{n(n-1)}{2}+1+(n-1)=n^2$$

as desired.

Fibonacci's triangle possesses yet another property: Since row n begins with $2t_{n-1}+1$, the various elements in row n are $2t_{n-1}+1, 2t_{n-1}+3, \cdots$, an $2t_{n-1}+(2n-1)$. So the sum S_n of the numbers in row n is given by:

$$\begin{aligned}S_n &= (2t_{n-1}+1)+(2t_{n-1}+3)+\cdots+[2t_{n-1}+(2n-1)]\\ &= 2nt_{n-1}+\sum_{i=1}^{n}(2i-1)=2n\cdot\frac{n(n-1)}{2}+n^2=n^3\end{aligned}$$

Thus, the row sum in row n equals n^3. See Figure 26.10.

Since each odd number is the difference of the squares of two consecutive integers, Fibonacci's triangle can be written in terms of differences of consecutive integers, as in Figure 26.11.

1^2-0^2

$2^2-1^2 \quad 3^2-2^2$

$4^2-3^2 \quad 5^2-4^2 \quad 6^2-5^2$

$7^2-6^2 \quad 8^2-7^2 \quad 9^2-8^2 \quad 10^2-9^2$

$11^2-10^2 \quad 12^2-11^2 \quad 13^2-12^2 \quad 14^2-13^2 \quad 15^2-14^2$

Figure 26.11

Consequently, the various elements in row n are

$$(t_{n-1}+1)^2-t_{n-1}^2, (t_{n-1}+2)^2-(t_{n-1}+1)^2, (t_{n-1}+3)^2-(t_{n-1}+2)^2,$$
$$\cdots, (t_{n-1}+n)^2-[t_{n-1}+(n-1)]^2$$

So S_n yields a telescopic sum:

$$S_n = [(t_{n-1}+1)^2 - t_{n-1}^2] + [(t_{n-1}+2)^2 - (t_{n-1}+1)^2] + \cdots + (t_{n-1}+n)^2 - [t_{n-1}+(n-1)]^2$$
$$= (t_{n-1}+n)^2 - t_{n-1}^2 = 2nt_{n-1} + n^2$$
$$= 2n \cdot \frac{n(n-1)}{2} + n^2 = n^3$$

A Byproduct

Since $t_{k-1} + k = t_k$, it follows that $t_k^2 - t_{k-1}^2 = k^3$. So $\sum_{k=1}^{n}(t_k^2 - t_{k-1}^2) = \sum_{k=1}^{n} k^3$; that is, $t_n^2 - t_0^2 = \sum_{k=1}^{n} k^3$. So $\sum_{k=1}^{n} k^3 = t_n^2 = \left[\frac{n(n+1)}{2}\right]^2$. Thus the cumulative row sum of the top n rows is given by

$$\sum_{k=1}^{n} k^3 = \left[\frac{n(n+1)}{2}\right]^2 = \left(\sum_{k=1}^{n} k\right)^2$$

Generating Functions for the Rising Diagonals

We can employ the generating functions for the columns of the array A in Figure 26.2 to find the generating function $g_k^*(x)$ for the columns of the left-justified Fibonacci's triangle B in Figure 26.12, as Hoggatt and Bicknell did in 1974.

1				
3	5			
7	9	11		
13	15	17	19	
21	23	25	27	29

Figure 26.12 Left-justified Fibonacci's Triangle

To this end, notice that every element in B is one less than the corresponding element in A. Consequently:

$$g_k^*(x) = \frac{2[t_{k+1} - (k+1)^2x + (t_k+1)x^2]}{(1-x)^3} - \frac{1}{1-x}$$
$$= \frac{2[t_{k+1} - (k+1)^2x + (t_k+1)x^2] - (1-2x+x^2)}{(1-x)^3}$$

$$= \frac{(2t_{k+1} - 1) - 2[(k+1)^2 - 1]x + (2t_k + 1)x^2}{(1-x)^3}$$

$$= \frac{(2t_{k+1} - 1) - 2[(k+1)^2 - 1]x + (2t_k + 1)x^2}{(1-x)^3}$$

The generating function $g_k^*(x)$ has an interesting byproduct. It can be used to find a generating function for the cubes of successive positive integers. To this end, we will need the following generating functions:

$$\sum_{k=0}^{\infty} x^k = \frac{1}{1-x} \qquad \sum_{k=0}^{\infty} t_k x^k = \frac{x}{(1-x)^3}$$

$$\sum_{k=0}^{\infty} t_{k+1} x^k = \frac{1}{(1-x)^3} \qquad \sum_{k=0}^{\infty} (k+1)^2 x^k = \sum_{k=0}^{\infty} (t_{k+1} + t_k) x^k = \frac{1+x}{(1-x)^3}$$

Since the row sums of Fibonacci's triangle are cubes of successive positive integers, we have:

$$\sum_{k=0}^{\infty} (k+1)^3 x^k = \sum_{k=0}^{\infty} x^k g_k^*(x)$$

$$= \frac{1}{(1-x)^3} \left[\sum_{k=0}^{\infty} (2k_{k+1} - 1)x^k - 2[(k+1)^2 - 1]x^{k+1} + (2t_k + 1)x^{k+2} \right]$$

$$= \frac{1}{(1-x)^3} \left[\frac{2}{(1-x)^3} - \frac{1}{1-x} - \frac{2x(1+x)}{(1-x)^3} + \frac{2x}{1-x} + \frac{2x^3}{(1-x)^3} + \frac{x^2}{1-x} \right]$$

$$= \frac{2 - (1-x)^2 - 2x(1+x) + 2x(1-x)^2 + 2x^3 + x^2(1-x)^2}{(1-x)^6}$$

$$= \frac{1 + 2x - 6x^2 + 2x^3 + x^4}{(1-x)^6} = \frac{(1+4x+x^2)(1-x)^2}{(1-x)^6}$$

$$= \frac{1+4x+x^2}{(1-x)^4}$$

This is the desired generating function.

Reversing the steps, it is fairly easy to show that the function $g(x) = \frac{1+4x+x^2}{(1-x)^4}$ does in fact generate the cubes. To confirm this, we will employ the following

Maclaurin series: $(1-x)^{-n} = \sum\limits_{r=0}^{n} \binom{n+r-1}{r} x^n$. Then:

$$g(x) = (1+4x+x^2)\sum_{r=0}^{\infty}\binom{r+3}{r}x^r$$

$$\text{Coefficient of } x^r = \binom{r+3}{r} + 4\binom{r+2}{r-1} + \binom{r+1}{r-2}$$

$$= \frac{(r+3)!}{r!3!} + 4\frac{(r+2)!}{(r-1)!3!} + \frac{(r+1)!}{(r-2)!3!}$$

$$= \frac{1}{6}[(r+3)(r+2)(r+1) + 4(r+2)(r+1)r + (r+1)r(r-1)]$$

$$= \frac{6r^3+18r^2+18r+6}{6} = (r+1)^3$$

Thus

$$\frac{1+4x+x^2}{(1-x)^4} = \sum_{r=0}^{\infty}(r+1)^3x^r$$

Fibonacci's triangle can be altered in a number of ways, as R. L. Ollerton of the University of Western Sydney, Australia, and A. G. Shannon of the University of New South Wales, Australia, did in 2003. For example, adding a 1 to each entry, we get the array in Figure 26.13.

	Row sums
2	$2 = 1^3 + 1$
4 6	$10 = 2^3 + 2$
8 10 12	$30 = 3^3 + 3$
14 16 18 20	$68 = 4^3 + 4$
22 24 26 28 30	$130 = 5^3 + 5$
32 34 36 38 40 42	$222 = 6^3 + 6$

Figure 26.13

Since row n contains n entries and the nth row sum in Fibonacci's triangle is n^3, it follows that the nth row sum of this array is $n^3 + n$.

Suppose we alter the shape of Fibonacci's triangle by increasing the size of each row by one more than the previous row. Figure 26.14 shows the resulting array left-justified. Notice that the nth row contains $2n-1$ elements, where $n \geq 1$.

1										
3	5	7								
9	11	13	15	17						
19	21	23	25	27	29	31				
33	35	37	39	41	43	45	47	49		
51	53	55	57	59	61	63	65	67	69	71

Figure 26.14

Let A_n denote the cumulative sum of the entries in the top n rows. Since $\sum_{j=1}^{n}(2j-1) = n^2$, it follows that

$$A_n = \sum_{j=1}^{n^2}(2j-1) = 2\sum_{j=1}^{n^2} j - \sum_{j=1}^{n^2} 1$$

$$= 2 \cdot \frac{n^2(n^2+1)}{2} - n^2 = n^4$$

Consequently, $A_{n-1} = (n-1)^4$.

Thus, the nth row sum S_n of this array is given by $S_n = A_n - A_{n-1} = n^4 - (n-1)^4$.

For example, $S_4 = 4^4 - 3^4 = 175 = 19 + 21 + 23 + 25 + 27 + 29 + 31$; see Figure 26.15.

									Row sums
1									$1 = 1^4 - 0^4$
3	5	7							$15 = 2^4 - 1^4$
9	11	13	15	17					$65 = 3^4 - 2^4$
19	21	23	25	27	29	31			$175 = 4^4 - 3^4$
33	35	37	39	41	43	45	47	49	$369 = 5^4 - 4^4$

Figure 26.15

Suppose the top row consists of two odd integers and the row lengths increase by two at each step. Figure 26.16 shows the resulting array.

The row sums manifest an intriguing pattern: The nth row sum is $2^2 \cdot n^3$.

More generally, suppose the top row contains the first m odd positive integers and the row lengths increase by m. The top $n-1$ rows contain

$$m + 2m + 3m + \cdots + (n-1)m = \frac{mn(n-1)}{2}$$

										Row sums
1	3									$4 = 2^2 \cdot 1^3$
5	7	9	11							$32 = 2^2 \cdot 2^3$
13	15	17	19	21	23					$108 = 2^2 \cdot 3^3$
25	27	29	31	33	35	37	39			$256 = 2^2 \cdot 4^3$
41	43	45	47	49	51	53	55	57	59	$500 = 2^2 \cdot 5^3$

Figure 26.16

elements. So the first element in row n is the $\left[\frac{mn(n-1)}{2}+1\right]$th odd integer: namely,

$$2\left[\frac{mn(n-1)}{2}+1\right]-1 = mn(n-1)+1$$

Since row n contains mn elements, the nth row sum of such an array is given by

$$\sum_{k=\frac{mn(n-1)}{2}+1}^{\frac{mn(n+1)}{2}} (2k-1) = \sum_{k=1}^{\frac{mn(n+1)}{2}} (2k-1) - \sum_{k=1}^{\frac{mn(n-1)}{2}} (2k-1)$$

$$= \left[\frac{mn(n+1)}{2}\right]^2 - \left[\frac{mn(n-1)}{2}\right]^2$$

$$= \frac{m^2n^2}{4}[(n+1)^2-(n-1)^2] = m^2n^3$$

1	3	5									
7	9	11	11	13	15	17					
19	21	23	25	27	29	31	33	35			
37	39	41				…				57	59

Figure 26.17

1				
3				
5	7			
9	11	13		
15	17	19	21	23

Figure 26.18

For example, the fourth row sum of the array in Figure 26.17 is given by $37+39+41+\cdots+59 = 3^2 \cdot 4^3 = 576$.

We can alter Fibonacci's triangle to construct an array with F_n number of entries in row n, as Ollerton and Shannon did in 2004. See Figure 26.18.

Cumulative Row Sums

Since the top $n-1$ rows contain $\sum_{i=1}^{n-1} F_i = F_{n+1} - 1$ elements, row n begins with the F_{n+1}-st odd number, namely, $2F_{n+1} - 1$, and it ends with the $(2F_{n+2} - 1) - 2 = (2F_{n+2} - 3)$rd odd number.

Since the top n rows contain $\sum_{i=1}^{n} F_i = F_{n+2} - 1$ elements, the cumulative sum S_n of the elements in the top n rows is given by:

$$
\begin{aligned}
S_n &= \sum_{i=1}^{F_{n+2}} (2i-1) \\
&= 2 \cdot \frac{(F_{n+2}-1)F_n}{2} - (F_{n+2} - 1) = (F_{n+2} - 1)^2
\end{aligned}
$$

Row Sums

Using the cumulative sum S_n, we can easily find the nth row sum R_n:

$$
\begin{aligned}
R_n &= S_n - S_{n-1} = (F_{n+2} - 1)^2 - (F_{n+1} - 1)^2 \\
&= \left(F_{n+2}^2 - F_{n+2}^2\right) - 2\left(F_{n+2} - F_{n+1}\right) \\
&= \left(F_{n+2} + F_{n+1}\right)\left(F_{n+2} - F_{n+1}\right) - 2F_n \\
&= F_{n+3}F_n - 2F_n = F_n(F_{n+3} - 2)
\end{aligned}
$$

Since $\sum_{i=1}^{n} R_i = S_n$, it follows that $\sum_{i=1}^{n} F_i(F_{i+3} - 2) = (F_{n+2} - 1)^2$; that is, $\sum_{i=1}^{n} F_iF_{i+3} = F_{n+2}^2 - 1$.

Next, we will study yet another triangular array investigated by Ollerton and Shannon; see Figure 26.19. Here also row n contains F_n odd numbers.

This array satisfies the following properties:

- Row n begins with $2F_{n+1} - 1$.
- Row n ends with $2F_{n+2} - 1$.
- The cumulative sum of the top n rows is $F_{n+2}^2 - 1$.
- The nth row sum is F_nF_{n+3}.

Consequently, $\sum_{i=1}^{n} F_iF_{i+3} = F_{n+2}^2 - 1$, as we obtained earlier. You can confirm these properties.

					Rows sums
3					$1 \cdot 3$
5					$1 \cdot 5$
7	9				$2 \cdot 8$
11	13	15			$3 \cdot 13$
17	19	21	23	25	$5 \cdot 21$

Figure 26.19

```
                0                  ← row 0
            1       2
        3       4       5
    6       7       8       9
10     11      12      13      14
```

Figure 26.20 Whole Number Triangle

Whole Number Triangle

The next triangle we investigate is the triangular array of consecutive whole numbers, studied by Hoggatt and Bicknell in 1974; they called it the *whole number triangle*. Figure 26.20 shows the array.

The leftmost southwest diagonal consists of triangular numbers t_n, where $n \geq 0$; that is, row n begins with t_n. Since row n contains $n + 1$ elements, the last element in row n is

$$t_n + n = [t_n + (n + 1)] - 1 = t_{n+1} - 1$$

So the last element in row n is one less than the next triangular number, as could be expected.

Thus, the various elements in row n are $t_n, t_n + 1, t_n + 2, \cdots$, and $t_n + n$. So the sum S_n of the elements in row n is given by:

$$\begin{aligned} S_n &= t_n + (t_n + 1) + (t_n + 2) + \cdots + (t_n + n) \\ &= (n + 1)t_n + \sum_{i=1}^{n} i = (n + 1)t_n + t_n \\ &= (n + 2)t_n = \frac{n(n + 1)(n + 2)}{2} \end{aligned}$$

Thus the nth row sum is $n + 2$ times t_n, where $n \geq 0$; see Figure 26.21.

```
                          0                              Row sums
                      1       2                      0 = (0 + 2) · 0
                  3       4       5                  3 = (1 + 2) · 1
              6       7       8       9             12 = (2 + 2) · 3
         10      11      12      13      14         30 = (3 + 2) · 6
     15      16      17      18      19      20     60 = (4 + 2) · 10
                                                   105 = (5 + 2) · 15
```

Figure 26.21

Since the last element in row n is $t_{n+1} - 1$, the various elements in row n can also be written as

$$t_{n+1} - (n+1), t_{n+1} - n, t_{n+1} - (n-1), \cdots, t_{n+1} - 3, t_{n+1} - 2, t_{n+1} - 1$$

So their sum is given by:

$$\begin{aligned} S_n &= [t_{n+1} - (n+1)] + [t_{n+1} - n] + [t_{n+1} - (n-1)] + \cdots \\ &\quad + (t_{n+1} - 2) + (t_{n+1} - 1) \\ &= (n+1)t_{n+1} - \sum_{i=1}^{n+1} i = (n+1)t_{n+1} - t_{n+1} \\ &= nt_{n+1} = (n+2)t_n \end{aligned}$$

An Interesting Dividend

Since $n(n+1)(n+2) \equiv 0 \pmod 3$, and $(2,3) = 1$, it follows that $3 \mid S_n$ for every n; see Figure 26.22.

											Row sums
					0						$0 = 3 \cdot 0$
				1		2					$3 = 3 \cdot 1$
			3		4		5				$12 = 3 \cdot 4$
		6		7		8		9			$30 = 3 \cdot 10$
	10		11		12		13		14		$60 = 3 \cdot 20$
15		16		17		18		19		20	$105 = 3 \cdot 35$

Figure 26.22

Interestingly, the remaining factors in S_n are the tetrahedral numbers 0, 1, 4, 10, 20, 35, $\cdots$. This is the case, since:

$$\begin{aligned} S_n = nt_{n+1} &= \frac{n(n+1)(n+2)}{2} \\ &= 3\left[\frac{n(n+1)(n+2)}{3!}\right] = 3\binom{n+2}{3} = 3T_n \end{aligned}$$

where T_n is the nth tetrahedral number.

Cumulative Row Sums

Since the rightmost element in row n is $t_{n+1} - 1$, the cumulative sum of the elements in the top n rows is given by:

$$\begin{aligned}\sum_{j=0}^{n} S_j &= 0 + 1 + 2 + \cdots + (t_{n+1} - 1)\\ &= t_{t_{n+1}-1} = \frac{(t_{n+1} - 1)t_{n+1}}{2}\\ &= \frac{\left[\frac{(n+1)(n+2)}{2} - 1\right]\left[\frac{(n+1)(n+2)}{2}\right]}{2}\\ &= \frac{n(n+1)(n+2)(n+3)}{8} = 3\binom{n+3}{4} \qquad (26.1)\end{aligned}$$

Since $S_j = jt_{j+1}$, it follows by equation (26.1) that

$$\sum_{j=0}^{n} jt_{j+1} = \sum_{j=0}^{n} (j+2)t_j = 3\binom{n+3}{4} = t_{t_{n+1}-1}$$

Lambda Array

The triangular array in Figure 26.23, called the *Lambda array*, was investigated in 1999 by J. Kappraff of New Jersey Institute of Technology, Newark, New Jersey. He showed that the array is related to the panels in the pavement of the Reading Room of the Laurentian Library on the second floor of the San Lorenzo Church complex in Florence, Italy; the library was designed by Michelangelo.

i \ j	0	1	2	3	4
0	1				
1	2	3			
2	4	6 ↓	9		
3	8	12 →	18	27	
4	16	24	36	54	81

Figure 26.23 Lambda Array

The array L, found in Plato's *Timaeus*, is defined recursively as follows:

$$L(0, 0) = 1$$

$$L(n, j) = L(n, j-1) + L(n-1, j-1), \quad n \geq j \geq 1$$

For example, $L(4, 2) = 12 + 24 = 36$; see Figure 26.26.

We now claim that $L(r, s) = 2^{r-s}3^s$, where $r, s \geq 0$. Since $L(0, 0) = 1 = 2^{0-0}3^0$, and

$$\begin{aligned} L(n,j) &= L(n,j-1) + L(n-1,j-1) \\ &= 2^{n-j+1}3^{j-1} + 2^{n-1-j+1}3^{j-1} = 2^{n-j+1}3^{j-1} + 2^{n-j}3^{j-1} \\ &= 2^{n-j}3^{j-1}(2+1) = 2^{n-j}3^j \end{aligned}$$

it follows by PMI that every element $L(r, s)$ in the array can be defined explicitly: $L(r, s) = 2^{r-s}3^s$.

Consequently, $L(n, 0) = 2^n$ and $L(n, n) = 3^n$. So every row begins with a power of 2 and ends in the same power of 3.

The nth row sum is given by:

$$\begin{aligned} \sum_{j=0}^{n} L(n,j) &= \sum_{j=0}^{n} 2^{n-j}3^j = 2^n \sum_{j=0}^{n} (3/2)^j \\ &= 2^n \cdot \frac{(3/2)^{n+1} - 1}{3/2 - 1} = 3^{n+1} - 2^{n+1}, \quad n \geq 0 \end{aligned}$$

A Triangular Array of Triangular Numbers

Next, we pursue the triangular array of triangular numbers t_n, shown in Figure 26.24.

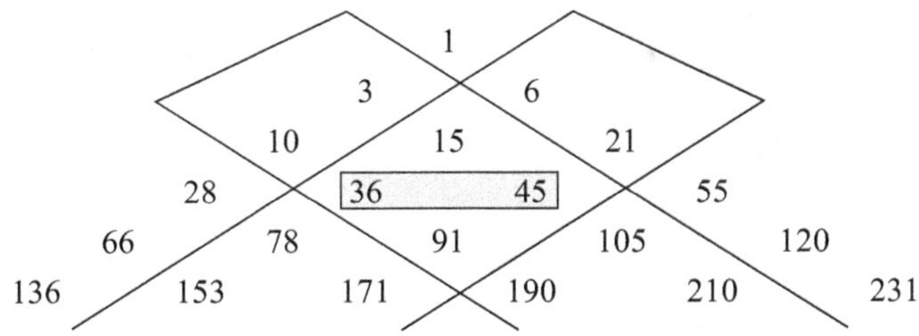

Figure 26.24 A Triangular Array of Triangular Numbers

Consider any band consisting of two adjacent northeast (or southeast) diagonals in the array; see Figure 26.24. The sum of the horizontal elements in each band is a square. This should not be a surprise, since $t_{n-1} + t_n = n^2$.

This result provides an interesting identity for triangular numbers whose subscripts are also triangular numbers: $t_{t_n} + t_{t_n - 1} = t_n^2$. This can be confirmed algebraically.

Cumulative Row Sums

Let B_n denote the cumulative row sum of the triangular numbers in the first n rows, where $n \geq 1$. Let $k = n(n+1)/2$. Then:

$$\begin{aligned}
B_n &= \sum_{i=1}^{k} t_i = \sum_{i=1}^{k} \frac{i(i+1)}{2} \\
&= \frac{1}{2}\left[\sum_{i=1}^{k} i^2 + \sum_{i=1}^{k} i\right] = \frac{k(k+1)(2k+1)}{12} + \frac{k(k+1)}{4} \\
&= \frac{k(k+1)}{12}[(2k+1)+3] = \frac{k(k+1)(k+2)}{6} \\
&= \frac{n(n+1)[n(n+1)+2][n(n+1)+4]}{48} = \frac{(n^2+n)(n^2+n+2)(n^2+n+4)}{48} \\
&= \frac{n^6+3n^5+9n^4+13n^3+14n^2+8n}{48}
\end{aligned}$$

Changing n to $-n$,

$$B_{n-1} = \frac{n^6-3n^5+9n^4-13n^3+14n^2-8n}{48}$$

So the sum S_n of the sum of the triangular numbers in row n is given by

$$S_n = B_n - B_{n-1} = \frac{6n^5+26n^3+16n}{48} = \frac{n(3n^4+13n^2+8)}{24}$$

(Finding a formula for S_n was proposed as a problem in 1988 by J. M. Howell of Los Angeles City College.)

Next, we derive an alternate formula for S_n. To this end, notice that:

$$\begin{aligned}
B_n &= \frac{k(k+1)(k+2)}{6} = \binom{k+2}{3} \\
&= \binom{(n^2+n+4)/2}{3}
\end{aligned}$$

B_{n-1} can be obtained by changing n to $-n$: $B_{n-1} = \binom{(n^2-n+4)/2}{3}$. Thus:

$$\begin{aligned}
S_n &= B_n - B_{n-1} \\
&= \binom{(n^2+n+4)/2}{3} - \binom{(n^2-n+4)/2}{3}
\end{aligned}$$

Palindromic Triangular Numbers in Base Nine

In 1973, C. W. Trigg of California found that there are 27 triangular numbers t_n that are palindromes,[†] where $n < 151340$. For example, 1, 3, 6, 55, 66, 171, 595, and 666 are eight palindromic triangular numbers < 1000.

$$\begin{array}{ccccc} 1 & & & & \\ 1 & 1 & & & \\ 1 & 1 & 1 & & \\ 1 & 1 & 1 & 1 & \\ 1 & 1 & 1 & 1 & 1 \\ & & \vdots & & \end{array}$$

Figure 26.25

Interestingly, every number in the array C in Figure 26.25 represents a palindromic triangular number in base nine. For example, $111_{\text{nine}} = 9^2+9+1 = 91 = t_{13}$ and $1111_{\text{nine}} = 9^3 + 9^2 + 9 + 1 = 820 = t_{40}$, where notice that $13 = 111_{\text{three}}$ and $40 = 1111_{\text{three}}$.

We now confirm that every number in the array C is a triangular number, as Hoggatt and Bicknell did. Let N be the decimal number in row n, so $N = 111\cdots 11_{\text{nine}}$, with n ones. Then:

$$\begin{aligned} N &= 111\cdots 11_{\text{nine}} \\ &= 9^{n-1} + 9^{n-1} + \cdots + 9 + 1 = \frac{9^n - 1}{8} \\ &= \frac{(3^n-1)(3^n+1)}{8} = \frac{1}{2}\left(\frac{3^n-1}{2}\right)\left(\frac{3^n+1}{2}\right) \\ &= t_m \end{aligned}$$

where $m = \frac{3^n-1}{2}$. Thus, N is a triangular number.

An observation: Notice that

$$m = \frac{3^n-1}{2} = 3^{n-1} + 3^{n-2} + \cdots + 3 + 1 = 111\cdots 11_{\text{three}} \qquad (n \text{ ones})$$

Thus, the ternary representation of the subscript m in t_m has the same number of 1s as the base-nine representation of N.

[†] A *palindrome* is a number or a word that reads the same backwards and forwards.

Besides, since

$$\begin{aligned} 9t_n + 1 &= \frac{9n(n+1)}{2} + 1 = \frac{9n^2 + 9n + 2}{2} \\ &= \frac{(3n+1)(3n+2)}{2} = t_{3n+1} \end{aligned}$$

it follows that appending any number of 1s to the right of t_n yields another triangular number. The subscript of the new triangular number is obtained by appending the same number of 1s to the ternary representation of n.

For example, let $N = 11111_{\text{nine}} = t_{21}$. We have $121 = 11111_{\text{three}}$. Suppose we append two 1s to N; let N' be the resulting number: $N' = 11111\underbrace{11}_{\text{nine}} = t_{1093}$. Notice that the ternary representation of $1093 = 11111\underbrace{11}_{\text{nine}}$ contains two more 1s, as expected.

Obviously, this discussion can be extended to palindromic triangular numbers in other non-decimal bases, as Hoggat and Bicknell have done.

Ensley's Triangle

The array E in Figure 26.26 was investigated in 2003 by D. Ensley of Shippensburg University, Shippensburg, Pennsylvania. The leftmost rising diagonal and the rightmost falling diagonal consist of Fibonacci numbers, and every internal element is computed using Pascal's recurrence relation.

$$\begin{array}{ccccccccccccc} &&&&&& 1 &&&&&& \leftarrow \text{row } 0 \\ &&&&& 1 && 1 &&&&& \\ &&&& 2 && 2 && 2 &&&& \\ &&& 3 && 4 && 4 && 3 &&& \\ && 5 && 7 && 8 && 7 && 5 && \\ & 8 && 12 && 15 && 15 && 12 && 8 & \\ 13 && 20 && 27 && 30 && 27 && 20 && 13 \end{array}$$

Figure 26.26 Ensley's Triangle

Accordingly, *Ensley's triangle E* can be defined recursively as follows:

$$E(n, 0) = F_{n+1} = E(n, n)$$

$$E(n, r) = E(n-1, r) + E(n-1, r-1), \quad 0 < r < n \tag{26.2}$$

Symmetry

We now prove by PMI that the array is symmetric about the vertical line through the apex; that is, $E(n, r) = E(n, n - r)$. Clearly, $E(1, 0) = E(1, 1 - 0)$.

Now, assume that it is true for all positive integers $< n$. Then, by the recurrence relation (26.2) and the inductive hypothesis, we have:

$$\begin{aligned} E(n, r) &= E(n-1, r) + E(n-1, r-1) \\ &= E(n-1, n-1-r) + E(n-1, r-1) \\ &= E(n-1, n-r) + E(n-1, n-r-1) = E(n, n-r) \end{aligned}$$

Thus, by the strong version of PMI, $E(n, r) = E(n, n - r)$ for every $n \geq 1$.

Parity of the Central Element

Since the array is symmetric,

$$\begin{aligned} E(2n, n) &= E(2n-1, n) + E(2n-1, n-1) \\ &= E(2n-1, n) + E(2n-1, n) = 2E(2n-1, n) \end{aligned}$$

Thus, the central elements in Ensley's triangle are even integers.

An Explicit Formula for $E(n, r)$

Next, we develop an explicit formula for $E(n, r)$:

$$E(n, r) = \binom{n}{r} + \sum_{i=1}^{n-r} \binom{n-i}{r} F_{i-1} + \sum_{j=1}^{r} \binom{n-j}{r-j} F_{j-1} \tag{26.3}$$

To establish this, it suffices to show that the RHS satisfies the three conditions in the recursive definition of $E(n, r)$. For convenience, we denote the RHS by $A(n, r)$. Then:

$$\begin{aligned} A(n, 0) &= \binom{n}{0} + \sum_{i=1}^{n} \binom{n-i}{0} F_{i-1} + \sum_{j=1}^{0} \binom{n-j}{-j} F_{j-1} \\ &= 1 + \sum_{i=1}^{n} F_{i-1} + 0 = 1 + \sum_{i=1}^{n-1} F_j \\ &= 1 + (F_{n+1} - 1) = F_{n+1} \end{aligned}$$

$$
\begin{aligned}
A(n,n) &= \binom{n}{n} + \sum_{i=1}^{0}\binom{n-i}{n}F_{i-1} + \sum_{j=1}^{n}\binom{n-j}{n-j}F_{j-1}\\
&= 1 + 0 + \sum_{j=1}^{n}F_{j-1} = 1 + \sum_{j=1}^{n-1}F_j\\
&= 1 + (F_{n+1} - 1) = F_{n+1}
\end{aligned}
$$

and

$$
\begin{aligned}
A(n-1,r)+A(n-1,r-1) = &\binom{n-1}{r} + \binom{n-1}{r-1} + \sum_{i=1}^{n-r-1}\binom{n-i-1}{r}F_{i-1} +\\
&\sum_{j=1}^{r}\binom{n-j-1}{r-j}F_{j-1} + \sum_{i=1}^{n-r}\binom{n-i-1}{r-1}F_{i-1} +\\
&\sum_{j=1}^{r-1}\binom{n-j-2}{r-j-1}F_{j-1}\\
= &\binom{n}{r} + \sum_{i=1}^{n-r}\left[\binom{n-i-1}{r} + \binom{n-i-1}{r-1}\right]F_{i-1} +\\
&\sum_{j=1}^{r}\left[\binom{n-j-1}{r-j} + \binom{n-j-1}{r-j-1}\right]F_{j-1}\\
= &\binom{n}{r} + \sum_{i=1}^{n-r}\binom{n-i}{r}F_{i-1} + \sum_{j=1}^{r}\binom{n-j}{r-j}F_{j-1}\\
= &A(n,r)
\end{aligned}
$$

Thus, $A(n, r)$ satisfies the recursive definition of $E(n, r)$, so $A(n, r) = E(n, r)$, as desired.

For example:

$$
\begin{aligned}
E(5,3) &= \binom{5}{3} + \sum_{i=1}^{2}\binom{5-i}{3}F_{i-1} + \sum_{j=1}^{3}\binom{5-j}{3-j}F_{j-1}\\
&= 10 + \left[\binom{4}{3}F_0 + \binom{3}{3}F_1\right] + \left[\binom{4}{2}F_0 + \binom{3}{1}F_1 + \binom{2}{0}F_2\right] = 15
\end{aligned}
$$

Thus, to compute $E(5, 3)$, it suffices to multiply the binomial coefficients along the diagonals intersecting at $\binom{5}{3}$ with weights $F_0, F_1, F_2, \cdots$ in the reverse order, and then add the products to $\binom{5}{3}$; see Figure 26.27.

```
                    1
               1         1
          2         2         2
     3         4         4         3
5         7         8         7         5
8    12        15       [15]       12        8
```

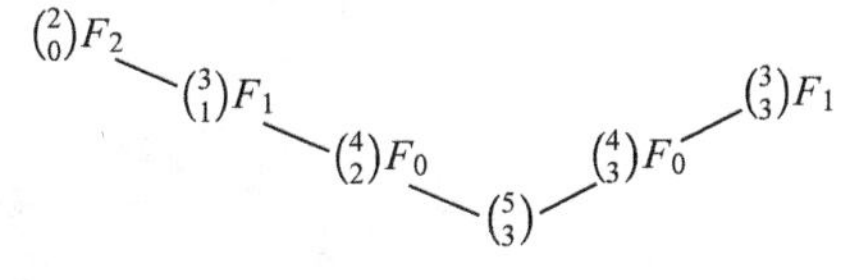

Figure 26.27

Central Elements

It follows from formula (26.3) that

$$\begin{aligned} E(2n, n) &= \binom{2n}{n} + \sum_{i=1}^{n} \binom{2n-i}{n} F_{i-1} + \sum_{i=1}^{n} \binom{2n-i}{n-i} F_{i-1} \\ &= \binom{2n}{n} + \sum_{i=1}^{n} \left[\binom{2n-i}{n} + \binom{2n-i}{n} \right] F_{i-1} \\ &= \binom{2n}{n} + 2\sum_{i=1}^{n} \binom{2n-i}{n} F_{i-1} \end{aligned}$$

This gives us an explicit formula for the central element in row $2n$. It also follows by the symmetry of the triangle.

Since $\binom{2n}{n}$ is even, it also follows that $E(2n, n)$ is even, where $n \geq 1$.

27

Tribinomial Triangles

This chapter presents a new class of numbers; it is obtained by extending the definition of the ordinary binomial coefficient to triangular numbers t_n. They were investigated by W. Hansell around 1974.

Tribinomial Coefficients

Corresponding to $n! = n(n-1)\cdots 2 \cdot 1$, we define $t_m^* = t_m t_{m-1} \cdots t_2 t_1$. For example, $t_3^* = t_3 t_2 t_1 = 6 \cdot 3 \cdot 1 = 18$. Corresponding to the binomial coefficient

$$\binom{n}{r} = \frac{n!}{r!(n-r)!} = \frac{n(n-1)\cdots(n-r+1)}{r(r-1)\cdots 2 \cdot 1}$$

we now define binomial coefficients for triangular numbers† (or *tribinomial coefficients* for short) $\begin{bmatrix} n \\ r \end{bmatrix}$:

$$\begin{bmatrix} n \\ r \end{bmatrix} = \frac{t_n^*}{t_r^* t_{n-r}^*} \tag{27.1}$$

$$= \frac{t_n t_{n-1} \cdots t_{n-r+1}}{t_r t_{r-1} \cdots t_2 t_1} \tag{27.2}$$

where $\begin{bmatrix} n \\ 0 \end{bmatrix} = \begin{bmatrix} n \\ n \end{bmatrix} = 1$. We will confirm later that they are are indeed integers.

†For a detailed discussion, see the author's *Calatalan Number with Applications*.

Since $t_k = \frac{k(k+1)}{2}$, it follows by ordinary algebra that

$$\begin{bmatrix} n \\ r \end{bmatrix} = \frac{1}{n-r+1}\binom{n+1}{r+1}\binom{n}{r} \tag{27.3}$$

For example, $\begin{bmatrix} 6 \\ 2 \end{bmatrix} = \frac{1}{5}\binom{7}{3}\binom{6}{2} = 105$, as expected.

The tribinomial coefficients $\begin{bmatrix} n \\ r \end{bmatrix}$ can be arranged to construct the *tribinomial array* in Figure 27.1.

$$\begin{array}{ccccccccccccc}
 & & & & & & \boxed{1} & & & & & & \leftarrow \text{row } 0 \\
 & & & & & 1 & & 1 & & & & & \\
 & & & & 1 & & \boxed{3} & & 1 & & & & \\
 & & & 1 & & 6 & & 6 & & 1 & & & \\
 & & 1 & & 10 & & \boxed{20} & & 10 & & 1 & & \\
 & 1 & & 15 & & 50 & & 50 & & 15 & & 1 & \\
1 & & 21 & & 105 & & \boxed{175} & & 105 & & 21 & & 1 \\
 & \nearrow & & & & & \uparrow & & & & & \nwarrow & \\
t_n & & & & & & \text{CTCs} & & & & & & t_n
\end{array}$$

Figure 27.1 Tribinomial Triangle

Figure 27.2 shows an interesting *Sierpiński-like* geometric representation of its binary version, where a black triangle represents an odd number and a light triangle an even number.

Figure 27.2 Tribinomial Array

The tribinomial array manifests two obvious properties that can easily be confirmed:

- The northeast diagonal 1 consists of triangular numbers: $\begin{bmatrix} n \\ 1 \end{bmatrix} = t_n$.
- The array is symmetric about the vertical line through the middle; that is, $\begin{bmatrix} n \\ r \end{bmatrix} = \begin{bmatrix} n \\ n-r \end{bmatrix}$.

Central Tribinomial Coefficients

The *central tribinomial coefficient* (CTC) is given by

$$\begin{aligned} \begin{bmatrix} 2n \\ n \end{bmatrix} &= \frac{1}{2n-n+1}\binom{2n+1}{n+1}\binom{2n}{n} \\ &= (2n+1)C_n^2 \end{aligned} \tag{27.4}$$

where C_n is the nth Catalan number.

Next, we turn to the parity of the CTCs.

Parity of the CTCs

It follows from property (27.4) $\begin{bmatrix} 2n \\ n \end{bmatrix}$ is odd if and only if C_n is odd. But C_n is odd if and only if either $n = 0$ or n is a Mersenne number.[†] Consequently, $\begin{bmatrix} 2n \\ n \end{bmatrix}$ is odd if and only if either $n = 0$ or n is a Mersenne number.

For example, let $n = 7 = 2^3 - 1$, a Mersenne number. Then $\begin{bmatrix} 2n \\ n \end{bmatrix} = \begin{bmatrix} 6 \\ 3 \end{bmatrix} = 7C_3^2 = 7 \cdot 25 = 175$ is odd.

Row Sums

The tribinomial array contains a hidden treasure: The nth row sum is C_{n+1}; that is,

$$\sum_{r=0}^{n} \begin{bmatrix} n \\ r \end{bmatrix} = C_{n+1} \tag{27.5}$$

[†] See T. Koshy and M. Salmassi, "Parity and Primality of Catalan Numbers,"*College Mathematics Journal* 37 (2006), 52–53.

n \ r	0	1	2	3	4	5	6	Row sums
0	1							1
1	1	1						2
2	1	3	1					5
3	1	6	6	1				14
4	1	10	20	10	1			42
5	1	15	50	50	15	1		132
6	1	21	105	175	105	21	1	429

Figure 27.3

For example, $\sum_{r=0}^{5}\begin{bmatrix}5\\r\end{bmatrix} = 1 + 15 + 50 + 50 + 15 + 1 = 132 = C_6$; see Figure 27.3.

In 2005, M. Salmassi established property (27.5)* using formula (27.3) and the following two combinatorial identities:†

$$\binom{a}{b} = \frac{a}{b}\binom{a-1}{b-1} \tag{27.6}$$

$$\sum_k \binom{r}{a+k}\binom{s}{b-k} = \binom{r+s}{a+b} \tag{27.7}$$

Tribinomial Triangle Revisited

The following theorem identifies the rows of the tribinomial array which consist of only odd entries.

Theorem 27.1 The tribinomial coefficient $\begin{bmatrix}n\\r\end{bmatrix}$ is odd for every r if and only if $n = M_k - 1$.

Proof: We have:

$$\begin{aligned}\begin{bmatrix}n\\r\end{bmatrix} &= \frac{1}{n-r+1}\binom{n}{r}\binom{n+1}{r+1}\\ &= \frac{n!}{r!(n-r+1)!}\binom{n+1}{r+1}\\ &= \frac{1}{n+1}\binom{n+1}{r}\binom{n+1}{r+1}\end{aligned} \tag{27.8}$$

*For a proof, see the author's *Catalan Numbers with Applications*.

†See R. L. Graham et al., *Concrete Mathematics*, Addison-Wesley, Reading, Massachusetts, 1990.

So

$$\begin{bmatrix} n \\ r+1 \end{bmatrix} = \frac{1}{n+1}\binom{n+1}{r+1}\binom{n+1}{r+2} \tag{27.9}$$

Thus

$$\begin{bmatrix} n \\ r \end{bmatrix} + \begin{bmatrix} n \\ r+1 \end{bmatrix} = \frac{1}{n+1}\binom{n+1}{r+1}\left[\binom{n+1}{r} + \binom{n+1}{r+2}\right] \tag{27.10}$$

Suppose $n = M_k - 1$, so $n + 1 = M_k$. Then $\binom{n+1}{r}$ is odd and $\binom{n+1}{r} + \binom{n+1}{r+2}$ even for every r. Hence the right-hand side of equation (27.10) is even.

Consequently, $\begin{bmatrix} n \\ r \end{bmatrix} + \begin{bmatrix} n \\ r+1 \end{bmatrix}$ is even for every r; in other words, $\begin{bmatrix} n \\ r \end{bmatrix}$ and $\begin{bmatrix} n \\ r+1 \end{bmatrix}$ have the same parity for every r. But $\begin{bmatrix} n \\ 0 \end{bmatrix} = 1$ is odd. Thus $\begin{bmatrix} n \\ r \end{bmatrix}$ is odd for every r.

On the other hand, suppose $\begin{bmatrix} n \\ r \end{bmatrix}$ is odd for every r. Then, by equation (27.8), $\frac{1}{n+1}\binom{n+1}{r}\binom{n+1}{r+1}$ is odd for very r. In particular, $\frac{1}{n+1}\binom{n+1}{0}\binom{n+1}{1} = n + 1$ must be odd. Therefore, $\binom{n+1}{r}\binom{n+1}{r+1}$ is odd for every r; so $\binom{n+1}{r}$ is odd for every r. Consequently, either $n + 1 = 0$ or $n + 1 = M_k$. But $n + 1$ cannot be zero, so $n = M_k - 1$.

Thus, $\begin{bmatrix} n \\ r \end{bmatrix}$ is odd for every r if and only if $n = M_k - 1$, as desired. ■

We now add that the oddness of $n + 1$, coupled with property (27.8), can be used to provide an interestingly different argument for the second half of the proof. Since $\begin{bmatrix} n \\ r \end{bmatrix}$ is odd for every r, the left-hand side (LHS) of property (4) is the sum of $n + 1$ odd numbers. But $n + 1$ is odd, so the LHS and hence C_{n+1} are odd. Therefore, either $n + 1 = 0$ or $n + 1 = M_k$. Again, $n + 1$ cannot be zero, so $n = M_k - 1$.

Finally, Hoggatt showed that tribinomial coefficients $a_{n,r}$ can be evaluated directly from Pascal's triangle. This follows from the fact that

$$\begin{vmatrix} \binom{n}{r} & \binom{n}{r+1} \\ \binom{n+1}{r} & \binom{n+1}{r+1} \end{vmatrix} = \frac{1}{n-r+1}\binom{n}{r}\binom{n+1}{r+1} = \begin{bmatrix} n \\ r \end{bmatrix} \tag{27.11}$$

Since binomial coefficients are integers, this result confirms the earlier observation that the tribinomial numbers are integers.

28

Fibonomial Triangles

Just as we extended the definition of ordinary binomial coefficients to triangular numbers, we can extend it to Fibonacci numbers as well. To this end, we first define F_n^* as the product of the first n Fibonacci numbers: $F_n^* = F_nF_{n-1}\cdots F_2F_1$, where $F_0^* = 1$. Recalling that the binomial coefficient $\binom{n}{r}$ is given by

$$\binom{n}{r} = \frac{n!}{r!(n-r)!} = \frac{n(n-1)\cdots(n-r+1)}{r(r-1)\cdots 2\cdot 1}$$

we define the nth *Fibonomial coefficient* (binomial coefficient for Fibonacci numbers) $\begin{bmatrix} n \\ r \end{bmatrix}$ as follows:

$$\begin{bmatrix} n \\ r \end{bmatrix} = \frac{F_n^*}{F_r^*F_{n-r}^*} = \frac{F_nF_{n-1}\cdots F_{n-r+1}}{F_rF_{r-1}\cdots F_1} \tag{28.1}$$

where $0 < r < n$. Clearly, $\begin{bmatrix} n \\ 0 \end{bmatrix} = 1 = \begin{bmatrix} n \\ n \end{bmatrix}$.

Fibonomial coefficients were originally studied in 1965 by S. Jerbic of then San Jose State College, and the bi-level notation was introduced the same year by Hoggatt. Five years later, D. Lind of the University of Virginia, Charlottesville, Virginia, showed that every Fibonomial coefficient is an integer.

Notice that $\begin{bmatrix} 5 \\ 3 \end{bmatrix} = \frac{F_5^*}{F_3^*F_2^*} = F_5F_4 = 15 = \begin{bmatrix} 5 \\ 2 \end{bmatrix}$. More generally, it follows from (28.1) that $\begin{bmatrix} n \\ r \end{bmatrix} = \begin{bmatrix} n \\ n-r \end{bmatrix}$, as in the case of binomial and tribinomial coefficients. So the Fibonomial triangle we will construct later is symmetric about the vertical line through the apex.

It follows from (28.1) that $\begin{bmatrix} n \\ 1 \end{bmatrix} = F_n = \begin{bmatrix} n \\ n-1 \end{bmatrix}$.

Fibonomial Recurrence Relations

Fibonomial numbers satisfy two Pascal-like Fibonomial recurrence relations:

$$\begin{bmatrix} n \\ r \end{bmatrix} = \begin{bmatrix} n-1 \\ r \end{bmatrix} F_{r+1} + \begin{bmatrix} n-1 \\ r-1 \end{bmatrix} F_{n-r-1} \tag{28.2}$$

$$\begin{bmatrix} n \\ r \end{bmatrix} = \begin{bmatrix} n-1 \\ r \end{bmatrix} F_{r-1} + \begin{bmatrix} n-1 \\ r-1 \end{bmatrix} F_{n-r+1} \tag{28.3}$$

Both can be confirmed algebraically using definition (28.1). They, coupled with the two initial conditions, can be used to compute Fibonomial coefficients.

Using the recurrence relation (28.2) and the two initial conditions, we can construct the Fibonomial triangle in Figure 28.1.

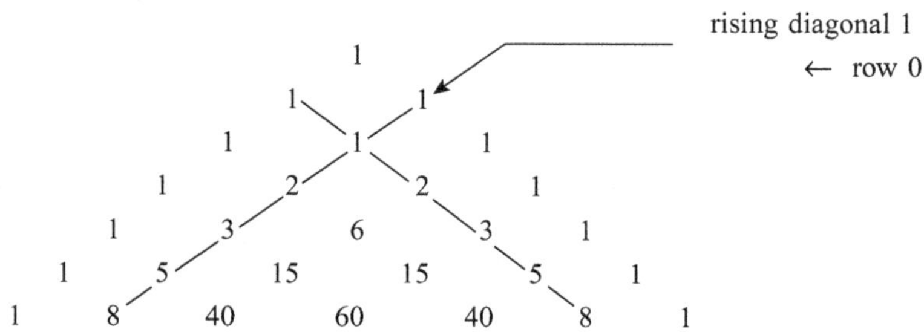

Figure 28.1 Fibonomial Triangle

It follows from identity (28.3) that

$$\begin{bmatrix} n \\ 1 \end{bmatrix} = \begin{bmatrix} n-1 \\ 1 \end{bmatrix} F_0 + \begin{bmatrix} n-1 \\ n-1 \end{bmatrix} F_n = 0 + 1 \cdot F_n = F_n = \begin{bmatrix} n \\ n-1 \end{bmatrix}$$

by symmetry. Thus, the rising diagonal 1 and hence the falling diagonal 1 consist of all Fibonacci numbers; see Figure 28.1.

An observation: $\begin{bmatrix} n \\ 2 \end{bmatrix} = \frac{F_n^*}{F_2^* F_{n-2}^*} = F_n F_{n-1} = \begin{bmatrix} n \\ 1 \end{bmatrix} \begin{bmatrix} n-1 \\ 1 \end{bmatrix}$. So, the elements on the rising diagonal 2 can be obtained by multiplying two consecutive elements on the rising diagonal 1. For instance, $15 = 5 \cdot 3$; see Figure 28.2.

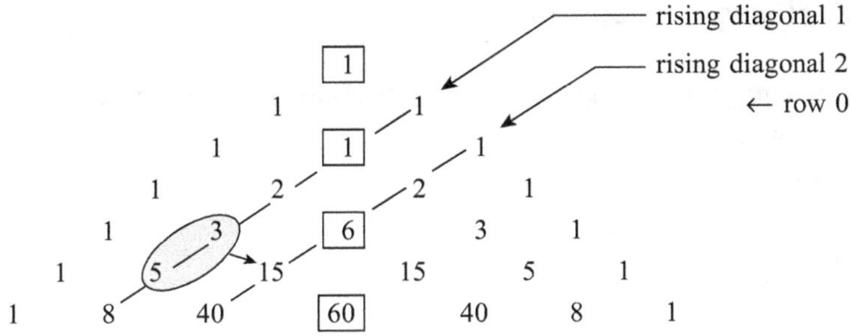

Figure 28.2

Hoggatt–Hansell Identity

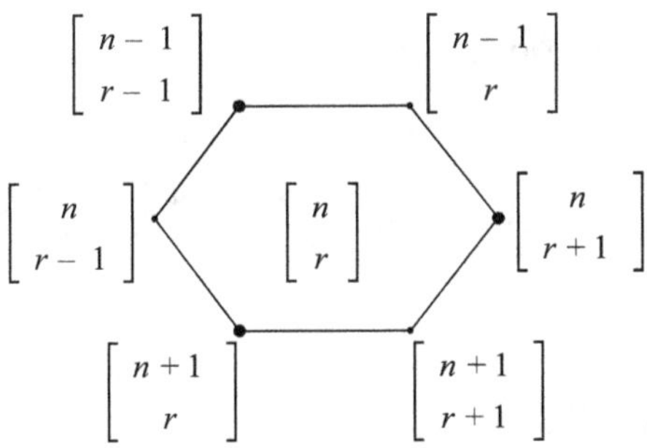

Figure 28.3

Interestingly, the Hoggatt–Hansell identity for binomial coefficients holds for Fibonomials also:

$$\begin{bmatrix} n-1 \\ r-1 \end{bmatrix}\begin{bmatrix} n \\ r+1 \end{bmatrix}\begin{bmatrix} n+1 \\ r \end{bmatrix} = \begin{bmatrix} n-1 \\ r \end{bmatrix}\begin{bmatrix} n+1 \\ r+1 \end{bmatrix}\begin{bmatrix} n \\ r-1 \end{bmatrix} \tag{28.4}$$

See Figure 28.3. This can be established algebraically.

For example, $\begin{bmatrix} 4 \\ 2 \end{bmatrix}\begin{bmatrix} 5 \\ 4 \end{bmatrix}\begin{bmatrix} 6 \\ 3 \end{bmatrix} = 6 \cdot 5 \cdot 60 = 3 \cdot 40 \cdot 15 = \begin{bmatrix} 4 \\ 3 \end{bmatrix}\begin{bmatrix} 6 \\ 4 \end{bmatrix}\begin{bmatrix} 5 \\ 2 \end{bmatrix}$; see Figure 28.4.

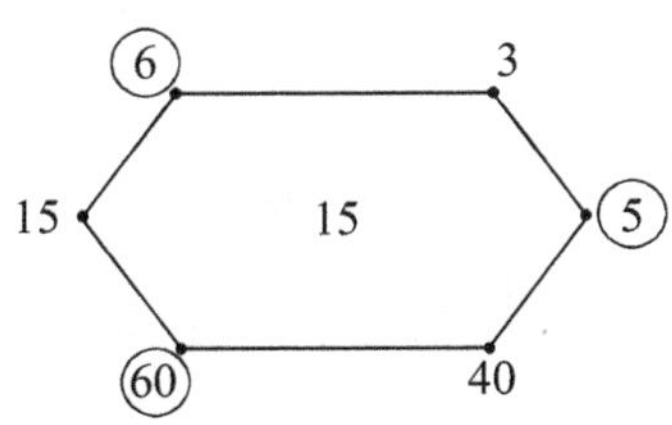

Figure 28.4

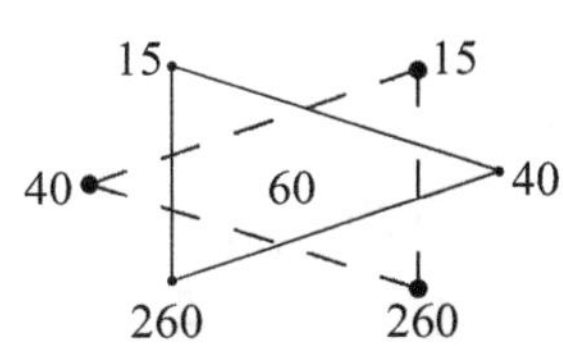

Figure 28.5

Gould's Hexagon Property

In 1971, Gould conjectured that the gcd of the three alternate vertices equals that of the remaining vertices:

$$\left(\begin{bmatrix} n-1 \\ r-1 \end{bmatrix}, \begin{bmatrix} n \\ r+1 \end{bmatrix}, \begin{bmatrix} n+1 \\ r \end{bmatrix}\right) = \left(\begin{bmatrix} n-1 \\ r \end{bmatrix}, \begin{bmatrix} n+1 \\ r+1 \end{bmatrix}, \begin{bmatrix} n \\ r-1 \end{bmatrix}\right)$$

This was confirmed in the following year by A. P. Hillman of the University of New Mexico, Albuquerque, and Hoggatt.

For example, consider Figure 28.5. Clearly, $(15, 40, 260) = (15, 260, 40)$.

Central Fibonomial Coefficients

The central Fibonomial coefficients (CFCs) $\begin{bmatrix} 2n \\ n \end{bmatrix}$, boxed in the Fibonomial triangle in Figure 28.2, also manifest a nice pattern. By the recurrence relation (28.2),

$$\begin{aligned}
\begin{bmatrix} 2n \\ n \end{bmatrix} &= \begin{bmatrix} 2n-1 \\ n \end{bmatrix} F_{n+1} + \begin{bmatrix} 2n-1 \\ n-1 \end{bmatrix} F_{n-1} \\
&= \begin{bmatrix} 2n-1 \\ n \end{bmatrix} F_{n+1} + \begin{bmatrix} 2n-1 \\ n \end{bmatrix} F_{n-1} \\
&= \begin{bmatrix} 2n-1 \\ n \end{bmatrix} (F_{n+1} + F_{n-1}) \\
&= \begin{bmatrix} 2n-1 \\ n \end{bmatrix} L_n = \begin{bmatrix} 2n-1 \\ n-1 \end{bmatrix} L_n
\end{aligned}$$

Consequently, we can obtain the CFC by multiplying its northeast (or northwest) neighbor by L_n.

For example, $\begin{bmatrix} 6 \\ 3 \end{bmatrix} = 60 = 15 \cdot 4 = \begin{bmatrix} 5 \\ 3 \end{bmatrix} L_3$; see Figure 28.2.

Since $L_n \mid \begin{bmatrix} 2n \\ n \end{bmatrix}$, $\frac{1}{L_n} \begin{bmatrix} 2n \\ n \end{bmatrix}$ is the integer $\begin{bmatrix} 2n-1 \\ n-1 \end{bmatrix}$, where $n \geq 1$. Such elements are $1, 2, 15, 260, \cdots$. They form the column on the right (or left) adjacent to the middle column.

Another Recurrence Relation for $\begin{bmatrix} n \\ r \end{bmatrix}$

In 1963, Brennan developed another recurrence relation for $\begin{bmatrix} n \\ r \end{bmatrix}$ using Lucas numbers:

$$2\begin{bmatrix} n \\ r \end{bmatrix} = \begin{bmatrix} n-1 \\ r \end{bmatrix} L_r + \begin{bmatrix} n-1 \\ r-1 \end{bmatrix} L_{n-r} \tag{28.5}$$

For example, $2\begin{bmatrix} 5 \\ 3 \end{bmatrix} = \begin{bmatrix} 4 \\ 3 \end{bmatrix} L_3 + \begin{bmatrix} 4 \\ 2 \end{bmatrix} L_2 = 3 \cdot 4 + 6 \cdot 3 = 30$; so $\begin{bmatrix} 5 \\ 3 \end{bmatrix} = 15$.

To establish the recurrence relation (28.5), we will use the identity[†]

$$2F_n = L_r F_{n-r} + L_{n-r} F_r$$

where $n \geq r \geq 0$, as J. L. Brown of Pennsylvania State University did in 1963. Multiplying both sides by $\begin{bmatrix} n \\ r \end{bmatrix}$, we get:

$$2\begin{bmatrix} n \\ r \end{bmatrix} F_n = L_r \begin{bmatrix} n \\ r \end{bmatrix} F_{n-r} + L_{n-r} \begin{bmatrix} n \\ r \end{bmatrix} F_r$$

Substituting for $\begin{bmatrix} n \\ r \end{bmatrix}$, this yields

$$2\begin{bmatrix} n \\ r \end{bmatrix} F_n = L_r \begin{bmatrix} n-1 \\ r \end{bmatrix} F_n + L_{n-r} \begin{bmatrix} n-1 \\ r-1 \end{bmatrix} F_n$$

That is,

$$2\begin{bmatrix} n \\ r \end{bmatrix} = \begin{bmatrix} n-1 \\ r \end{bmatrix} L_r + \begin{bmatrix} n-1 \\ r-1 \end{bmatrix} L_{n-r}$$

as desired.

Using the fact that $L_m = F_{m+1} + F_{m-1}$, and then adding formulas (28.2) and (28.3), we can again get formula (28.5). We invite you to try this approach.

Binet's formula also can be used to derive the recurrence relation, again as Brown did. To see this approach, we have:

$$\begin{aligned} F_n &= \frac{\alpha^n - \beta^n}{\alpha - \beta} = \frac{(\alpha^r - \beta^r)\alpha^{n-r} + (\alpha^{n-r} - \beta^{n-r})\beta^r}{\alpha - \beta} \\ &= \alpha^{n-r} F_r + \beta^r F_{n-r} \end{aligned}$$

[†] See the author's *Fibonacci and Lucas Numbers with Applications*.

where $n \geq r$. Then:

$$\begin{bmatrix} n \\ r \end{bmatrix} = \frac{1}{F_n}\begin{bmatrix} n \\ r \end{bmatrix} \cdot F_n = \frac{1}{F_n}\begin{bmatrix} n \\ r \end{bmatrix}[\alpha^{n-r}F_r + \beta^r F_{n-r}]$$

$$= \alpha^{n-r}\begin{bmatrix} n-1 \\ r-1 \end{bmatrix} + \beta^r \begin{bmatrix} n-1 \\ r \end{bmatrix} \tag{28.6}$$

Replacing r with $n-r$, this yields

$$\begin{bmatrix} n \\ n-r \end{bmatrix} = \alpha^r \begin{bmatrix} n-1 \\ n-r-1 \end{bmatrix} + \beta^{n-r}\begin{bmatrix} n-1 \\ n-r \end{bmatrix}$$

$$\begin{bmatrix} n \\ r \end{bmatrix} = \alpha^r \begin{bmatrix} n-1 \\ r \end{bmatrix} + \beta^{n-r}\begin{bmatrix} n-1 \\ r-1 \end{bmatrix} \tag{28.7}$$

Adding equations (28.6) and (28.7),

$$2\begin{bmatrix} n \\ r \end{bmatrix} = (\alpha^r + \beta^r)\begin{bmatrix} n-1 \\ r \end{bmatrix} + (\alpha^{n-r} + \beta^{n-r})\begin{bmatrix} n-1 \\ r-1 \end{bmatrix}$$

$$= \begin{bmatrix} n-1 \\ r \end{bmatrix} L_r + \begin{bmatrix} n-1 \\ r-1 \end{bmatrix} L_{n-r}$$

as desired.

Jerbic's Formula for Odd-Numbered Row Sums

The row sums S_{2n-1} satisfy the following pattern:

$$S_1 = 1 + 1 = 2 = L_0$$

$$S_3 = 1 + 2 + 2 + 1 = 6 = L_0 L_2$$

$$S_5 = 1 + 5 + 15 + 15 + 5 + 1 = 42 = L_0 L_2 L_4$$

More generally, S. Jerbic discovered in 1965 that

$$\sum_{r=0}^{2n-1}\begin{bmatrix} 2n-1 \\ r \end{bmatrix} = \prod_{i=0}^{n-1} L_{2i} \quad n \geq 1 \tag{28.8}$$

Hermite's Analogous Divisibility Properties for Fibonacci Numbers

Hermite's divisibility properties, and the fact that $(n+1)|\binom{2n}{n}$ for every $n \geq 0$, have analogous results for Fibonomials also, as Gould discovered.

The first analogous divisibility property is

$$\frac{F_m}{(F_m, F_n)} \mid \begin{bmatrix} m \\ n \end{bmatrix} \tag{28.9}$$

The proof follows on similar lines, as in Theorem 2.4. Let $d = (F_m, F_n)$. By the Euclidean algorithm, there are integers A and B such that $d = AF_m + BF_n$. Then:

$$\begin{aligned} d\begin{bmatrix} m \\ n \end{bmatrix} &= A\begin{bmatrix} m \\ n \end{bmatrix} F_m + B\begin{bmatrix} m \\ n \end{bmatrix} F_n \\ &= A\begin{bmatrix} m \\ n \end{bmatrix} F_m + B\begin{bmatrix} m-1 \\ n-1 \end{bmatrix} F_m \\ &= \left(A\begin{bmatrix} m \\ n \end{bmatrix} + B\begin{bmatrix} m-1 \\ n-1 \end{bmatrix}\right) F_m \end{aligned}$$

So $F_m | d\begin{bmatrix} m \\ n \end{bmatrix}$; that is, $\frac{F_m}{(F_m, F_n)} | \begin{bmatrix} m \\ n \end{bmatrix}$, as desired.

For example, let $m = 8$ and $n = 4$. Then $F_m = 21, F_n = 3, (F_m, F_n) = (21, 3) = 3$, $\frac{F_m}{(F_m, F_n)} = \frac{21}{3} = 7$, and $\begin{bmatrix} 8 \\ 4 \end{bmatrix} = 1820$. Clearly, $7 | 1820$.

A Byproduct

In particular, let $m = 2n$. Then, by property (28.9), $\frac{F_{2n}}{(F_{2n}, F_n)} | \begin{bmatrix} 2n \\ n \end{bmatrix}$. But $(F_a, F_b) = F_{(a,b)}$[†]. So $(F_{2n}, F_n) = F_{(2n,n)} = F_n$. Thus $\frac{F_{2n}}{F_n} | \begin{bmatrix} 2n \\ n \end{bmatrix}$. But $F_{2n} = F_n L_n$. Consequently, this implies that $L_n | \begin{bmatrix} 2n \\ n \end{bmatrix}$. Thus L_n is a factor of every CFC, as observed earlier.

Hermite's second divisibility property also has a Fibonomial analog, as Gould discovered:

$$\frac{F_{m-n+1}}{(F_{m+1}, F_n)} | \begin{bmatrix} m \\ n \end{bmatrix} \tag{28.10}$$

[†] See the author's *Fibonacci and Lucas Numbers with Applications*.

Its proof uses the facts that $(a, b) = (a - b, b)$ and $(F_a, F_b) = F_{(a,b)}$. Let $d = (F_{m+1}, F_n)$. Then:

$$\begin{aligned} d &= (F_{m+1}, F_n) = F_{(m+1,n)} \\ &= F_{(m-n+1,n)} = (F_{m-n+1}, F_n) \\ &= AF_{m-n+1} + BF_n \end{aligned}$$

for some integers A and B. Therefore:

$$\begin{aligned} d\begin{bmatrix} m \\ n \end{bmatrix} &= A\begin{bmatrix} m \\ n \end{bmatrix} F_{m-n+1} + B\begin{bmatrix} m \\ n \end{bmatrix} F_n \\ &= A\begin{bmatrix} m \\ n \end{bmatrix} F_{m-n+1} + B\begin{bmatrix} m \\ n-1 \end{bmatrix} F_{m-n+1} \\ &= \left(A\begin{bmatrix} m \\ n \end{bmatrix} + B\begin{bmatrix} m \\ n-1 \end{bmatrix}\right) F_{m-n+1} \end{aligned}$$

So $F_{m-n+1} | d\begin{bmatrix} m \\ n \end{bmatrix}$; that is, $\frac{F_{m-n+1}}{(F_{m+1}, F_n)} | \begin{bmatrix} m \\ n \end{bmatrix}$, again as desired.

For example, let $m = 8$ and $n = 3$. Then $F_{m-n+1} = 8$, $(F_{m+1}, F_n) = (34, 2) = 2$, and $\begin{bmatrix} 8 \\ 3 \end{bmatrix} = 1092$. Clearly, $4|1092$.

A Byproduct

Property (28.10) also has a valuable corollary. When $m = 2n$, it yields $\frac{F_{n+1}}{(F_{2n+1}, F_n)} | \begin{bmatrix} 2n \\ n \end{bmatrix}$. But $(F_{2n+1}, F_n) = F_{(2n+1,n)} = F_1 = 1$. So $F_{n+1} | \begin{bmatrix} 2n \\ n \end{bmatrix}$. That is, $\frac{1}{F_{n+1}} \begin{bmatrix} 2n \\ n \end{bmatrix}$ is an integer, and it corresponds to the Catalan number $\frac{1}{n+1}\binom{2n}{n}$.

Gould called the numbers $\frac{1}{F_{n+1}} \begin{bmatrix} 2n \\ n \end{bmatrix}$ *Fibonomial Catalan numbers*. The various Fibonomial Catalan numbers are

$$1, 1, 3, 20, 364, 17017, 4194036, \cdots$$

Fibonomial Identities

Interestingly, the first four powers of F_n can be expressed as linear combinations of Fibonomials:

$$F_n = \begin{bmatrix} n \\ 1 \end{bmatrix} - \begin{bmatrix} n-1 \\ 1 \end{bmatrix}$$

$$F_n^2 = \begin{bmatrix} n+1 \\ 2 \end{bmatrix} - \begin{bmatrix} n \\ 2 \end{bmatrix}$$

$$F_n^3 = \begin{bmatrix} n+2 \\ 3 \end{bmatrix} - 2\begin{bmatrix} n+1 \\ 3 \end{bmatrix} - \begin{bmatrix} n \\ 3 \end{bmatrix} \qquad (28.11)$$

$$F_n^4 = \begin{bmatrix} n+3 \\ 4 \end{bmatrix} - 4\begin{bmatrix} n+2 \\ 4 \end{bmatrix} - 4\begin{bmatrix} n+1 \\ 4 \end{bmatrix} + \begin{bmatrix} n \\ 4 \end{bmatrix} \qquad (28.12)$$

Identities (28.11) and (28.12) were discovered in 1969 by P. Mana of the University of New Mexico. These can be confirmed algebraically. It is *not* known if similar formulas exist for higher powers of F_n.

Wall's Identities

Next, we establish two Fibonomial identities, discovered in 1966 by C. R. Wall of Harker Heights, Texas:

$$\prod_{r=1}^{n} \sqrt{5}F_r\alpha^{-r} = 1 + \sum_{r=1}^{n}(-1)^{r(r-1)/2}\begin{bmatrix} n \\ r \end{bmatrix}\alpha^{-r(n+1)} \qquad (28.13)$$

$$\prod_{r=1}^{n} L_r\alpha^{-r} = 1 + \sum_{r=1}^{n}(-1)^{r(r+1)/2}\begin{bmatrix} n \\ r \end{bmatrix}\alpha^{-r(n+1)} \qquad (28.14)$$

where α denotes the golden ratio.

To establish them, we use the identity

$$\prod_{r=0}^{n-1}(1 - q^r x) = \sum_{r=0}^{n}(-1)^r q^{r(r-1)/2}\begin{Bmatrix} n \\ r \end{Bmatrix}x^n \qquad (28.15)$$

where $\begin{Bmatrix} n \\ r \end{Bmatrix}$ denotes the Gaussian binomial coefficient (see Chapter 3), as D. Lind did in 1968. We have $\sqrt{5}F_r\alpha^{-r} = 1 - (\beta/\alpha)^r$, where $\beta = (1-\sqrt{5})/2$. Letting $q = \beta/\alpha$,

$$\begin{Bmatrix} n \\ r \end{Bmatrix} = \alpha^{r^2 - nr}\begin{bmatrix} n \\ r \end{bmatrix}$$

Substituting $x = q$ in (28.15), we get:

$$\prod_{r=1}^{n} \sqrt{5}F_r\alpha^{-r} = \prod_{r=1}^{n}(1-q^r)$$

$$= \sum_{r=0}^{n}(-1)^r \begin{bmatrix} n \\ r \end{bmatrix} q^{n(n+1)/2}\alpha^{r^2-nr}$$

$$= \sum_{r=0}^{n}(-1)^{r(r-1)/2} \begin{bmatrix} n \\ r \end{bmatrix} \alpha^{-r(n+1)}$$

where we have used the fact that $\alpha\beta = -1$.

To prove identity (28.13), we have $L_r\alpha^{-r} = 1 - (\beta/\alpha)^r$. Letting $q = \beta/\alpha$ and $x = -q$ in (28.15), it yields:

$$\prod_{r=1}^{n} L_r\alpha^{-r} = \prod_{r=1}^{n}(1+q^r)$$

$$= \sum_{r=0}^{n}(-1)^r \begin{bmatrix} n \\ r \end{bmatrix} q^{r(r-1)/2}\alpha^{r^2-nr}(-q)^r$$

$$= \sum_{r=0}^{n} \begin{bmatrix} n \\ r \end{bmatrix} q^{r(r+1)/2}\alpha^{r^2-nr}$$

$$= \sum_{r=0}^{n}(-1)^{r(r+1)/2} \begin{bmatrix} n \\ r \end{bmatrix} \alpha^{-r(n+1)}$$

as desired.

Generalized Fibonomial Coefficients

Using the Fibonacci numbers $U_n = F_{nk}$, we can generalize Fibonomial coefficients as follows:

$$\begin{bmatrix} n \\ r \end{bmatrix}_k = \frac{U_n^*}{U_r^* U_{n-r}^*} = \frac{U_n U_{n-1} \cdots U_{n-r+1}}{U_r U_{r-1} \cdots U_1}$$

where k is a positive integer, $0 < r < n$, and $\begin{bmatrix} n \\ 0 \end{bmatrix}_k = 1 = \begin{bmatrix} n \\ n \end{bmatrix}_k$.

Notice that $\begin{bmatrix} n \\ r \end{bmatrix}_1 = \begin{bmatrix} n \\ r \end{bmatrix}$ and $\begin{bmatrix} n \\ r \end{bmatrix}_k = \begin{bmatrix} n \\ n-r \end{bmatrix}_k$.

As before, it can be shown that

$$\begin{bmatrix} n \\ r \end{bmatrix}_k = \begin{bmatrix} n-1 \\ r \end{bmatrix}_k F_{kr+1} + \begin{bmatrix} n-1 \\ r-1 \end{bmatrix}_k F_{k(n-r)-1} \tag{28.16}$$

$$\begin{bmatrix} n \\ r \end{bmatrix}_k = \begin{bmatrix} n-1 \\ r \end{bmatrix}_k F_{kr-1} + \begin{bmatrix} n-1 \\ r-1 \end{bmatrix}_k F_{k(n-r)+1}$$

and

$$2\begin{bmatrix} n \\ r \end{bmatrix}_k = \begin{bmatrix} n-1 \\ r \end{bmatrix}_k L_{kr} + \begin{bmatrix} n-1 \\ r-1 \end{bmatrix}_k L_{k(n-r)}$$

These are generalizations of identities (28.2), (28.3), and (28.5), respectively. They can be used to construct a triangular array for each value of $k \geq 1$.

In particular, let $k = 2$. Then:

$$\begin{bmatrix} n \\ 0 \end{bmatrix}_2 = 1 = \begin{bmatrix} n \\ n \end{bmatrix}_2$$

$$\begin{bmatrix} n \\ r \end{bmatrix}_2 = \begin{bmatrix} n-1 \\ r \end{bmatrix}_2 F_{2r-1} + \begin{bmatrix} n-1 \\ r-1 \end{bmatrix}_2 F_{2(n-r)+1} \tag{28.17}$$

For example, $\begin{bmatrix} 3 \\ 2 \end{bmatrix}_2 = 8 = \begin{bmatrix} 3 \\ 1 \end{bmatrix}_2$. So $\begin{bmatrix} 4 \\ 2 \end{bmatrix}_2 = \begin{bmatrix} 3 \\ 2 \end{bmatrix}_2 F_3 + \begin{bmatrix} 3 \\ 1 \end{bmatrix}_2 F_5 = 8 \cdot 2 + 8 \cdot 5 = 56$.

The various values of $\begin{bmatrix} n \\ r \end{bmatrix}_2$ can be used to construct the triangular array in Figure 28.6.

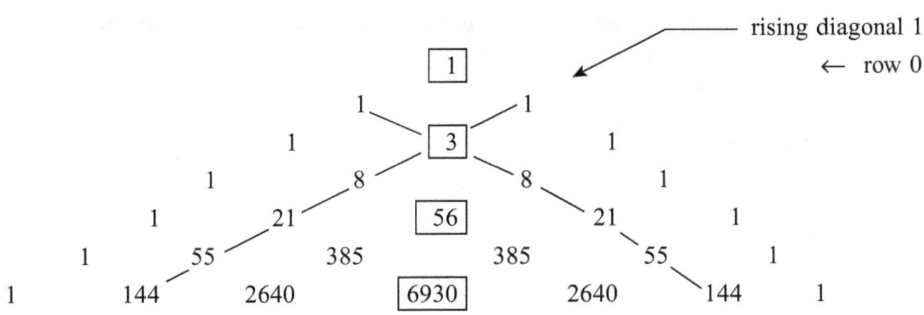

Figure 28.6

Interestingly, the rising diagonal 1 consists of the Fibonacci numbers F_{2n}. This is so, since $\begin{bmatrix} n \\ 1 \end{bmatrix}_2 = \frac{F_{2n}F_{2n-2}\cdots F_2}{F_2(F_{2n-2}\cdots F_2)} = F_{2n}$; see Figure 28.6. More generally, $\begin{bmatrix} n \\ 1 \end{bmatrix}_k = \frac{F_{kn}F_{kn-k}\cdots F_k}{F_k(F_{kn-k}\cdots F_k)} = \frac{F_{kn}}{F_k}$.

A Formula for $\begin{bmatrix} n \\ r \end{bmatrix}_{2k}$

The coefficient $\begin{bmatrix} n \\ r \end{bmatrix}_{2k}$ can be computed using the corresponding Fibonomial coefficient $\begin{bmatrix} n \\ r \end{bmatrix}_k$ and Lucas numbers. Using the fact that $F_{2m} = F_m L_m$, we have:

$$\begin{aligned}
\begin{bmatrix} n \\ r \end{bmatrix}_{2k} &= \frac{U_n^*}{U_r^* U_{n-r}^*} = \frac{F_{2nk}^*}{F_{2rk}^* F_{2(n-r)k}^*} \\
&= \frac{F_{nk}^*}{F_{rk}^* F_{(n-r)k}^*} \cdot \frac{L_{nk}^*}{L_{rk}^* L_{(n-r)k}^*} \\
&= \begin{bmatrix} n \\ r \end{bmatrix}_k \cdot \frac{L_{nk} L_{(n-1)k} \cdots L_{(n-r+1)k}}{L_{rk} L_{(r-1)k} \cdots L_k} \qquad (28.18)
\end{aligned}$$

For example, $\begin{bmatrix} 6 \\ 3 \end{bmatrix}_2 = \begin{bmatrix} 6 \\ 3 \end{bmatrix} \cdot \frac{L_6 L_5 L_4}{L_3 L_2 L_1} = 60 \cdot \frac{18 \cdot 11 \cdot 7}{4 \cdot 3 \cdot 1} = 6{,}930.$

(Notice that $\frac{L_n^*}{L_r^* L_{n-r}^*}$ need not be an integer. For example, $\frac{L_6^*}{L_3^* L_3^*} = \frac{L_6 L_5 L_4}{L_3 L_2 L_1} = \frac{18 \cdot 11 \cdot 7}{4 \cdot 3 \cdot 1}$ is not an integer; but $\frac{L_6^*}{L_2^* L_4^*} = \frac{L_6 L_5}{L_2 L_1} = \frac{18 \cdot 11}{3 \cdot 1}$ is an integer.)

Odd-Numbered Row Sums

As in the case of the Fibonomial triangle in Figure 28.2, the odd-numbered row sums in Figure 28.6 also follow a pattern:

$$\begin{aligned}
\sum_{r=0}^{1} \begin{bmatrix} 1 \\ r \end{bmatrix}_2 &= 2 \\
\sum_{r=0}^{3} \begin{bmatrix} 3 \\ r \end{bmatrix}_2 &= 18 = 2 \cdot 3^2 = 2\prod_{i=1}^{1} L_{2i}^2 \\
\sum_{r=0}^{5} \begin{bmatrix} 5 \\ r \end{bmatrix}_2 &= 882 = 2(3 \cdot 7)^2 = 2\prod_{i=1}^{2} L_{2i}^2 \\
\sum_{r=0}^{7} \begin{bmatrix} 7 \\ r \end{bmatrix}_2 &= 285{,}768 = 2(3 \cdot 7 \cdot 18)^2 = 2\prod_{i=1}^{3} L_{2i}^2
\end{aligned}$$

More generally, we conjecture that

$$\sum_{r=0}^{2n-1} \begin{bmatrix} 2n-1 \\ r \end{bmatrix}_2 = 2\prod_{i=1}^{n-1} L_{2i}^2 \qquad (28.19)$$

Central Elements $\begin{bmatrix} 2n \\ n \end{bmatrix}_2$

The central elements $\begin{bmatrix} 2n \\ n \end{bmatrix}_2$ in Figure 28.6 also possess an interesting property. To see this, by the recurrence relation (28.9), we have:

$$\begin{aligned}
\begin{bmatrix} 2n \\ n \end{bmatrix}_2 &= \begin{bmatrix} 2n-1 \\ n \end{bmatrix}_2 F_{2n-1} + \begin{bmatrix} 2n-1 \\ n-1 \end{bmatrix}_2 F_{2n+1} \\
&= \begin{bmatrix} 2n-1 \\ n \end{bmatrix}_2 F_{2n-1} + \begin{bmatrix} 2n-1 \\ n \end{bmatrix}_2 F_{2n+1} \\
&= \begin{bmatrix} 2n-1 \\ n \end{bmatrix}_2 (F_{2n-1} + F_{2n+1}) \\
&= \begin{bmatrix} 2n-1 \\ n \end{bmatrix}_2 L_{2n} = \begin{bmatrix} 2n-1 \\ n-1 \end{bmatrix}_2 L_{2n} \qquad (28.20)
\end{aligned}$$

Thus each central element $\begin{bmatrix} 2n \\ n \end{bmatrix}_2$ in Figure 28.6 can be obtained by multiplying its northeast (or northwest) neighbor by L_{2n}.

For example, $\begin{bmatrix} 6 \\ 3 \end{bmatrix}_2 = 6930 = 385 \cdot 18 = 385 \cdot L_6 = \begin{bmatrix} 5 \\ 3 \end{bmatrix}_2$; see Figure 28.6.

More generally, by the recurrence relation (28.16),

$$\begin{aligned}
\begin{bmatrix} 2n \\ n \end{bmatrix}_k &= \begin{bmatrix} 2n-1 \\ n \end{bmatrix}_k F_{kn-1} + \begin{bmatrix} 2n-1 \\ n-1 \end{bmatrix}_k F_{kn+1} \\
&= \begin{bmatrix} 2n-1 \\ n \end{bmatrix}_2 F_{kn-1} + \begin{bmatrix} 2n-1 \\ n \end{bmatrix}_k F_{kn+1} \\
&= \begin{bmatrix} 2n-1 \\ n \end{bmatrix}_k (F_{kn-1} + F_{kn+1}) \\
&= \begin{bmatrix} 2n-1 \\ n \end{bmatrix}_k L_{kn} = \begin{bmatrix} 2n-1 \\ n-1 \end{bmatrix}_k L_{kn}
\end{aligned}$$

Next, we turn to the array in Figure 28.7. It is obtained by changing the sign of every element on each odd-numbered northeast diagonal. The elements in row $m+1$ provide the coefficients in the recurrence relation for F_{2n}^m.

For example, using row 5, $F_{2n+10}^4 - 55F_{2n+8}^4 + 385F_{2n+6}^4 - 385F_{2n+4}^4 + 55F_{2n+2}^4 - F_{2n}^4 = 0$; that is, $F_{2n+10}^4 = 55F_{2n+8}^4 - 385F_{2n+6}^4 + 385F_{2n+4}^4 - 55F_{2n+2}^4 + F_{2n}^4$.

					1					
				1		−1				
			1		−3		1			
		1		−8		8		−1		
	1		−21		56		−21		1	
1		−55		385		−385		55		−1

Figure 28.7

Hermite's Analogous Divisibility Properties

Finally, Hermite's divisibility properties have analogous results for the generalized Fibonomial coefficient $\begin{bmatrix} n \\ r \end{bmatrix}_k$ also:

$$\frac{U_m}{(U_m, U_n)} \mid \begin{bmatrix} m \\ n \end{bmatrix}_k \tag{28.21}$$

$$\frac{U_{m-n+1}}{(U_{m+1}, U_n)} \mid \begin{bmatrix} m \\ n \end{bmatrix}_k \tag{28.22}$$

To establish (28.21), let $d = (U_m, U_n) = AU_m + BU_n$ for some integers A and B. Then:

$$\begin{aligned} d\begin{bmatrix} m \\ n \end{bmatrix}_k &= A\begin{bmatrix} m \\ n \end{bmatrix}_k U_m + B\begin{bmatrix} m \\ n \end{bmatrix}_k U_n \\ &= A\begin{bmatrix} m \\ n \end{bmatrix}_k U_m + B\begin{bmatrix} m-1 \\ n-1 \end{bmatrix}_k U_m \\ &= \left(A\begin{bmatrix} m \\ n \end{bmatrix}_k U_m + B\begin{bmatrix} m-1 \\ n-1 \end{bmatrix}_k\right) U_m \end{aligned}$$

So $U_m | d\begin{bmatrix} m \\ n \end{bmatrix}_k$; that is, $\frac{U_m}{(U_m, U_n)} | \begin{bmatrix} m \\ n \end{bmatrix}_k$, as desired.

For example, let $m = 6$, $n = 4$, and $k = 2$. Then $U_m = F_{12}$, $U_n = F_8$, and $(U_m, U_n) = (F_{12}, F_8) = F_{(12,8)} = F_4 = 3$; so $\frac{U_m}{(U_m, U_n)} = 48$; and $\begin{bmatrix} m \\ n \end{bmatrix}_2 = \begin{bmatrix} 6 \\ 4 \end{bmatrix}_2 =$ 2,640. Clearly, $48|2640$.

A Byproduct

In particular, let $m = 2n$. Then, by property (28.21),

$$\frac{U_m}{(U_m, U_n)} = \frac{F_{4n}}{(F_{4n}, F_{2n})} = \frac{F_{4n}}{F_{(4n,2n)}} = \frac{F_{2n}L_{2n}}{F_{2n}} = L_{2n}$$

So $L_{2n} | \begin{bmatrix} 2n \\ n \end{bmatrix}_2$. Thus $\frac{1}{L_{2n}} \begin{bmatrix} 2n \\ n \end{bmatrix}_2$ is an integer.

It follows from (28.21) that

$$\frac{1}{L_{2n}} \begin{bmatrix} 2n \\ n \end{bmatrix}_2 = \begin{bmatrix} 2n-1 \\ n-1 \end{bmatrix}_2 \tag{28.23}$$

The numbers $\frac{1}{L_{2n}} \begin{bmatrix} 2n \\ n \end{bmatrix}_2$ correspond to Fibonomial Catalan numbers, where $n \geq 1$. The first five Fibonomial Catalan numbers are 1, 8, 385, 124410, and 73331470368. They are the northwest neighbors of the central elements in Figure 28.7, and form the adjacent column to the left of the central column.

To establish property (28.22), we use the facts that $(a, b) = (a - b, b)$ and $(F_a, F_b) = F_{(a,b)}$. Let:

$$\begin{aligned} d &= (U_{m+1}, U_n) \\ &= (F_{mk+k}, F_{nk}) = F_{(mk+k,nk)} \\ &= F_{(mk+k-nk,nk)} = (F_{mk-nk+k}, F_{nk}) \\ &= (U_{m-n+1}, U_n) = CU_{m-n+1} + DU_n \end{aligned}$$

for some integers C and D. Then:

$$\begin{aligned} d\begin{bmatrix} m \\ n \end{bmatrix}_k &= C\begin{bmatrix} m \\ n \end{bmatrix}_k U_{m-n+1} + D\begin{bmatrix} m \\ n \end{bmatrix}_k U_n \\ &= C\begin{bmatrix} m \\ n \end{bmatrix}_k U_{m-n+1} + D\begin{bmatrix} m \\ n-1 \end{bmatrix}_k U_{m-n+1} \\ &= \left(C\begin{bmatrix} m \\ n \end{bmatrix}_k + D\begin{bmatrix} m \\ n-1 \end{bmatrix}_k\right) U_{m-n+1} \end{aligned}$$

So $U_{m-n+1} | d\begin{bmatrix} m \\ n \end{bmatrix}_k$; that is, $\frac{U_{m-n+1}}{(U_{m+1}, U_n)} | \begin{bmatrix} m \\ n \end{bmatrix}_k$, again as required.

In particular, when $m = 2n$, this yields $\frac{U_{n+1}}{(U_{2n+1}, U_n)} | \begin{bmatrix} 2n \\ n \end{bmatrix}_k$. But $(2n+1, n) = 1$; so

$$(U_{2n+1}, U_n) = (F_{(2n+1)k}, F_{nk}) = F_{((2n+1)k,nk)} = F_{(2n+1,n)k} = F_k$$

Thus $\frac{F_{nk+k}}{F_k} \mid \begin{bmatrix} 2n \\ n \end{bmatrix}_k$.

Clearly, $(U_{2n+1}, U_n) = 1$ if and only if $k = 1$ or $k = 2$. So $F_{nk+k} \mid \begin{bmatrix} 2n \\ n \end{bmatrix}_k$ if and only if $k = 1$ or $k = 2$. So $F_{n+1} \mid \begin{bmatrix} 2n \\ n \end{bmatrix}$ and $F_{2n+2} \mid \begin{bmatrix} 2n \\ n \end{bmatrix}_2$, where $n \geq 1$. Thus, every CFC in Figure 28.2 is divisible by a Fibonacci number and a Lucas number. Also, every CFC in Figure 28.8 is divisible by an even-subscripted Fibonacci number and an even-numbered Lucas number. Since $F_{2m} = F_m L_m$, it follows that $F_{n+1}L_{n+1} \mid \begin{bmatrix} 2n \\ n \end{bmatrix}_2$ for every $n \geq 0$.

For example, let $n = 5$. We have $F_6 = 8, L_6 = 18$, and $F_{12} = 144$, and each divides $\begin{bmatrix} 10 \\ 5 \end{bmatrix}_2 = 73,331,470,368$.

The term $\frac{1}{F_{2n+2}} \begin{bmatrix} 2n \\ n \end{bmatrix}_2$ generates the sequence $1, 7, 330, 106314, \cdots$.

Brennan Matrix Revisited

The fact that $\frac{F_{nk+k}}{F_k} \mid \begin{bmatrix} 2n \\ n \end{bmatrix}_k$ presents a nice relationship between the $(n+1) \times (n+1)$ Brennan matrix P_n from Chapter 25 and the CFC $\begin{bmatrix} 2n \\ n \end{bmatrix}_k$. In 1965, Carlitz showed that the trace of P_n^k is $\frac{F_{nk+k}}{F_k}$: $tr(P_n^k) = \frac{F_{nk+k}}{F_k}$. Thus $tr(P_n^k) \mid \begin{bmatrix} 2n \\ n \end{bmatrix}_k$.

Hoggat–Hansell Identity for $\begin{bmatrix} n \\ r \end{bmatrix}_k$

The Hoggat–Hansell identity (28.5) can be extended to $U_n = \begin{bmatrix} n \\ r \end{bmatrix}_k$ also:

$$\begin{aligned}
\begin{bmatrix} n-1 \\ r-1 \end{bmatrix}_k \begin{bmatrix} n \\ r+1 \end{bmatrix}_k \begin{bmatrix} n+1 \\ r \end{bmatrix}_k &= \frac{U_{n-1}^*}{U_{r-1}^* U_{n-r}^*} \cdot \frac{U_n^*}{U_{r+1}^* U_{n-r-1}^*} \cdot \frac{U_{n+1}^*}{U_r^* U_{n-r+1}^*} \\
&= \frac{U_{n-1}^*}{U_r^* U_{n-r-1}^*} \cdot \frac{U_{n+1}^*}{U_{r+1}^* U_{n-r}^*} \cdot \frac{U_n^*}{U_{r-1}^* U_{n-r+1}^*} \\
&= \begin{bmatrix} n-1 \\ r \end{bmatrix}_k \begin{bmatrix} n+1 \\ r+1 \end{bmatrix}_k \begin{bmatrix} n \\ r-1 \end{bmatrix}_k \qquad (28.24)
\end{aligned}$$

See Figure 28.8. So the product of the six Fibonomial coefficients is a square.

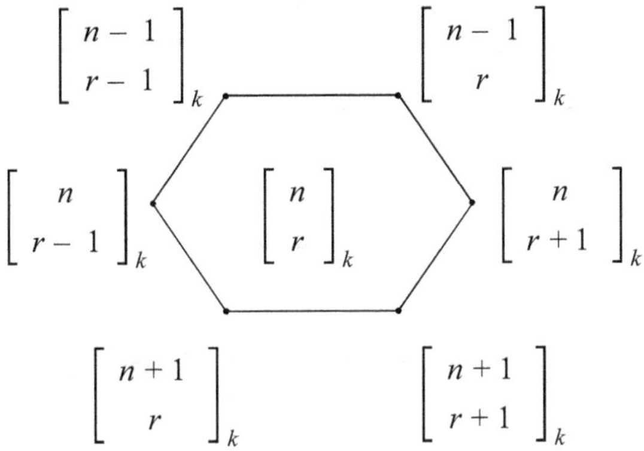

Figure 28.8

For example, consider Figure 28.9, where $k = 2$. Clearly, $56 \cdot 55 \cdot 6930 = 21 \cdot 2640 \cdot 385$, and $56 \cdot 21 \cdot 55 \cdot 2640 \cdot 6930 \cdot 385 = 21{,}344{,}400^2$.

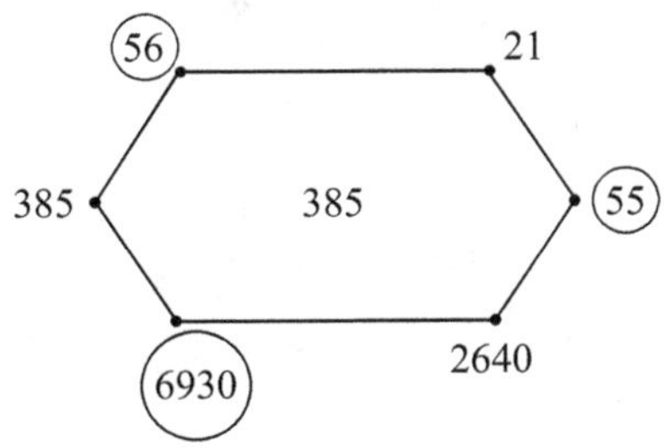

Figure 28.9

Just as Gould generalized the Hoggatt–Hansell identity for binomial coefficients, identity (28.24) can be generalized as follows, where a is a positive integer:

$$\begin{bmatrix} n-a \\ r-a \end{bmatrix}_k \begin{bmatrix} n \\ r+a \end{bmatrix}_k \begin{bmatrix} n+a \\ r \end{bmatrix}_k = \begin{bmatrix} n-a \\ r \end{bmatrix}_k \begin{bmatrix} n+a \\ r+a \end{bmatrix}_k \begin{bmatrix} n \\ r-a \end{bmatrix}_k \tag{28.25}$$

This too can be confirmed algebraically without much effort.

For example, when $n = 5$, $a = 2$, $r = 3$, and $k = 2$:

$$\begin{aligned} \text{LHS} &= \begin{bmatrix} 3 \\ 1 \end{bmatrix}_2 \begin{bmatrix} 5 \\ 5 \end{bmatrix}_2 \begin{bmatrix} 7 \\ 3 \end{bmatrix}_2 = 8 \cdot 1 \cdot 124410 \\ &= 1 \cdot 18096 \cdot 55 = \begin{bmatrix} 3 \\ 3 \end{bmatrix}_2 \begin{bmatrix} 7 \\ 5 \end{bmatrix}_2 \begin{bmatrix} 5 \\ 1 \end{bmatrix}_2 = \text{RHS} \end{aligned}$$

See Figure 28.10.

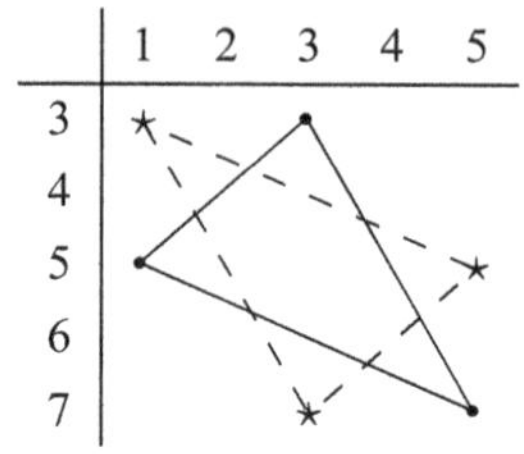

Figure 28.10

Ironically, Gould's hexagon property does not hold for the Fibonomial coefficients in identity (28.25). The case $n = 5, a = 2, r = 3$, and $k = 2$ provides a counterexample: $(8, 1, 124410) = 2 \neq 1 = (1, 18096, 55)$.

Although Gould's hexagon property holds for $\begin{bmatrix} n \\ r \end{bmatrix}$, it is not known if it holds for $k > 1$. However, using Figure 28.9, we find that $(56, 55, 6930) = (21, 2640, 385)$; also $(55, 6930, 18096) = (385, 124850, 144)$; see Figure 28.11.

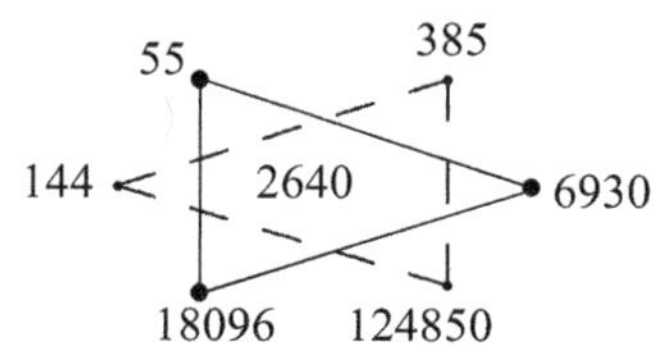

Figure 28.11

For the curious-minded, Figure 28.12 shows the Fibonomial triangle for $k = 3$. As expected, it satisfies the Hoggatt–Hansell identity. For example, $72 \cdot 5490 \cdot 98515 = 306 \cdot 417240 \cdot 305$. It appears that this array also satisfies Gould's hexagon property. For instance, $(72, 5490, 98515) = (306, 417240, 305)$.

```
                    1
               1         1
          1         4         1
     1        17        17         1
  1      72       306        72        1
1    305     5490      5490      305      1
```

Figure 28.12

29

Multinomial Arrays

This chapter extends the binomial theorem to more than two variables. To this end, first notice that the binomial theorem can be expressed in a slightly different way:

$$(x+y)^n = \sum_{i=0}^{n} \binom{n}{i} x^{n-i} y^i = \sum_{i=0}^{n} \frac{n!}{i!(n-i)!} x^{n-i} y^i$$

$$= \sum_{\substack{i,j \ge 0 \\ i+j=n}} \frac{n!}{i!j!} x^i y^j$$

The binomial coefficient $\frac{n!}{i!j!}$ represents the number of ways of partitioning an n-element set into two disjoint subsets of sizes i and $j = n - i$.

Trinomial Theorem

The binomial theorem can be extended to a finite number of monomials. To this end, we begin with the trinomial expansion of $(x+y+z)^n$. We have

$$(x+y+z)^n = \underbrace{(x+y+z)(x+y+z)\cdots(x+y+z)}_{n \text{ factors}} \tag{29.1}$$

Every term in the expansion is of the form $Cx^i y^j z^k$, where i, j, and k are nonnegative integers and $i+j+k=n$, and C is the coefficient to be determined.

The coefficient C is the number of ways i x's, j y's, and k z's can be selected from the n factors on the RHS of equation (29.1). The i x's can be selected in $\binom{n}{i}$ ways. Now, $n - i$ factors are left; so the j y's can be selected in $\binom{n-i}{j}$ ways. Now,

$n-i-j$ factors are left; so the k z's can be selected in $\binom{n-i-j}{k}$ ways. Thus, by the multiplication principle,

$$C = \binom{n}{i}\binom{n-i}{j}\binom{n-i-j}{k}$$

$$= \frac{n!}{i!(n-i)!} \cdot \frac{(n-i)!}{j!(n-i-j)!} \cdot \frac{(n-i-j)!}{k!(n-i-j-k)!}$$

$$= \frac{n!}{i!j!k!0!} = \frac{n!}{i!j!k!}$$

Thus, we have the following theorem.

Theorem 29.1 (Trinomial Theorem) Let x, y, and z be any real numbers, and n any nonnegative integer. Then

$$(x+y+z)^n = \sum_{\substack{i,j,k\geq 0\\ i+j+k=n}} \frac{n!}{i!j!k!} x^i y^j z^k \tag{29.2}$$ ■

[The *trinomial coefficient* $\frac{n}{i!j!k!}$ is often written as $\binom{n}{i,j,k}$ to correspond to the binomial coefficient $\binom{n}{r}$.]

For example,

$$(x+y+z)^2 = \sum_{\substack{i,j,k\geq 0\\ i+j+k=2}} \frac{2!}{i!j!k!} x^i y^j z^k$$

$$= x^2 + y^2 + z^2 + 2xy + 2yz + 2zx$$

In particular, let $y = x^2$ and $z = 1$ in the trinomial theorem. Then

$$(1+x+x^2)^n = \sum_{\substack{i,j,k\geq 0\\ i+j+k=n}} \frac{n!}{i!j!k!} x^i x^{2j}$$

$$= \sum_{\substack{i,j,k\geq 0\\ i+j+k=n}} \frac{n!}{i!j!k!} x^{i+2j} \tag{29.3}$$

This expansion is often written as

$$(1+x+x^2)^n = \sum_{r=0}^{2n} \binom{n, 3}{r} x^r \tag{29.4}$$

where

$$\binom{n,3}{r} = \sum_{\substack{i,j,k\geq 0\\ i+2j=r\\ i+j+k=n}} \frac{n!}{i!j!k!}$$

denotes the rth *trinomial coefficient* and $0 \leq r \leq 2n$.

For example,

$$(1+x+x^2)^0 = 1 \qquad (1+x+x^2)^2 = 1+2x+3x^2+2x^3+x^4$$
$$(1+x+x^2)^1 = 1+x+x^2 \qquad (1+x+x^2)^3 = 1+3x+6x^2+7x^3+6x^4+3x^5+x^6$$

The trinomial coefficients $\binom{n,3}{r}$ can be used to construct a Pascal-like array, as in Figure 29.1, where the top row is row 0.

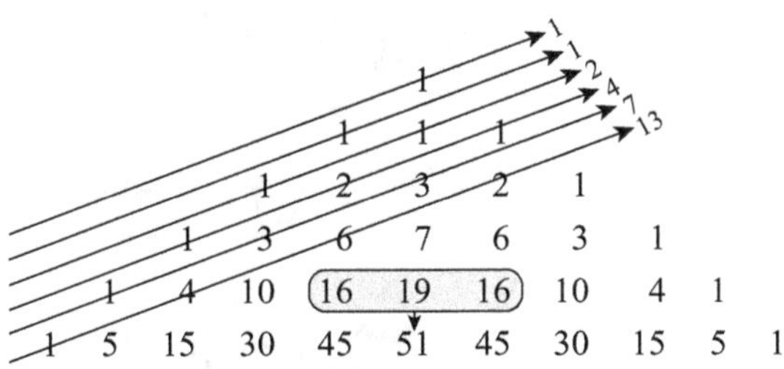

Figure 29.1 Trinomial Triangle

As in the case of Pascal's triangle, the trinomial triangle also satisfies a number of interesting properties:

1) Each row begins with and ends in a 1.
2) Each row, except row 0, contains two more elements than the preceding row.
3) The trinomial array is symmetric about the vertical axis through the apex.
4) Each internal entry is the sum of its northwestern, northern, and northeastern neighbors. That is,

$$\binom{n,3}{r} = \binom{n-1,3}{r-1} + \binom{n-1,3}{r} + \binom{n-1,3}{r+1}$$

For example, $51 = 16 + 19 + 16$; see Figure 29.2.

5) Row n contains $2n+1$ entries.
6) The northeast diagonal 1 consists of positive integers and diagonal 2 consists of triangular numbers. See Figure 29.2. The latter property follows by

induction and the fact that

$$\binom{n,3}{2} = \binom{n-1,3}{0} + \binom{n-1,3}{1} + \binom{n-1,3}{2}$$
$$= 1 + (n-1) + \frac{n(n-1)}{2} = t_n$$

7) The sum of the numbers in row n equals 3^n, where $n \geq 0$; that is, $\sum_{r=0}^{2n} \binom{n,3}{r} = 3^n$. This follows by letting $x = 1$ in the trinomial expansion (29.4). For example,

$$1 = 3^0 \qquad 1+2+3+2+1 = 3^2$$
$$1+1+1 = 3^1 \qquad 1+3+6+7+6+3+1 = 3^3$$

8) Suppose we let $x = -1$ in formula (29.4). Then

$$(1-1+1)^n = \sum_{r=0}^{2n} \binom{n,3}{r}(-1)^r$$
$$1 = \sum_{r=0}^{2n} \binom{n,3}{r}(-1)^r$$

Accordingly, the sum of the numbers in row n with alternate signs equals 1. For example,

$$1 = 1 \qquad 1-2+3-2+1 = 1$$
$$1-1+1 = 1 \qquad 1-3+6-7+6-3+1 = 1$$

9) The northeast diagonal sums yield the *tribonacci numbers* T_n: $\sum_{r\geq 0} \binom{n-r,3}{r} = T_n$.

10) Let $a_n = \sum_{r=0}^{\lfloor n/3 \rfloor} \binom{n-2r}{r}$, $b_{2n} = \sum_{r=0}^{\lfloor n/2 \rfloor} \binom{n-2r,3}{2r}$, and $b_{2n+1} = \sum_{r=0}^{\lfloor (n-1)/2 \rfloor} \binom{n-r,3}{2r+1}$. In 1977, G. Berzsenyi of Lamar University, Beaumont, Texas, established that $a_n = b_n + b_{n-1}$, where $n \geq 1$.

11) Unlike Pascal's triangle, the trinomial array has a *central trinomial coefficient* in every row; see the circled numbers in Figure 29.2. They are the coefficients of x^n. Using expansion (29.4), it is given by $\binom{n,3}{n} = \sum_{\substack{i,j,k\geq 0 \\ i+j+k=n}} \frac{n!}{i!j!k!}$, where $i+2j = n$. Since $i+j+k = i+2j$, it follows that $j = k$. So, the central trinomial coefficient is given by $\binom{n,3}{n} = \sum_{\substack{i,j\geq 0 \\ i+2j=n}} \frac{n!}{i!j!j!}$.

$$\begin{array}{ccccccccccc}
 & & & & & \boxed{1} & & & & & \\
 & & & & 1 & \boxed{1} & 1 & & & & \\
 & & & 1 & 2 & \boxed{3} & 2 & 1 & & & \\
 & & 1 & 3 & 6 & \boxed{7} & 6 & 3 & 1 & & \\
 & 1 & 4 & 10 & 16 & \boxed{19} & 16 & 10 & 4 & 1 & \\
1 & 5 & 15 & 30 & 45 & \boxed{51} & 45 & 30 & 15 & 5 & 1
\end{array}$$

Figure 29.2

For example, the central trinomial coefficient in row 4 is given by

$$\sum_{\substack{i,j\geq 0\\ i+2j=4}} \frac{4!}{i!j!j!} = \frac{4!}{4!0!0!} + \frac{4!}{2!1!1!} + \frac{4!}{0!2!2!} = 1 + 12 + 6 = 19$$

Similarly, the central trinomial coefficient in row 3 is given by

$$\sum_{\substack{i,j\geq 0\\ i+2j=3}} \frac{3!}{i!j!j!} = \frac{3!}{1!1!1!} + \frac{3!}{3!0!0!} = 6 + 1 = 7$$

12) The sum of the squares of the entries in each row is a central trinomial coefficient; that is,

$$\sum_{r=0}^{n} \binom{n,3}{r}^2 = \binom{2n,3}{n}$$

This corresponds to Lagrange's identity for binomial coefficients, and follows from the facts that $\binom{n,3}{r} = \binom{n,3}{n-r}$ and $(1 + x + x^2)^{2n} = (1 + x + x^2)^n(1 + x + x^2)^n$.

For example,

$$1^2 = 1 \qquad 1^2 + 2^2 + 3^2 + 2^2 + 1^2 = 19$$

$$1^2 + 1^2 + 1^2 = 3 \qquad 1^2 + 3^2 + 6^2 + 7^2 + 6^2 + 3^2 + 1^2 = 141$$

See Figure 29.2.

Left-justified Trinomial Triangle

Next, we turn to the column generators of the left-justified trinomial triangle in Figure 29.3.

Let $g_j(x)$ denote the column generator of column j, where $j \geq 0$. Then:

$$g_0(x) = \frac{1}{1-x}, \quad g_1(x) = \frac{x}{(1-x)^2}, \quad g_2(x) = \frac{x}{(1-x)^3},$$

$$\begin{array}{ccccccccccccl}
1 & & & & & & & & & & & & \leftarrow \text{row } 0 \\
1 & 1 & 1 & & & & & & & & & & \\
1 & 2 & 3 & 2 & 1 & & & & & & & & \\
1 & 3 & 6 & 7 & 6 & 3 & 1 & & & & & & \\
1 & 4 & 10 & 16 & 19 & 16 & 10 & 4 & 1 & & & & \\
1 & 5 & 15 & 30 & 45 & 51 & 45 & 30 & 15 & 5 & 1 & & \\
\uparrow & & & & & & & & & & & & \\
\text{column } 0 & & & & & & & & & & & &
\end{array}$$

Figure 29.3

and in general

$$g_n(x) = \frac{x}{1-x}[g_{n-1}(x) + g_{n-2}(x)], \quad n \geq 2$$

Next, we turn to an application of the trinomial theorem.

Odd Factorials

In 1975, G. B. Huff proposed the following problem: Solve the recurrence relation

$$a_n = \sum_{r=0}^{\lfloor n/2 \rfloor} \binom{n}{2r}(n-2r)!a_r^2$$

where $a_0 = 1$.

Notice that:

$$a_1 = \sum_{r=0}^{0} \binom{1}{2r}(1-2r)!a_r^2 = 1$$

$$a_2 = \sum_{r=0}^{1} \binom{2}{2r}(2-2r)!a_r^2$$

$$= \binom{2}{0}2!a_0^2 + \binom{2}{2}(2-2)!a_1^2 = 2 + 1 = 3$$

The solution we now present was given in the following year by A. Stenger, then a student at Pennsylvania State University. We will show by PMI that

$$a_n = 1 \cdot 3 \cdot 5 \cdots (2n-1) \tag{29.5}$$

This product is often denoted by $(2n-1)!!$, where we define $(-1)!! = 1$.

First, notice that:

$$\frac{[(2r-1)!!]^2}{[(2r)!]^2} = \frac{[1\cdot 3\cdot 5\cdots(2r-1)][1\cdot 3\cdot 5\cdots(2r-1)]}{[1\cdot 2\cdot 3\cdots(2r)][1\cdot 2\cdot 3\cdots(2r)]}$$

$$= \frac{1}{[2\cdot 4\cdots(2r)][2\cdot 4\cdots(2r)]}$$

$$= \frac{1}{(2^r\, r!)(2^r\, r!)} = \frac{2^{-2r}}{(r!)^2} \tag{29.6}$$

Second, by the trinomial theorem,

$$(x + \frac{1}{x} + 2)^n = \sum_{\substack{i,j,k\geq 0\\ i+j+k=n}} \frac{n!}{i!j!k!} x^{i-j} 2^k$$

So the constant term C in the expansion is

$$C = \sum_{\substack{i,k\geq 0\\ 2i+k=n}} \frac{n!2^k}{(i!)^2 k!} = \sum_{i=0}^{\lfloor n/2\rfloor} \frac{n!2^{n-2i}}{(i!)^2(n-2i)!}$$

$$\frac{C}{n!2^n} = \sum_{i=0}^{\lfloor n/2\rfloor} \frac{2^{-2i}}{(i!)^2(n-2i)!}$$

But $(x+\frac{1}{x}+2)^n = x^{-n}(1+x)^{2n}$ and the constant term in the expansion of $x^{-n}(1+x)^{2n}$ is $\binom{2n}{n}$. So $C = \binom{2n}{n}$. Thus

$$\frac{1}{n!2^n}\binom{2n}{n} = \sum_{i=0}^{\lfloor n/2\rfloor} \frac{2^{-2i}}{(i!)^2(n-2i)!} \tag{29.7}$$

Since $a_0 = (-1)!! = 1$, formula (29.5) is true when $n = 0$. Now, assume that it is true for all nonnegative integers $k < n$. Then, by the recurrence relation, and equations (29.6) and (29.7),

$$a_n = \sum_{r=0}^{\lfloor n/2\rfloor} \binom{n}{2r}^2 (n-2r)![(2r-1)!!]^2$$

$$= \sum_{r=0}^{\lfloor n/2\rfloor} \frac{(n!)^2[(2r-1)!!]^2}{(n-2r)![(2r)!]^2} = \sum_{r=0}^{\lfloor n/2\rfloor} \frac{(n!)^2}{(n-2r)!} \cdot \frac{2^{-2r}}{(r!)^2}$$

$$\frac{a_n}{(n!)^2} = \sum_{r=0}^{\lfloor n/2\rfloor} \cdot \frac{2^{-2r}}{(r!)^2(n-2r)!} = \frac{1}{n!2^n}\binom{2n}{n}$$

$$a_n = \frac{n!}{2^n}\binom{2n}{n} = (2n-1)!!$$

Thus, by PMI, $a_n = (2n-1)!!$ for every $n \geq 0$, as desired.

A Generating Function for Central Trinomial Coefficients

Next, we will find a generating function for the central trinomial coefficient $b_n = \binom{n,3}{n}$. To this end, we will employ the same technique used by Brown and Hoggatt in 1978, and the generating function $A(x)$ for the CBC; namely, $A(x) = \frac{1}{\sqrt{1-4x}} = \sum_{n=0}^{\infty} \binom{2n}{n} x^n$.

Let

$$(1 + x + x^2)^n = \sum_{i=0}^{2n} b_i x^i$$

$$[(1 + x) + x^2]^n = \sum_{k=0}^{n} \binom{n}{k} x^{2k}(1 + x)^{n-k}$$

$$= \binom{n}{0}(1 + x)^n + \binom{n}{1}x^2(1 + x)^{n-1} + \cdots + \binom{n}{k}x^{2k}(1 + x)^{n-k} + \cdots + \binom{n}{n}x^{2n}(1 + x)^0$$

So b_n is given by $b_n = \binom{n}{0}\binom{n}{n} + \binom{n}{1}\binom{n-1}{n-2} + \cdots + \binom{n}{r}\binom{n-r}{n-2r}$, where $r = \lfloor n/2 \rfloor$.

Since

$$\binom{n}{k}\binom{n-k}{k} = \frac{n!}{k!(n-k)!} \cdot \frac{(n-k)!}{k!(n-2k)!}$$

$$= \frac{n!(2k)!}{(2k)!(n-2k)!k!k!} = \binom{n}{2k}\binom{2k}{k}$$

b_n can also be written as

$$b_n = \sum_{k=0}^{\lfloor n/2 \rfloor} \binom{n}{k}\binom{n-k}{n-2k} = \sum_{k=0}^{\lfloor n/2 \rfloor} \binom{n}{k}\binom{n-k}{k}$$

$$= \sum_{k=0}^{\lfloor n/2 \rfloor} \binom{n}{2k}\binom{2k}{k}$$

So

$$B(x) = \sum_{n=0}^{\infty} b_n x^n = \sum_{n=0}^{\infty} \left[\sum_{k=0}^{\lfloor n/2 \rfloor} \binom{n}{2k}\binom{2k}{k} \right] x^n$$

$$= \sum_{k=0}^{\infty} \binom{2k}{k} \left[\sum_{n=2k}^{\infty} \binom{n}{2k} x^n \right]$$

since $\binom{n}{2k} = 0$ for $0 \leq n < 2k$. Thus:

$$B(x) = \sum_{k=0}^{\infty} \binom{2k}{k} \sum_{n=0}^{\infty} \binom{n}{2k} x^n = \sum_{k=0}^{\infty} \binom{2k}{k} \left[\frac{x^{2k}}{(1-x)^{2k+1}} \right]$$

$$= \frac{1}{1-x} \sum_{k=0}^{\infty} \binom{2k}{k} \left(\frac{x}{1-x} \right)^{2k}$$

But

$$A(t) = \frac{1}{\sqrt{1-4t}} = \sum_{k=0}^{\infty} \binom{2k}{k} t^k$$

So

$$B(x) = \frac{1}{1-x} \cdot A\left[\frac{x^2}{(1-x)^2} \right]$$

$$= \frac{1}{1-x} \cdot \frac{1}{\sqrt{1 - \left(\frac{x}{1-x}\right)^2}} = \frac{1}{\sqrt{1-2x-3x^2}}$$

Thus

$$\frac{1}{\sqrt{1-2x-3x^2}} = \sum_{n=0}^{\infty} \binom{n,3}{n} x^n$$

is the desired generating function.

As a byproduct, it follows that the generating function $B(x)$ is related to $A(x)$ by a transformation:

$$B(x) = \frac{1}{1-x} A\left[\frac{x^2}{(1-x)^2} \right]$$

Trinomial Catalan Numbers

By successively finding the differences of consecutive Catalan numbers C_n from the Catalan sequence $1, 2, 5, 14, 42, 132, 429, \cdots$, *trinomial Catalan numbers* C_n^* can be obtained. C_n^* is the first term in the nth sequence of successive differences; see Figure 29.4.

Recall from Chapter 17 that the generating function $C(x)$ for Catalan numbers is

$$C(x) = \sum_{n=0}^{\infty} C_n x^n = \frac{1 - \sqrt{1-4x}}{2x}$$

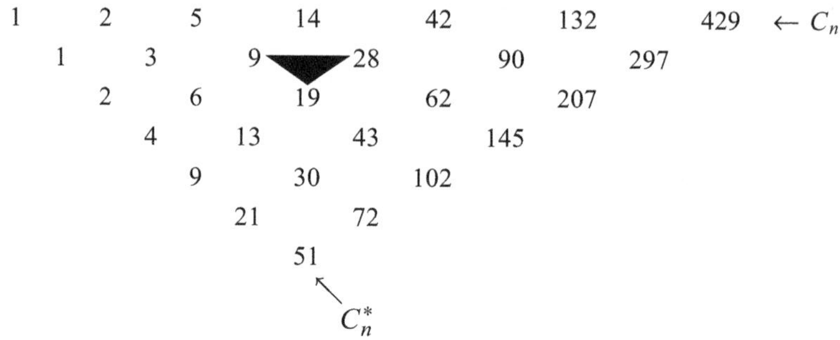

Figure 29.4

So

$$C^2(x) = \sum_{n=0}^{\infty} C_{n+1}x^n = \frac{C(x) - 1}{x}$$

$$= \frac{1 - 2x - \sqrt{1 - 4x}}{2x^2}$$

Let $C^*(x)$ be the generating function for trinomial Catalan numbers. Then:

$$C^*(x) = \frac{1}{1+x}C^2\left(\frac{x}{1+x}\right) = \frac{1}{1+x}\left[\frac{1 - \frac{2x}{1+x} - \sqrt{1 - \frac{4x}{1+x}}}{\frac{2x^2}{(1+x)^2}}\right]$$

$$= \frac{1 - x - \sqrt{(1+x)(1-3x)}}{2x^2} = \frac{1 - x - \sqrt{1 - 2x - 3x^2}}{2x^2}$$

Trinomial Catalan numbers can also be generated by inserting a zero between every two consecutive Catalan numbers and then by taking their successive sums; see Figure 29.5.

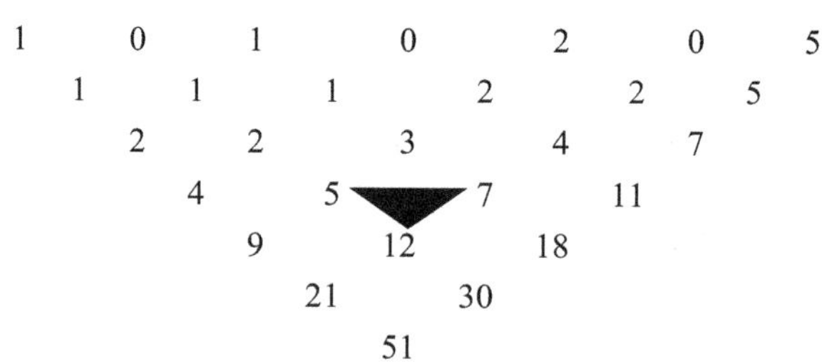

Figure 29.5

The numbers in the top row in Figure 29.5 are generated by $C(x^2) = \frac{1-\sqrt{1-4x^2}}{2x^2}$. So

$$C^*(x) = \frac{1}{1-x}C\left[\left(\frac{x}{1-x}\right)^2\right] = \frac{1}{1-x} \cdot \frac{1 - \sqrt{1 - \frac{4x^2}{(1-x)^2}}}{2\left(\frac{x}{1-x}\right)^2}$$

$$= \frac{1 - x - \sqrt{1 - 2x - 3x^2}}{2x^2}$$

as obtained earlier.

Motzkin Numbers

Trinomial Catalan numbers are often called *Motzkin numbers*, after T.S. Motzkin. He discovered them in his investigation of partitions of polygons in 1948, while he was at the University of California at Los Angeles. They have applications to combinatorics, geometry, and number theory.

In 1977, R. Donaghey of the City University of New York and L. W. Shapiro of Howard University, Washington, D.C. developed 14 different interpretations of Motzkin numbers M_n.[†] For example, they count the number of lattice paths on the cartesian plane from the origin to the point $(n, 0)$ using the steps (1,0), (1,1), and $(1, -1)$, without going below the x-axis. Figure 29.6 shows such paths for $1 \leq n \leq 4$.

Motzkin numbers also count the number of different ways a subset of n people, seated around a round table, can shake hands without their hands crossing; see Figure 29.7.

Next, we present two identities involving both Catalan and Motzkin numbers.

Catalan–Motzkin Identities

Suppose there are n people seated around a round table. There are exactly C_k ways $2k$ people can shake hands without their hands crossing[†] and the $2k$ people can be selected in $\binom{n}{2k}$ ways, where $0 \leq 2k \leq n$. So, by the multiplication and addition

[†]For additional occurrences, see R. P. Stanley, *Enumerative Combinatorics*, Vol. 2, Cambridge University Press, New York, 1999.

[†]See the author's *Catalan Numbers with Applications*.

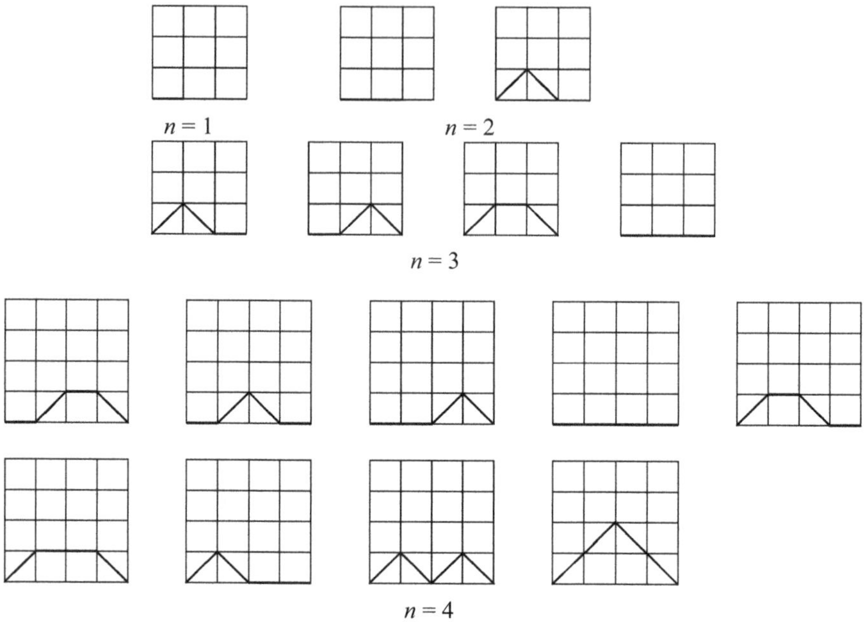

Figure 29.6

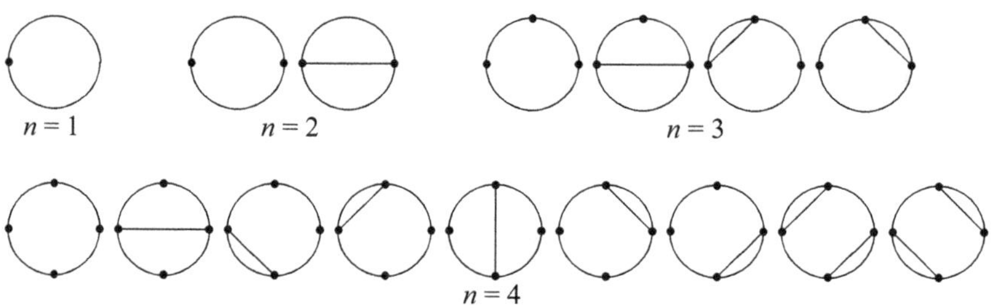

Figure 29.7

principles, the total number of ways a subset of n people can shake hands without their hands crossing equals $\sum_{k=0}^{\lfloor n/2 \rfloor} \binom{n}{2k} C_k$. Thus

$$M_n = \sum_{k=0}^{\lfloor n/2 \rfloor} \binom{n}{2k} C_k \tag{29.8}$$

For example,

$$M_5 = \sum_{k=0}^{2} \binom{5}{2k} C_k = \binom{5}{0} C_0 + \binom{5}{2} C_1 + \binom{5}{4} C_2 = 1 \cdot 1 + 10 \cdot 1 + 5 \cdot 2 = 21$$

as expected.

While identity (29.8) expresses Motzkin numbers in terms of Catalan numbers, the following identity expresses them in the reverse order. It was developed by Donaghey in 1977:†

$$C_{n+1} = \sum_{k=0}^{n} \binom{n}{k} M_k \tag{29.9}$$

For example,

$$C_4 = \sum_{k=0}^{3} \binom{3}{k} M_k = \binom{3}{0} M_0 + \binom{3}{1} M_1 + \binom{3}{2} M_2 + \binom{3}{3} M_3$$
$$= 1 \cdot 1 + 3 \cdot 1 + 3 \cdot 2 + 1 \cdot 4 = 14$$

Using identities (2.4), (29.8), and (29.9), and Corollary 3.1, we can now derive a new formula for C_{n+1}. It was developed by J. Touchard of France in 1924.

Touchard's Catalan Formula

Substituting for M_k from (29.8) into (29.9), we get:

$$C_{n+1} = \sum_{k=0}^{n} \binom{n}{k} \left[\sum_{j=0}^{\lfloor k/2 \rfloor} \binom{k}{2j} C_j \right] = \sum_{j=0}^{\lfloor n/2 \rfloor} C_j \left[\sum_{k=2j}^{n} \binom{n}{k} \binom{k}{2j} \right]$$
$$= \sum_{j=0}^{\lfloor n/2 \rfloor} C_j \left[\sum_{k=2j}^{n} \binom{n}{2j} \binom{n-2j}{k-2j} \right] = \sum_{j=0}^{\lfloor n/2 \rfloor} \binom{n}{2j} C_j \left[\sum_{k=2j}^{n} \binom{n-2j}{k-2j} \right]$$
$$= \sum_{j=0}^{\lfloor n/2 \rfloor} \binom{n}{2j} C_j \left[\sum_{\ell=0}^{n-2j} \binom{n-2j}{\ell} \right] = \sum_{j=0}^{\lfloor n/2 \rfloor} \binom{n}{2j} 2^{n-2j} C_j \tag{29.10}$$

For example,

$$C_6 = \sum_{j=0}^{2} \binom{5}{2j} 2^{5-2j} C_j = \binom{5}{0} 2^5 \cdot 1 + \binom{5}{2} 2^3 \cdot 1 + \binom{5}{4} 2^1 \cdot 2$$
$$= 1 \cdot 32 + 10 \cdot 8 + 5 \cdot 4 = 132$$

Touchard's formula implies that Catalan numbers C_{n+1} can be computed as a weighted sum of the entries $\binom{n}{2j}$ in row n of Pascal's triangle and weights $2^{n-2j} C_j$, where $0 \leq j \leq \lfloor n/2 \rfloor$.

†See R. Donaghey, "Restricted Plane Tree Representations of four Motzkin–Catalan Equations," *J. Combinatorial Theory, Series B*, 22 (1977), 114–121.

A Generating Function for Motzkin Numbers

Finally, we add that Motzkin numbers are defined by the generating function $M(x) = C^*(x)$:

$$\begin{aligned} M(x) &= \frac{1 - x - \sqrt{1 - 2x - 3x^2}}{2x^2} \\ &= 1 + x + 2x^2 + 4x^3 + 9x^4 + 21x^5 + 51x^6 + 127x^7 + \cdots \end{aligned}$$

Notice that the generating function $M(x)$ satisfies the quadratic equation $M(x) = 1 + xM(x) + x^2M^2(x)$; so Motzkin numbers can be defined by this equation also.

A Recursive Definition for the Central Trinomial Coefficients

Next, we will derive a recurrence relation for the central trinomial coefficients, following the technique used by Brown and Hoggatt.

To this end, we will employ the well-known *Legendre polynomials* $P_n(x)$, defined recursively as follows:

$$\begin{aligned} P_0(x) &= 1, \qquad P_1(x) = x \\ nP_n(x) &= (2n - 1)xP_{n-1}(x) - (n - 1)P_{n-2}(x), \qquad n \geq 2 \end{aligned}$$

They are generated by the function†

$$\frac{1}{\sqrt{1 - 2xt + x^2}} = \sum_{n=0}^{\infty} P_n(t)x^n$$

We now introduce a phantom parameter t in the generating function $B(x)$ of the central trinomial coefficients b_n:

$$B(x, t) = \frac{1}{\sqrt{1 - 2xt - 3x^2}} = \sum_{n=0}^{\infty} M_n(t)x^n \qquad \text{(say)}$$

where $B(x, 1) = B(x)$ and $M_n(1) = b_n$.

† See J. Riordan, *An Introduction to Combinatorial Analysis*, Princeton University Press, Princeton, New Jersey, 1980.

Let $x_1 = -\sqrt{3}xi$ and $t_1 = \frac{ti}{\sqrt{3}}$, where $i = \sqrt{-1}$. Then

$$\sum_{n=0}^{\infty} M_n(t)x^n = \frac{1}{\sqrt{1-2xt-3x^2}} = \frac{1}{\sqrt{1-2x_1t_1-x_1^2}}$$

$$= \sum_{n=0}^{\infty} P_n(t_1)x_1^n = \sum_{n=0}^{\infty} P_n\left(\frac{ti}{\sqrt{3}}\right)(-\sqrt{3}xi)^n$$

So $b_n = M_n(1) = (-\sqrt{3}i)^n P_n\left(\frac{i}{\sqrt{3}}\right)$. Since $P_0(x) = 1$ and $P_1(x) = x$, it follows that $b_0 = (-\sqrt{3}i)^0 P_0\left(\frac{i}{\sqrt{3}}\right) = 1$ and $b_1 = (-\sqrt{3}i)P_1\left(\frac{i}{\sqrt{3}}\right) = (-\sqrt{3}i)\left(\frac{i}{\sqrt{3}}\right) = 1$.

Substituting $x = \frac{i}{\sqrt{3}}$ in the recurrence relation for $P_n(x)$, we get

$$nP_n\left(\frac{i}{\sqrt{3}}\right) = (2n-1)\frac{i}{\sqrt{3}}P_{n-1}\left(\frac{i}{\sqrt{3}}\right) - (n-1)P_{n-2}\left(\frac{i}{\sqrt{3}}\right)$$

Multiplying both sides with $(-\sqrt{3}i)^n$, this yields

$$n(-\sqrt{3}i)^n P_n\left(\frac{i}{\sqrt{3}}\right) = (2n-1)(-\sqrt{3}i)^n \frac{i}{\sqrt{3}}P_{n-1}\left(\frac{i}{\sqrt{3}}\right) - (n-1)(-\sqrt{3}i)^n P_{n-2}\left(\frac{i}{\sqrt{3}}\right)$$

That is,

$$nb_n = (2n-1)b_{n-1} + 3(n-1)b_{n-2} \tag{29.11}$$

Thus, b_n can be defined recursively as follows:

$$b_0 = 1 = b_1$$
$$nb_n = (2n-1)b_{n-1} + 3(n-1)b_{n-2}, \qquad n \geq 2$$

An Interesting Byproduct

The recurrence relation (29.11) can be used to compute $\lim_{n\to\infty} \frac{b_n}{b_{n-1}}$. To this end, let ℓ be the limit. Then

$$\frac{b_n}{b_{n-1}} = \frac{2n-1}{n} + \frac{3(n-1)}{n}\frac{b_{n-2}}{b_{n-1}}$$

$$\lim_{n\to\infty} \frac{b_n}{b_{n-1}} = 2 + 3\lim_{n\to\infty} \frac{b_{n-2}}{b_{n-1}}$$

$$\ell = 2 + \frac{3}{\ell}$$

$$\ell^2 - 2\ell - 3 = 0$$

$$(\ell - 3)(\ell + 1) = 0$$

Since ℓ is positive, this implies that $\lim\limits_{n\to\infty} \frac{b_n}{b_{n-1}} = 3$.

From Recurrence Relation to Generating Function

Next, we will derive the generating function $B(x)$ from the recurrence relation (29.11). Again, we let $B(x) = \sum\limits_{n=0}^{\infty} b_n x^n$. Then

$$xB'(x) = \sum_{n=0}^{\infty} nb_n x^n$$

$$3xB'(x) + 3B(x) = \sum_{n=0}^{\infty} 3(n+1)b_n x^n$$

Besides,

$$xB'(x) - 0 \cdot b_0 - xb_1 = \sum_{n=2}^{\infty} nb_n x^n$$

That is,

$$\frac{B'(x) - 1}{x} = \sum_{n=0}^{\infty} (n+2)b_{n+2} x^n$$

We also have

$$B(x) = 1 + x\sum_{n=0}^{\infty} b_{n+1} x^n$$

and

$$B'(x) = \sum_{n=0}^{\infty} b_{n+1} x^n + \sum_{n=0}^{\infty} nb_{n+1} x^n$$

So

$$\frac{B(x) - 1}{x} = \sum_{n=0}^{\infty} b_{n+1} x^n$$

and

$$2B'(x) + \frac{B(x) - 1}{x} = \sum_{n=0}^{\infty}(2n+3)b_{n+1}x^n$$

By the recurrence relation (29.11), we have:

$$(n+2)b_{n+2} = (2n+3)b_{n+1} + 3(n+1)b_n$$

This yields

$$\frac{B'(x) - 1}{x} = 2B'(x) + \frac{B(x) - 1}{x} + 3xB'(x) + 3B(x)$$

That is,

$$(1 - 2x - 3x^2)B'(x) = (3x+1)B(x)$$

$$\frac{B'(x)}{B(x)} = \frac{3x+1}{1 - 2x - 3x^2}$$

Integrating both sides with respect to x, we get:

$$\ln B(x) = -\frac{1}{2}\ln(1 - 2x - 3x^2) + \ln C$$

$$B(x) = \frac{C}{\sqrt{1 - 2x - 3x^2}}$$

Since $B(0) = b_0 = 1$, $C = 1$. Thus, $B(x) = \frac{1}{\sqrt{1-2x-3x^2}}$, as expected.

The trinomial theorem can be generalized to any finite number of variables $x_1, x_2, \cdots, x_k$, as the following theorem shows.

Theorem 29.2 (**Multinomial Theorem**) Let $x_1, x_2, \cdots, x_k$ be any k real variables, and n any nonnegative integer. Then

$$(x_1 + x_2 + \cdots + x_n)^n = \sum \frac{n!}{i_1!i_2!\cdots i_k!}x_1^{i_1}x_2^{i_2}\cdots x_k^{i_k} \tag{29.12}$$

where $i_1, i_2, \cdots, i_k \geq 0$ and $i_1 + i_2 + \cdots + i_k = n$. ∎

The *multinomial coefficient* $\frac{n!}{i_1!i_2!\cdots i_k!}$ is denoted by $\binom{n}{i_1, i_2, \cdots, i_k}$.

For example, the multinomial coefficient of $x_1x_2^2x_3^3x_4^4x_5^5$ in the expansion of $(x_1 + x_2^2 + x_3^3 + x_4^4 + x_5^5)^{15}$ is $\frac{15!}{1!2!3!4!5!} = 37{,}837{,}800$.

In particular,

$$(1 + x + x^2 + x^3)^n = \sum_{\substack{i,j,k,l \geq 0 \\ i+j+k+l=n}} \frac{n!}{i!j!k!l!}x^{i+2j+3k} \tag{29.13}$$

This expansion is often written as

$$(1+x+x^2+x^3)^n = \sum_{r=0}^{3n} \binom{n,4}{r} x^r \tag{29.14}$$

For example,

$$(1+x+x^2+x^3)^2 = \sum_{r=0}^{6} \binom{n,4}{r} x^r = 1+2x+3x^2+4x^3+3x^4+2x^5+x^6$$

The *quadrinomial coefficients* $\binom{n,4}{r}$ in the *quadrinomial expansion* of $(1+x+x^2+x^3)^n$ can be constructed to form the *quadrinomial triangle* in Figure 29.8.

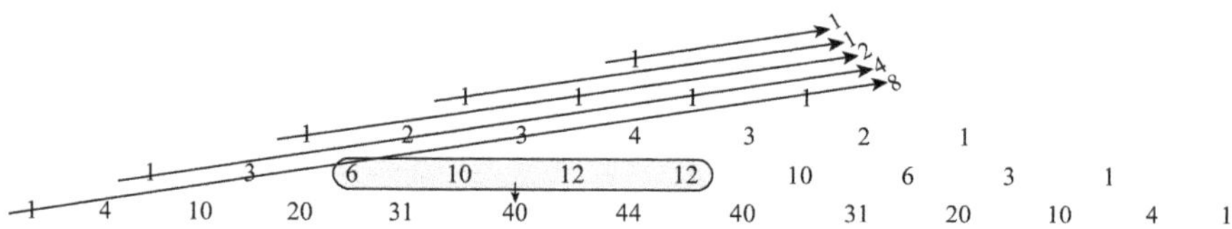

Figure 29.8 Quadrinomial Triangle

The quadrinomial array also has several interesting properties:

1) Every row begins with and ends in a 1.
2) Every row, except row 0, contains three more elements than the previous row.
3) Row n contains $3n+1$ elements, where $n \geq 0$.
4) The array is symmetric about the vertical line through the apex; that is, $\binom{n,4}{r} = \binom{n,4}{n-r}$.
5) Each internal entry can be obtained by adding the two neighbors to its left in the preceding row to the neighbors to its right in the same preceding row; that is,

$$\binom{n,4}{r} = \binom{n-1,4}{r-3} + \binom{n-1,4}{r-2} + \binom{n-1,4}{r-1}\binom{n-1,4}{r}$$

For example, $44 = 10+12+12+10$ and $155 = 31+40+44+40$.

6) The sum of the entries in row n is 4^n, where $n \geq 0$. This follows from the fact that $\sum_{r \geq 0} \binom{n,4}{r} = 4^n$, obtained by letting $x = 1$ in the expansion (29.13).

For example,

$$1 = 4^0 \qquad 1+2+3+4+3+2+1 = 4^2$$
$$1+1+1+1 = 4^1 \qquad 1+3+6+\cdots+6+3+1 = 4^3$$

7) Suppose we let $x = -1$ in the expansion (29.13), where $n \geq 1$. Then $0 = \sum_{r\geq 0} \binom{n,4}{r}(-1)^r$. Consequently, the sum of the neighbors in row n, when we alternately change their signs, equals 0, where $n \geq 1$.
8) The element $\binom{n,4}{2}$ in position 2 in each row n is the triangular number t_{n-1}. This follows by induction and the fact that

$$\binom{n,4}{2} = \binom{n-1,4}{-1} + \binom{n-1,4}{0} + \binom{n-1,4}{1} + \binom{n-1,4}{2}$$

$$= 0 + 1 + (n-1) + \binom{n-1,4}{2} = n + \binom{n-1,4}{2}$$

9) The *central quadrinomial coefficient* in row n equals the coefficient of $x^{\lfloor 3n/2 \rfloor}$, namely, $\binom{n,4}{\lfloor 3n/2 \rfloor}$, where n is even.
10) The sum of the squares of the entries in each row is a central quadrinomial coefficient:

$$\sum_{r=0}^{3n+1} \binom{n,4}{r}^2 = \binom{2n,4}{\lfloor 3n/2 \rfloor}$$

For example:

$$1^2 = 1 \qquad 1^2 + 2^2 + 3^2 + 4^2 + 3^2 + 2^2 + 1^2 = 44$$
$$1^2 + 1^2 + 1^2 + 1^2 = 4 \qquad 1^2 + 3^2 + 6^2 + \cdots + 6^2 + 3^2 + 1^2 = 580$$

11) The northeast diagonal sums yield the various *quadrinacci numbers* 1, 1, 2, 4, 8, 15, 29, $\cdots$; see Figure 29.8. They can be defined recursively, as follows:

$$Q_1 = Q_2 = 1, \quad Q_3 = 2, \quad Q_4 = 4$$
$$Q_n = Q_{n-1} + Q_{n-2} + Q_{n-3} + Q_{n-4}, \qquad n \geq 5$$

Thus $\sum_{r\geq 0} \binom{n-r,4}{r} = Q_n$.

Left-justified Quadrinomial Triangle

Next, we develop the column generators of the left-justified quadrinomial triangle in Figure 29.9.

Let $g_j(x)$ denote the column generator of column j, where $j \geq 0$. Then:

$$g_0(x) = \frac{1}{1-x}, \quad g_1(x) = \frac{x}{(1-x)^2}, \quad g_2(x) = \frac{x}{(1-x)^3},$$

and in general $g_n(x) = \frac{x}{1-x}[g_{n-1}(x) + g_{n-2}(x) + g_{n-3}(x)]$, where $n \geq 3$.

1													← row 0
1	1	1	1										
1	2	3	4	3	2	1							
1	3	6	10	12	12	10	6	3	1				
1	4	10	20	31	40	44	40	31	20	10	4	1	
↑ column 0													

Figure 29.9

Next, we compute the number of terms in the multinomial expansion (29.12).

Number of Terms in the Multinomial Expansion

In 1904, S. Epsteen looked for the number of terms in the multinomial expansion of $(x_1 + x_2 + \cdots + x_m)^m$. In the following year, F. Owens *et al* generalized it further, by finding the number of terms in the expansion of $(x_1 + x_2 + \cdots + x_m)^n$.

To see this, let $U(m, n)$ denote the number of terms in the expansion. Since

$$(x_1 + x_2 + \cdots + x_m)^n = [x_1 + (x_2 + x_3 + \cdots + x_m)]^n$$

it follows that

$$\begin{aligned} U(m,n) &= 1 + U(m-1,1) + U(m-1,2) + \cdots + U(m-1,n) \\ &= [1 + U(m-1,1) + U(m-1,2) + \cdots + U(m-1,n-1)] + U(m-1,n) \\ &= U(m,n-1) + U(m-1,n) \end{aligned}$$

This recurrence relation with the initial conditions $U(1,0) = U(1,1) = U(2,0) = 1$ yields the rectangular array in Figure 29.10.

$m \backslash n$	0	1	2	3	4	5	6
0	1	1	1	1	1	1	1
1	1	2	3	4	5	6	7
2	1	3	6	10	15	21	28
3	1	4	10	20	35	56	84
4	1	5	15	35	70	126	210
5	1	6	21	56	126	252	462

Figure 29.10 Tartaglia's Rectangular Array

Clearly, it is Tartaglia's array, which we studied in Chapter 25. Consequently,

$$U(m,n) = \binom{m+n-1}{m}$$

In particular,

$$U(n,n) = \binom{2n-1}{n} = \frac{1}{2}\binom{2n}{n}$$

Thus the expansion of $(x_1 + x_2 + \cdots + x_n)^n$ contains $\frac{1}{2}\binom{2n}{n}$ terms.

For example, the expansion of $(x+y+z)^3$ contains $\frac{1}{2}\binom{6}{3} = 10$ terms:

$$(x+y+z)^3 = x^3+y^3+z^3+3x^2y+3xy^2+3y^2z+3yz^2+3x^2z+3xz^2+6xyz$$

J. Scheffer of Hagerstown, Maryland, solved the same problem using a totally different approach. Since $\frac{1}{1-a_ix} = \sum\limits_{n=0}^{\infty} a_i^n x^n$,

$$\frac{1}{1-a_1x} \cdot \frac{1}{1-a_2x} \cdots \frac{1}{1-a_nx} = 1 + \sum_{n\geq 1}^{\infty} s_n x^n \tag{29.15}$$

where

$$s_1 = a_1 + a_2 + \cdots + a_n$$
$$s_2 = a_1^2 + a_2^2 + \cdots + a_n^2 + a_1a_2 + a_1a_3 + \cdots + a_2a_3 + \cdots$$
$$s_3 = a_1^3 + a_2^3 + \cdots + a_n^3 + a_1^2a_2 + a_1^2a_3 + \cdots + a_1a_2a_3 + \cdots$$
$$\vdots$$

Letting $a_1 = a_2 = \cdots = a_n = 1$, (29.15) becomes $\frac{1}{(1-x)^n} = 1 + \sum\limits_{n\geq 1}^{\infty} U(n,n)x^n$.

Consequently,

$$U(n,n) = \text{the coefficient of } x^n \text{ in the Maclaurin expansion of } (1-x)^{-n}$$

$$= \binom{2n-1}{n} = \frac{1}{2}\binom{2n}{n}, \text{ as desired}$$

References

Aggarwala, R. and Lamoureux, M. P., "Inverting the Pascal Matrix Plus One" *American Mathematical Monthly* 109 (2002), 371–377.

Alladi, K. and Hoggatt, V. E. Jr, "Compositions With Ones and Twos," *Fibonacci Quarterly* 13 (1975), 233–239.

Althoen, S. C. and Lacampagne, C. B., "Tetrahedral Numbers as Sums of Square Numbers," *Mathematics Magazine* 64 (1991), 104–108.

Andrews, G. E. et al., *Special Functions*, Cambridge University Press, New York, 1999.

Anning, N., "The Pascal Triangle," *American Mathematical Monthly* 37 (1930), 22–25.

Anselone, P., Solution to Problem E 839, *American Mathematical Monthly* 56 (1949), 342.

Aude, H. T. R., Solution to Problem 3468, *American Mathematical Monthly* 38 (1931), 355.

Barlaz, J., Solution to Problem 4322, *American Mathematical Monthly* 57 (1950), 347–348.

Bardell, R. H., Problem E349, *American Mathematical Monthly* 45 (1938), 629.

Basil, M., "Pascal's Pyramid," *Mathematics Teacher* 61 (1968), 19–21.

Bateman, P., Solution to Problem E706, *American Mathematical Monthly* 53 (1946), 533.

Beiler, A. H., *Recreations in the Theory of Numbers,* 2nd edition, Dover, New York, 1966.

Benjamin, A. T. and Quinn, J. J., *Proofs that Really Count: The Art of Combinatorial Proof*, Mathematical Association of America, Washington, D.C., 2003.

Bicknell, M., "A Primer for the Fibonacci Numbers: Part VII," *Fibonacci Quarterly* 8 (1970), 407–420.

Bicknell, M., "A Primer on the Pell Sequence and Related Sequences," *Fibonacci Quarterly* 13 (1975), 345–349.

Bidwell, J. K., "Pascal's Triangle Revisited," *Mathematics Teacher* 66 (1973), 448–452.

Biggs, N. L., "The Roots of Combinatorics," *Historia Mathematica* 6 (1979), 109–136.

Boas, R., "Snowfalls and Elephants, Pop Bottles and π," *Two-Year College Mathematics Journal* 11 (1980), 82–89.

Boscarol, M., "A Property of Binomial Coefficients," *Fibonacci Quarterly* 20 (1982), 249–251.

Boyd, A. V. and Glencross, M. J., "Dissecting a Circle by Chords through n Points," *Mathematics Teacher* 84 (1991), 318–319.

Boyer, C. B., "Cardan and the Pascal Triangle," *American Mathematical Monthly* 57 (1950), 387–390.

Brady, W. G., Solution to Problem B-155, *Fibonacci Quarterly* 7 (1969), 547.

Brady, W. G., Problem H-273, *Fibonacci Quarterly* 15 (1977), 185.

Brennan, T. A., "Fibonacci Powers and Pascal's Triangle in a Matrix," *Fibonacci Quarterly* 2 (1967), 93–103.

Breusch, R., Solution to Problem E2686, *American Mathematical Monthly* 86 (1979), 131.

Brezsenyi, G., Solution to Problem H-257, *Fibonacci Quarterly* 15 (1977), 283.

Broomhead, W. A., "PASCAL (MOD *p*)," *Mathematical Gazette* 56 (1972), 267–271.

Brown, A. L., "Geometrical Patterns in Pascal's Triangle," *Scripta Mathematica* 22 (1956), 273–274.

Brown, J. L. Jr, and Hoggatt, V. E. Jr, "A Primer for the Fibonacci Numbers, Part XVI: The Central Column Sequence," *Fibonacci Quarterly* 16 (1978), 41–46.

Browne, D. H., Problem E706, *American Mathematical Monthly* 53 (1946), 97.

Brualdi, R. A., *Introductory Combinatorics,* 3rd edition, Prentice Hall, Upper Saddle River, New Jersey, 1999.

Bruckman, P. S., Solution to Problem H-201, *Fibonacci Quarterly* 12 (1974), 218–219.

Bruckman, P. S., Solution to Problem H-217, *Fibonacci Quarterly* 13 (1975), 91–92.

Bruckman, P. S., Solution to Problem H-218, *Fibonacci Quarterly* 13 (1975), 92–93.

Busbridge, I. W., "On Note 2023," *Mathematical Gazette* 33 (1949), 294–295.

Buschman, R. G., Problem H-18, *Fibonacci Quarterly* 2 (1964), 126–127.

Call, G. S., and Velleman, D. J., "Pascal's Matrices," *American Mathematical Monthly* 100 (1993), 372–376.

Callan, D. Problem 498, *College Mathematics Journal* 24 (1993), 184.

Cameron, P. J., *Combinatorics: Topics, Techniques, Algorithms*, Cambridge University Press, New York, 1994.

Cannon, L. O., "Locating Multiples of Primes in Pascal's Triangle," *College Mathematics Journal* 20 (1989), 324–328.

Carlitz, L., "The Number of Binomial Coefficients Divisible by a Fixed Power of a Prime," *Rendiconti del Circolo Mathematico* 16 (1967), 299–320.

Carlitz, L., Problem H-131, *Fibonacci Quarterly* 7 (1969), 285–286.

Carlitz, L., Solution to Problem H-195, *Fibonacci Quarterly* 11 (1973), 502–506.

Carlitz, L., Problem H-229, *Fibonacci Quarterly* 11 (1973), 502.

Carlitz, L., Solution to Problem H-229, *Fibonacci Quarterly* 13 (1975), 371–372.

Carlitz, L., "Degenerate Stirling, Bernoulli and Eulerian Numbers," *Utilitas Mathematica* 15 (1979), 51–88.

Carlitz, L. and Gould, H. W., "Bracket Function Congruences For Binomial Coefficients," *Mathematics Magazine* 37 (1964), 91–93.

Chakravarti, G., "Growth and Development of Permutations and Combinations in India," *Bulletin of the Calcutta Mathematical Society* 24 (1932), 79–88.

Chapman, R. J., Solution to Problem 10570, *American Mathematical Monthly* 106 (1999), 73.

Chernoff, P. R., Solution to Problem E1896, *American Mathematical Monthly* 74 (1967), 1007.

The Chico Problem Group, Solution to Problem 1330, *Mathematics Magazine* 63 (1990), 277–278.

Clarke, L. E., Problem 4704, *American Mathematical Monthly* 63 (1956), 584.

Clarke, J. E., "Clark's Triangle," *Mathematics Student* 26 (1978), 4.

Clarke, R. J., Letter to the Editor, *Mathematics Magazine* 61 (1988), 269.

Cohn, D. I. A., Solution to Problem E1600, *American Mathematical Monthly* (1964), 324–325.

Cohn, D. I. A., *Basic Techniques of Combinatorial Theory*, Wiley, New York, 1978.

Comtet, L., *Advanced Combinatorics*, Dreidel, Boston, Massachusetts, 1974.

Coolidge, J. L., "The Story of the Binomial Theorem," *American Mathematical Monthly* 56 (1949), 147–157.

Cortada, J. W., *Historical Dictionary of Data Processing,* Greenwood Press, Westport, Connecticut, 1987.

Cortada, J. W., (ed.), *American Men and Women of Science*, 20th edition, R. R. Bowker, New Providence, New Jersey, 1998–1999.

Daus, P. H., "Note on Sums Involving Binomial Coefficients," *National Mathematics Magazine* 10 (1936), 165–166.

Deo, N. and Quinn, M. J., "Pascal Graphs and Their Properties," *Fibonacci Quarterly* 21 (1983), 203–214.

Deutsch, E. Problem 10570, *American Mathematical Monthly,* 104 (1997), 69.

Deutsch, E. and Gessel, I. M. Problem 1494, *Mathematics Magazine* 69 (1996), 143.

Dickson, L. E. *History of the Theory of Numbers*, Vol. 1, Chelsea, New York, 1952.

DiDomenico, A. *Private Communications,* 1990–2009.

Donaghey, R., and Shapiro, L. W., "Motzkin Numbers," *Journal of Combinatorial Theory, Series A* 23 (1977), 291–301.

Doster, D., Problem 1318, *Mathematics Magazine* 62 (1989), 136.

Dudley, U., Problem 1277, *Mathematics Magazine* 60 (1987), 328.

Duncan, D. C., "The Diagonals of Pascal's Triangle," *American Mathematical Monthly* 72 (1965), 1115.

Duncan, D. R. and Litwitter, B. H., "A Pattern: Pascal's Triangle and Prime Numbers," *Mathematics Teacher* 68 (1975), 23–26.

Eaton, C. F., "Perfect Numbers in Terms of Triangular Numbers," *Mathematics Magazine* 68 (1995), 307.

Ecklund, E. F. Jr, "On Prime Divisors of the Binomial Coefficients," *Pacific Journal of Mathematics* 29 (1969), 267–270.

Edwards, A. W. F., "Sums of Powers of Integers: A Little of the History," *Mathematical Gazette* 66 (1982), 22–28.

Edwards, A. W. F., "Pascal's Triangle – and Bernoulli's and Vieta's," *Mathematical Spectrum* 20 (1987–1988), 33–37.

Edwards, A. W. F., "Patterns and Primes in Bernoulli's Triangle," *Mathematical Spectrum* 23 (1990–1991), 105–109.

Edwards, A. W. F., *Pascal's Arithmetical Triangle,* The Johns Hopkins University Press, Baltimore, Maryland, 2002.

Eisner, M. P., Solution to Problem 498, *College Mathematics Journal* 24 (1994), 161.

Encyclopaedia Britannica, Chicago, Illinois, 1973.

Enison, R. L., Solution to Problem E2325, *American Mathematical Monthly* 79 (1972), 1136.

Ensley, D., "Fibonacci's Triangle and Other Abominations," *MAA Focus*, Mathematical Association of America, Washington, D.C. (Sept. 2003), 10–14.

Epsteen, S., Problem 219, *American Mathematical Monthly* 11 (1904), 240.

Erdös, P. and Szekeres, G., "Some Number-Theoretic Problems on Binomial Coefficients," *Australian Mathematical Society Gazette* 5 (1978), 97–99.

Eves, H., Problem E579, *American Mathematical Monthly* 50 (1943), 386.

Eves, H., *An Introduction to the History of Mathematics,* 3rd edition, Holt, Rinehart and Winston, New York, 1969.

Fearnehough, A., "π from Pascal's Triangle," *Mathematical Spectrum* 22 (1989–1990), 62–63.

Feemster, H. C., Problem 172, *American Mathematical Monthly* 17 (1910), 75.

Feemster, H. C., Problem 175, *American Mathematical Monthly* 17 (1910), 150.

Feinberg, M., "A Lucas Triangle," *Fibonacci Quarterly* 5 (1967), 486–490.

Fine, N. J., "Binomial Coefficients Modulo a Prime," *American Mathematical Monthly* 54 (1947), 589–592.

Fomin, D. and Kirichenko, A., *Leningrad Mathematical Olympiads 1987–1991*, MathPro Press, Mansfield, Ohio, 1994.

Forder, H. G., "Some Problems in Combinatorics," *Mathematical Gazette* 45 (1961), 199–201.

Frame, J. S., Problem 5113, *American Mathematical Monthly* 70 (1963), 672.

Fry, T. C., Solution to Problem E349, *American Mathematical Monthly* 46 (1939), 512–513.

Gandhi, J. M., Problem E1346, *American Mathematical Monthly* 66 (1959), 61.

Gardner, M., "The Multiple Charms of Pascal's Triangle," *Scientific American* 215 (1966), 128–132.

Gardner, M., "Catalan Numbers: an Integer Sequence that Materializes in Unexpected Places," *Scientific American* 234 (1976), 120–125.

Geldzahler, B., "Powers of Eleven by Pascal's Triangle," *Journal of Recreational Mathematics* 3 (1970), 188–189.

Gillispie, C. C., (ed.), *Dictionary of Scientific Biography,* Charles Scribner's Sons, New York, 1973.

Ginsburg, J., "A Geometrical Interpretation of Binomial Coefficients," *Scripta Mathematica* 5 (1938), 141–142.

Ginsburg, J., "Visualizing Binomial Coefficients," *Scripta Mathematica* 8 (1941), 78.

Ginsburg, J., Problem 136, *Scritpa Mathematica* 13 (1947), 16.

Ginsburg, J., "Trinomial Coefficients," *Scritpa Mathematica* 22 (1956), 136.

Golomb, S. W., "Iterated Binomial Coefficients," *American Mathematical Monthly* 87 (1980), 719–727.

Gould, H. W., "Some Generalizations of Vandermonde's Convolution," *American Mathematical Monthly* 63 (1956), 84–91.

Gould, H. W., "Stirling Number Representation Problems," *Proceedings of the AMS* 11 (1960), 447–451.

Gould, H. W., "A Variant of Pascal's Triangle," *Fibonacci Quarterly* 3 (1965), 257–271.

Gould, H. W., "Fibonomial Catalan Numbers: Arithmetic Properties and a Table of the First Fifty Numbers," *Notices of the AMS* 18 (1971), 938.

Gould, H. W., "A New Greatest Common Divisor Property of the Binomial Coefficients," *Fibonacci Quarterly* 10 (1972), 579–584, 628.

Gould, H. W., "Euler's Formula for the *n*th Differences of Powers," *American Mathematical Monthly* 85 (1978), 450–467.

Graham, R. L. et al., *Concrete Mathematics*, 2nd edition, Addison-Wesley, Reading, Massachusetts, 1994.

Grassl, R., "The Squares Do Fit!," *Mathematical Gazette* 79 (1995), 361–363.

Greenwood, G. W., Solution to Problem 219, *American Mathematical Monthly* 12 (1905), 69.

Greenwood, G. W. and Armstrong, H. M., Solution to Problem 218, *American Mathematical Monthly* 12 (1905), 47.
Grimaldi, R. P., *Discrete and Combinatorial Mathematics,* 5th edition, Pearson/Addison-Wesley, Boston, Massachusetts, 2004.
Gross, O. A., "Preferential Arrangements," *American Mathematical Monthly* 69 (1962), 4–8.
Grossman, H. D., "Binomial Coefficients," *Scripta Mathematica* 12 (1946), 160–161.
Grossman, H. D., Problem 117, *Mathematics Magazine* 25 (1951), 108.
Gunderson, N. G., Solution to Problem 1346, *American Mathematical Monthly* 66 (1959), 30591.
Guy, R. K., Letter to the Editor, *Mathematics Magazine* 61 (1988), 269.
Häggmark, P., "Pascal + Fermat→ Gauss," *Mathematical Gazette* 61 (1977), 285–288.
Harborth, H., "Divisibility of Binomial Coefficients by Their Row Number," *American Mathematical Monthly* 84 (1977), 35–36.
Hardy, G. H. and Wright, E. M., *An Introduction to the Theory of Numbers,* 5th edition, Oxford University Press, Oxford, 1979.
Hathaway, D. K. and Brown, S. L., "Fibonacci Powers and a Fascinating Triangle," *College Mathematics Journal* 28 (1997), 124–128.
Heller, S., Problem 2313, *American Mathematical Monthly* 78 (1971), 904.
Hill, C., Solution to Problem 1318, *Mathemaitcs Magazine* 63 (1990), 128.
Hillman, A. P., The William Lowell Putnam Mathematical Competition, *American Mathematical Monthly*, 82 (1975), 905–912.
Hillman, A. P. et al., The William Lowell Putnam Mathematical Competition, *American Mathematical Monthly*, 85 (1978), 26–33.
Hillman, A. P. and Hoggatt, V. E. Jr, "A Proof of Gould's Pascal Hexagon Conjecture," *Fibonacci Quarterly* 10 (1972), 565–568, 598.
Hilton, P. and Pedersen, J., "Looking into Pascal's Triangle: Combinatorics, Arithmetic, and Geometry," *Mathematics Magazine* 60 (1987), 305–316.
Hodges, J. H., Solution to Problem 1346, *American Mathematical Monthly* 66 (1959), 590.
Hoggatt, V. E. Jr, "A Variant of Pascal's Triangle," *Fibonacci Quarterly* 3 (1965), 257–271.
Hoggatt, V. E. Jr, Problem H-73, *Fibonacci Quarterly* 3 (1965), 300.
Hoggatt, V. E. Jr, Problem H-76, *Fibonacci Quarterly* 3 (1965), 300–301.
Hoggatt, V. E. Jr, Problem H-82, *Fibonacci Quarterly* 4 (1966), 57.
Hoggatt, V. E. Jr, "Fibonacci Numbers and Generalized Binomial Coefficients," *Fibonacci Quarterly* 5 (1967), 383–400.
Hoggatt, V. E. Jr, Problem H-131, *Fibonacci Quarterly* 6 (1968), 142.
Hoggatt, V. E. Jr, "A New Angle on Pascal's Triangle," *Fibonacci Quarterly* 6 (1968), 221–234.
Hoggatt, V. E. Jr, "Convolution Triangle for Generalized Fibonacci Numbers," *Fibonacci Quarterly* 8 (1970), 158–171.
Hoggatt, V. E. Jr, "An Application of the Lucas Triangle," *Fibonacci Quarterly* 8 (1970), 360–364, 427.
Hoggatt, V. E. Jr, Problem H-183, *Fibonacci Quarterly* 9 (1971), 389.
Hoggatt, V. E. Jr, "Some Special Fibonacci and Lucas Generating Functions," *Fibonacci Quarterly* 9 (1971), 121–133.
Hoggatt, V. E. Jr, Problem H-195, *Fibonacci Quarterly* 10 (1972), 413.
Hoggatt, V. E. Jr, Solution to Problem H-183, *Fibonacci Quarterly* 10 (1972), 288–289.
Hoggatt, V. E. Jr, Problem H-201, *Fibonacci Quarterly* 10 (1972), 630.
Hoggatt, V. E. Jr, Problem H-217, *Fibonacci Quarterly* 11 (1973), 185.

Hoggatt, V. E. Jr, Problem H-218, *Fibonacci Quarterly* 11 (1973), 185.

Hoggatt, V. E. Jr, Problem H-257, *Fibonacci Quarterly* 15 (1977), 283.

Hoggatt, V. E. Jr, and Alexander, G. L., "A Property of Multinomial Coefficients," *Fibonacci Quarterly* 9 (1971) 351–356.

Hoggatt, V. E. Jr, and Bicknell, M., "Diagonal Sums of Generalized Pascal Triangles," *Fibonacci Quarterly* 7 (1969), 341–358.

Hoggatt, V. E. Jr, and Bicknell, M., "Generalized Fibonacci Polynomials," *Fibonacci Quarterly* 11 (1973), 457–465.

Hoggatt, V. E. Jr, and Bicknell, M., "Triangular Numbers," *Fibonacci Quarterly* 12 (1974), 221–230.

Hoggatt, V. E. Jr, and Hansell, W., "The Hidden Hexagon Squares," *Fibonacci Quarterly* 9 (1971), 120, 133.

Huff, G. B., Problem E2520, *American Mathematical Monthly* 82 (1975), 169.

Honsberger, R., *Mathematical Gems II*, Mathematical Association of America, Washington, D.C., 1976.

Honsberger, R., *Mathematical Delights*, Mathematical Association of America, Washington, D.C., 2004.

Howard, F. T., "The Number of Binomial Coefficients Divisible by a Fixed Power of 2," *Proceedings of the American Mathematical Society* 29 (1971), 236–242.

Howell, J. M. and Horton, R. E., "Number Bases and Binomial Coefficients," *Mathematics Magazine* 35 (1962), 177–179.

Hsu, L. C., Problem E995, *American Mathematical Monthly* 58 (1951), 700.

http://en.wikipedia.org/wiki/Motzkin-number.

http://en.wikipedia.org/wiki/Pronic-number.

http://mathworld.wolfram.com/MotzkinNumber.html.

http://numbers.computation.free.fr/Constants/Gamma/gamma.html.

http://www.absoluteastronmy.com/topics.Halayudha.

http://www.absoluteastronmy.com/topics/Pingala.

http://binomial.csuhaward.edu/India.html.

http://en.wikipedia.org/wiki/Gaussian-binomial.

Huff, G. B., Problem E2520, *American Mathematical Monthly* 82 (1975), 169.

Ivanoff, V. F., Problem E21, *American Mathematical Monthly* 40 (1933), 110.

Jerbic, S., Problem H-63, *Fibonacci Quarterly* 3 (1965), 116.

Johnston, L. S., Solution to Problem E21, *American Mathematical Monthly* 40 (1933), 362–363.

Jones, C. H., "Generalized Hockey Stick Identities and *n*-dimensional Blockwalking," *Fibonacci Quarterly* 34 (1996), 280–288.

Josef, S., "Lucas Triangle," *Fibonacci Quarterly* 21 (1983), 192–195.

Kirkpatrick, T. B. Jr, "Fibonacci Sequences and Additive Triangles of Higher Order and Degree," *Fibonacci Quarterly* 15 (1977), 319–322.

Klamkin, M. and Liu, A., Problem 488, *College Mathematics Journal* 23 (1992), 435–436.

Klavžar, S. et al., "Combinatorics of Topmost Discs of Multi-peg Tower of Hanoi Problem," *Ars Combinatoria* 59 (2001), 55–64.

Klavžar, S. et al., "Hanoi Graphs and Some Classical Numbers," *Expositiones Mathematicae* 23 (2005), 371–378.

Knebelman, M. S., Solution to Problem E706 *American Mathematical Monthly* 53 (1946), 534.

Knuth, D. E., *The Art of Computer Programming: Fundamental Algorithms*, Vol. 1, Addison-Wesley, Reading, Massachusetts, 1973.

Knuth, D. E., Solution to Problem 6480, *American Mathematical Monthly* 93 (1986), 220–221.

Koshy, T., *Fibonacci and Lucas Numbers with Applications*, Wiley, New York, 2001.

Koshy, T., "A Generalized Fibonacci Triangular Array," *Journal of Recreational Mathematics* 32 (2003–2004), 286–289.

Koshy, T., *Discrete Mathematics with Applications,* Elsevier/Academic Press, Boston, Massachusetts, 2004.

Koshy, T., *Elementary Number Theory with Applications,* 2nd edition, Academic Press, Boston, Massachusetts, 2007.

Koshy, T., "The Central Elements in Josef's Triangle," *Mathematical Gazette* 91 (2007), 63–68.

Koshy, T., *Catalan Numbers with Applications*, Oxford University Press, New York, 2009.

Koshy, T., "Lobb's Generalization of Catalan's Parenthesization Problem," *College Mathematics Journal* 40 (2009), 99–107.

Koshy, T. and Salmassi, M., "Parity and Primality of Catalan Numbers," *College Mathematics Journal* 37 (2006), 52–53.

Koshy, T. and Salmassi, M., "Tribinomial Coefficients and Catalan Numbers," *Mathematical Gazette* 91 (2009), 449–455.

Krishnapriyan, H. K., "Eulerian Polynomials and Faulhaber's Result on Sums of Powers of Integers," *College Mathematics Journal* 26 (1995), 118–123.

Kruse, R. L., *Data Structures and Program Design*, 3rd edition, Prentice Hall, Englewood Cliffs, New Jersey, 1994.

Kung, S. H. L., "Parity Triangles of Pascal's Triangle," *Fibonacci Quarterly* 14 (1976), 54.

Larcombe, P. J., "The 18th Century Chinese Discovery of the Catalan Numbers," *Mathematical Spectrum* 32 (1999), 5–7.

Larsen, M. E., "The Eternal Triangle – A History of a Counting Problem," *College Mathematics Journal* 20 (1989), 370–384.

Larson, L., 51st Annual William Lowell Putnam Mathematical Competition, *Mathematics Magazine* 64 (1991), 141–142.

Lehmer, D. H., Problem E621, *American Mathematical Monthly* 43 (1936), 307.

Lehmer, D. H., Problem 3801, *American Mathematical Monthly* 43 (1936), 580.

Lehmer, D. H., Problem 3802, *American Mathematical Monthly* 43 (1936), 580.

Lehmer, D. H., "Interesting Series Involving the Central Binomial Coefficients," *American Mathematical Monthly* 92 (1985), 449–457.

Lih, K., Problem 2775, *American Mathematical Monthly* 86 (1979), 393.

Lilley, G., "Pascal's Arithmetical Triangle," *American Mathematical Monthly* 1 (1894), 426.

Lind, D., Solution to Problem H-63, *Fibonacci Quarterly* 5 (1967), 256.

Lind, D., Solution to Problem H-73, *Fibonacci Quarterly* 5 (1967), 256.

Lind, D., Problem H-140, *Fibonacci Quarterly* 6 (1968), 251.

Lind, D., Solution to Problem H-99, *Fibonacci Quarterly* 6 (1966), 257–258.

Lind, D., Solution to Problem H-140, *Fibonacci Quarterly* 8 (1970), 81–82.

Lind, D., Solution to Problem B-161, *Fibonacci Quarterly* 8 (1970), 107–108.

Little, T. M., Solution to Problem E1346, *American Mathematical Monthly* 66 (1959), 590.

Lobb, A., "Deriving the *n*th Catalan Number," *Mathematical Gazette* 83 (1999), 109–110.

Long, C. T., *Elementary Introduction to the Theory of Numbers,* 2nd edition, Heath, Lexington, Massachusetts, 1972.

Long, C. T., "Pascal's Triangle Modulo p," *Fibonacci Quarterly* 19 (1981), 458–463.

Lu, X. M., "Towers of Hanoi Graphs," *International Journal of Computational Mathematics* 19 (1986), 23–38.

Lunnon, W. F., "The Pascal Matrix," *Fibonacci Quarterly* 15 (1977), 201–204.

Josef, S., "Lucas Triangle," *Fibonacci Quarterly* 21 (1983), 192–195.

Mana, P., Problems B-136, B-137, *Fibonacci Quarterly* 6 (1968), 185.

Mann, H. B. and Shanks, D., "A Necessary and Sufficient Condition for Primality and Its Source," *Journal of Combinatorial Theory,* Part A, 13 (1972), 131–134.

Martson, H. M., Problem 1494, *Mathematics Magazine* 70 (1997), 143.

Mauldon, J. G., Solution to Problem 2775, *American Mathematical Monthly* 87 (1980), 578–579.

Mauldon, J. G., Solution to Problem 1277, *Mathematics Magazine* 61 (1988), 316.

Mays, M., Solution to Problem 2775, *American Mathematical Monthly* 87 (1980), 579–580.

McAnaney, K. L., Problem 1330, *Mathematics Magazine* 62 (1989), 275.

McKay, J. H., The William Lowell Putnam Mathematical Competition, *American Mathematical Monthly* 80 (1973), 1017–1028.

Mendelson, E., "Races with Ties," *Mathematics Magazine* 55 (1982), 170–175.

Mendelsohn, N. S., Problem E864, *American Mathematical Monthly* 56 (1949), 263.

Mendelsohn, N. S., Problem E2227, *American Mathematical Monthly* 77 (1970), 308.

Moen, C., "Infinite Series with Binomial Coefficients," *Mathematics Magazine* 64 (1991), 53–55.

Montgomery, P. L., Problem E2686, *American Mathematical Monthly* 84 (1977), 820.

Morgan-Voyce, A. M., "Ladder Network Analysis using Fibonacci Numbers," *IRE Transactions on Circuit Theory* 6 (1959), 321–322.

Moser, L., Solution to Problem E864, *American Mathematical Monthly* 57 (1950), 38.

Moser, L., Problem 60, *Mathematics Magazine* 26 (1952), 225.

Moser, L., Solution to Problem 1346, *American Mathematical Monthly* 66 (1959), 591.

Motzkin, T. S., "Arithmetical Progressions in the Binomial Triangle," *Scripta Mathematica* 12 (1946), 14.

Motzkin, T. S., "Relations between Hypersurface Cross Ratios, and a Combinatorial Formula for Partitions of a Polygon, for Permanent Preponderance, and for Non-associative Products," *Bulletin of the American Mathematical Society* 54 (1948), 352–360.

Mubayi, D., Problem 523, *College Mathematics Journal* 26 (1995), 160–161.

Needham, J., "Mathematics," *Science and Civilization in China* 3 (1959), 137–139.

Niven, I., "Formal Power Series," *American Mathematical Monthly* 76 (1969), 871–889.

Noltie, S., Solution to Problem 1494, *Mathematics Magazine* 70 (1997), 143–144.

Olds, C. D., Problem 4428, *American Mathematical Monthly* 58 (9151), 194.

Ollerton, R. L. and Shannon, A. G., "Some Useful Integer Arrays for the Teaching of Generalization Skills," *International Journal of Mathematical Education in Science and Technology* 34 (2003), 135–143.

Ollerton, R. L. and Shannon, A. G., "Fibonacci's Odd Number Array," *Journal of Recreational Mathematics* 32 (2003–2004), 198–204.

Osler, T. J., "Variations of a Theme from Pascal's Triangle," *College Mathematics Journal* 34 (2003), 216–223.

Piza, P. A., Problem 4322, *American Mathematical Monthly* 55 (1948), 642.

Piza, P. A., "Kummer Numbers," *Mathematics Magazine* 21 (1948), 257–260.

Piza, P. A., Problem 92, *Mathematics Magazine* 24 (1951), 225.

Polya, G., *Mathematical Discovery,* Vol. 1, Wiley, New York, 1966.

Poole, D. G., "The Towers and Triangles of Professor Claus (or, Pascal Knows Hanoi)," *Mathematics Magazine* 67 (1994), 323–344.

Richards, I. M., "π from Pascal's Triangle," *Mathematical Spectrum* 22 (1989–1990), 104–105.

Ringenberg, L. A., Solution to Problem 236, *Mathematics Magazine* 29 (1956), 164.

Riordan, J., *Combinatorial Identities,* Wiley, New York, 1968.

Riordan, J., *An Introduction to Combinatorial Analysis,* Princeton University Press, Princeton, New Jersey, 1980.

Roberts, J. B., "On Binomial Coefficient Residues," *Canadian Journal of Mathematics* 9 (1957), 363–370.

Rogers, D. G., "Pascal's Triangles, Catalan Numbers, and Renewal Arrays," *Discrete Mathematics* 22 (1978), 301–310.

Rosencrans, S. I., Problem E2325, *American Mathematical Monthly* 78 (1971), 1137.

Rupp, C. A., Problem 3468, *American Mathematical Monthly* 37 (1930), 552.

Schell, E. D., Problem E839, *American Mathematical Monthly* 55 (1948), 576.

Schmitt, F. G., Solution to Problem 2313, *American Mathematical Monthly* 79 (1972), 906–907.

Schneider, I. E., Problem E1600, *American Mathematical Monthly* (1963), 569.

Shafer, R. E., Problem 6480, *American Mathematical Monthly* 91 (1984), 651.

Shannon, A. G., Solution to Problem H-273, *Fibonacci Quarterly* 16 (1978), 568–569.

Sharp, R. T., Problem 52, *Pi Mu Epsilon Journal* 1 (1954), 411.

Shapiro, L. W., "A Catalan Triangle," *Discrete Mathematics* 14 (1976), 83–90.

Shapiro, L. W., *American Mathematical Monthly*, 106 (1999), 777.

Shaumyan, A., Solution to Problem 19, *Math Horizons* (April 1995), 33.

Shimshoni, M., Solution to Problem 2313, *American Mathematical Monthly* 79 (1972), 906.

Singer, D., Problem E1896, *American Mathematical Monthly* 73 (1966), 665.

Singh, P., "The So-called Fibonacci Numbers in Ancient and Medieval India," *Historia Mathematica* 12 (1985), 229–244.

Smiley, L. M., "A Quick Solution of Triangle Counting," *Mathematics Magazine* 66 (1993), 40.

Spiegel, M. R., Solution to Problem E995, *American Mathematical Monthly* 59 (1952), 411.

Stanley, R. P., *Enumerative Combinatorics*, Vol. 1, Wadsworth & Brooks/Cole, Monterey, California, 1986.

Stanley, R. P., *Enumerative Combinatorics*, Vol. 2, Cambridge University Press, New York, 1999.

Starke, E. P., Solution to Problem E706 *American Mathematical Monthly* 53 (1946), 533.

Starke, E. P., Solution to Problem 117, *Mathematics Magazine* 26 (1952), 104–107.

Starke, E. P., Solution to Problem 92, *Mathematics Magazine* 26 (1952), 216.

Stenger, A., Solution to Problem E2520, *American Mathematical Monthly* 83 (1976), 383.

St. Olaf College Students, Solution to Problem E2227, *American Mathematical Monthly* 78 (1971), 201–202.

Struyk, A., "Two Notes on Binomial Coefficients" *Mathematics Teacher* 49 (1956), 192–194.

Sumner, J. S. and Dove, K. L., Solution to Problem 488, *College Mathematics Journal* 24 (1993), 477.

Takács, L., Letter to the Editor, *Mathematics Magazine* 61 (1988), 269.

Talbot, R. F., "Pascal's Semi-infinite Plane," *Mathematical Spectrum* 21 (1988–1989), 25–26.

Tanton, J., "Sums of Powers," *Math Horizons* (Sept. 2003), 15–20.

Trigg, C. W., Solution to Problem 52, *Pi Mu Epsilon Journal* 1 (1954), 411–412.

Trigg, C. W., "Geometric Proof of a Result of Lehmer's," *Fibonacci Quarterly* 11 (1973), 593–540.

Trost, E., Solution to Problem 4428, *American Mathematical Monthly* 59 (1952), 470.

Tucker, A., *Applied Combinatorics*, 4th edition, Wiley, New York, 2002.

van Lint, J. H. and Wilson, R. M., *A Course in Combinatorics,* Cambridge University Press, New York, 1992.

Vun, I. and Belcher, P., "Catalan Numbers," *Mathematical Spectrum* 29 (1996–1997), 3–5.

Umansky, H. L., Letter to the Editor, *Fibonacci Quarterly* 8 (1970), 89.

Usiskin, Z., "Perfect Square Patterns in the Pascal Triangle," *Mathematics Magazine* 46 (1973), 203–208.

Uspenskii, V. A., *Pascal's Triangle*, The University of Chicago Press, Chicago, Illinois, 1974.

Wahlin, G. E., Solution to Problem 175, *American Mathematical Monthly* 18 (1911), 41–43.

Wall, C. R., Problem H-99, *Fibonacci Quarterly* 4 (1966), 333.

Wayne, A., Solution to Problem E579, *American Mathematical Monthly* 51 (1944), 165.

Wang, E. T. H., Problem 523, *College Mathematics Journal* 25 (1994), 157.

Weinmann, A., Solution to Problem 5113, *American Mathematical Monthly* 71 (1964), 691–692.

Weisstein, E. W., "Binomial Coefficient," Mathworld.wolfram.com/*q*-Binomial Coefficient.html.

Weisstein, E. W., "Hanoi Graph," Mathworld.wolfram.com/HanoiGraph.html.

Wells, D., *The Penguin Dictionary of Curious and Interesting Numbers,* Penguin Books, London, England, 1987.

Wong, C. K. and Maddocks, T. W., "A Generalized Pascal's Triangle," *Fibonacci Quarterly* 13 (1975), 134–136.

Wright, E. M., Solution to Problem 4704, *American Mathematical Monthly* 64 (1957), 597–598.

Wright, T., "Pascal's Triangle Gets Its Genes from Stirling Numbers of the First Kind," *College Mathematics Journal* 26 (1995), 368–371.

Wong, C. K. and Maddocks, T. W., "A Generalized Pascal's Triangle," *Fibonacci Quarterly* 13 (1975), 134–136.

Young, J. W., Problem 106, *American Mathematical Monthly* 9 (1902), 86.

Zeitlin, D., Solution to Problem E1346, *American Mathematical Monthly* 66 (1959), 590.

Zerr, G. B. M., Solution to Problem 172, *American Mathematical Monthly* 17 (1910), 147.

Zerr, G. B. M. and Walker, L. C., Solution to Problem 106, *American Mathematical Monthly* 9 (1902), 86.

Zhiqing, L., "Pascal's Pyramid," *Mathematical Spectrum* 16 (1983–1084), 1–3.

Index

G

H

U

V

W

Z

www.ingramcontent.com/pod-product-compliance
Lightning Source LLC
LaVergne TN
LVHW011904200926
848484LV00002B/520